MONOGRAPHS ON STATISTICS AND APPLIED PROBABILITY

General Editors

V. Isham, N. Keiding, T. Louis, N. Reid, R. Tibshirani, and H. Tong

1 Stochastic Population Models in Ecology and Epidemiology *M.S. Barlett* (1960)
2 Queues *D.R. Cox and W.L. Smith* (1961)
3 Monte Carlo Methods *J.M. Hammersley and D.C. Handscomb* (1964)
4 The Statistical Analysis of Series of Events *D.R. Cox and P.A.W. Lewis* (1966)
5 Population Genetics *W.J. Ewens* (1969)
6 Probability, Statistics and Time *M.S. Barlett* (1975)
7 Statistical Inference *S.D. Silvey* (1975)
8 The Analysis of Contingency Tables *B.S. Everitt* (1977)
9 Multivariate Analysis in Behavioural Research *A.E. Maxwell* (1977)
10 Stochastic Abundance Models *S. Engen* (1978)
11 Some Basic Theory for Statistical Inference *E.J.G. Pitman* (1979)
12 Point Processes *D.R. Cox and V. Isham* (1980)
13 Identification of Outliers *D.M. Hawkins* (1980)
14 Optimal Design *S.D. Silvey* (1980)
15 Finite Mixture Distributions *B.S. Everitt and D.J. Hand* (1981)
16 Classification *A.D. Gordon* (1981)
17 Distribution-Free Statistical Methods, 2nd edition *J.S. Maritz* (1995)
18 Residuals and Influence in Regression *R.D. Cook and S. Weisberg* (1982)
19 Applications of Queueing Theory, 2nd edition *G.F. Newell* (1982)
20 Risk Theory, 3rd edition *R.E. Beard, T. Pentikäinen and E. Pesonen* (1984)
21 Analysis of Survival Data *D.R. Cox and D. Oakes* (1984)
22 An Introduction to Latent Variable Models *B.S. Everitt* (1984)
23 Bandit Problems *D.A. Berry and B. Fristedt* (1985)
24 Stochastic Modelling and Control *M.H.A. Davis and R. Vinter* (1985)
25 The Statistical Analysis of Composition Data *J. Aitchison* (1986)
26 Density Estimation for Statistics and Data Analysis *B.W. Silverman* (1986)
27 Regression Analysis with Applications *G.B. Wetherill* (1986)
28 Sequential Methods in Statistics, 3rd edition
G.B. Wetherill and K.D. Glazebrook (1986)
29 Tensor Methods in Statistics *P. McCullagh* (1987)
30 Transformation and Weighting in Regression
R.J. Carroll and D. Ruppert (1988)
31 Asymptotic Techniques for Use in Statistics
O.E. Bandorff-Nielsen and D.R. Cox (1989)
32 Analysis of Binary Data, 2nd edition *D.R. Cox and E.J. Snell* (1989)
33 Analysis of Infectious Disease Data *N.G. Becker* (1989)
34 Design and Analysis of Cross-Over Trials *B. Jones and M.G. Kenward* (1989)

35 Empirical Bayes Methods, 2nd edition *J.S. Maritz and T. Lwin* (1989)

36 Symmetric Multivariate and Related Distributions
K.T. Fang, S. Kotz and K.W. Ng (1990)

37 Generalized Linear Models, 2nd edition *P. McCullagh and J.A. Nelder* (1989)

38 Cyclic and Computer Generated Designs, 2nd edition
J.A. John and E.R. Williams (1995)

39 Analog Estimation Methods in Econometrics *C.F. Manski* (1988)

40 Subset Selection in Regression *A.J. Miller* (1990)

41 Analysis of Repeated Measures *M.J. Crowder and D.J. Hand* (1990)

42 Statistical Reasoning with Imprecise Probabilities *P. Walley* (1991)

43 Generalized Additive Models *T.J. Hastie and R.J. Tibshirani* (1990)

44 Inspection Errors for Attributes in Quality Control
N.L. Johnson, S. Kotz and X. Wu (1991)

45 The Analysis of Contingency Tables, 2nd edition *B.S. Everitt* (1992)

46 The Analysis of Quantal Response Data *B.J.T. Morgan* (1992)

47 Longitudinal Data with Serial Correlation—A state-space approach
R.H. Jones (1993)

48 Differential Geometry and Statistics *M.K. Murray and J.W. Rice* (1993)

49 Markov Models and Optimization *M.H.A. Davis* (1993)

50 Networks and Chaos—Statistical and probabilistic aspects
O.E. Barndorff-Nielsen, J.L. Jensen and W.S. Kendall (1993)

51 Number-Theoretic Methods in Statistics *K.-T. Fang and Y. Wang* (1994)

52 Inference and Asymptotics *O.E. Barndorff-Nielsen and D.R. Cox* (1994)

53 Practical Risk Theory for Actuaries
C.D. Daykin, T. Pentikäinen and M. Pesonen (1994)

54 Biplots *J.C. Gower and D.J. Hand* (1996)

55 Predictive Inference—An introduction *S. Geisser* (1993)

56 Model-Free Curve Estimation *M.E. Tarter and M.D. Lock* (1993)

57 An Introduction to the Bootstrap *B. Efron and R.J. Tibshirani* (1993)

58 Nonparametric Regression and Generalized Linear Models
P.J. Green and B.W. Silverman (1994)

59 Multidimensional Scaling *T.F. Cox and M.A.A. Cox* (1994)

60 Kernel Smoothing *M.P. Wand and M.C. Jones* (1995)

61 Statistics for Long Memory Processes *J. Beran* (1995)

62 Nonlinear Models for Repeated Measurement Data
M. Davidian and D.M. Giltinan (1995)

63 Measurement Error in Nonlinear Models
R.J. Carroll, D. Rupert and L.A. Stefanski (1995)

64 Analyzing and Modeling Rank Data *J.J. Marden* (1995)

65 Time Series Models—In econometrics, finance and other fields
D.R. Cox, D.V. Hinkley and O.E. Barndorff-Nielsen (1996)

66 Local Polynomial Modeling and its Applications *J. Fan and I. Gijbels* (1996)

67 Multivariate Dependencies—Models, analysis and interpretation
D.R. Cox and N. Wermuth (1996)

68 Statistical Inference—Based on the likelihood *A. Azzalini* (1996)

69 Bayes and Empirical Bayes Methods for Data Analysis
B.P. Carlin and T.A Louis (1996)

70 Hidden Markov and Other Models for Discrete-Valued Time Series
I.L. Macdonald and W. Zucchini (1997)

71 Statistical Evidence—A likelihood paradigm *R. Royall* (1997)

72 Analysis of Incomplete Multivariate Data *J.L. Schafer* (1997)

73 Multivariate Models and Dependence Concepts *H. Joe* (1997)

74 Theory of Sample Surveys *M.E. Thompson* (1997)

75 Retrial Queues *G. Falin and J.G.C. Templeton* (1997)

76 Theory of Dispersion Models *B. Jørgensen* (1997)

77 Mixed Poisson Processes *J. Grandell* (1997)

78 Variance Components Estimation—Mixed models, methodologies and applications
P.S.R.S. Rao (1997)

79 Bayesian Methods for Finite Population Sampling
G. Meeden and M. Ghosh (1997)

80 Stochastic Geometry—Likelihood and computation
O.E. Barndorff-Nielsen, W.S. Kendall and M.N.M. van Lieshout (1998)

81 Computer-Assisted Analysis of Mixtures and Applications—
Meta-analysis, disease mapping and others *D. Böhning* (1999)

82 Classification, 2nd edition *A.D. Gordon* (1999)

83 Semimartingales and their Statistical Inference *B.L.S. Prakasa Rao* (1999)

84 Statistical Aspects of BSE and vCJD—Models for Epidemics
C.A. Donnelly and N.M. Ferguson (1999)

85 Set-Indexed Martingales *G. Ivanoff and E. Merzbach* (2000)

86 The Theory of the Design of Experiments *D.R. Cox and N. Reid* (2000)

87 Complex Stochastic Systems
O.E. Barndorff-Nielsen, D.R. Cox and C. Klüppelberg (2001)

88 Multidimensional Scaling, 2nd edition *T.F. Cox and M.A.A. Cox* (2001)

89 Algebraic Statistics—Computational Commutative Algebra in Statistics
G. Pistone, E. Riccomagno and H.P. Wynn (2001)

90 Analysis of Time Series Structure—SSA and Related Techniques
N. Golyandina, V. Nekrutkin and A.A. Zhigljavsky (2001)

91 Subjective Probability Models for Lifetimes
Fabio Spizzichino (2001)

92 Empirical Likelihood *Art B. Owen (2001)*

93 Statistics in the 21st Century *Adrian E. Raftery, Martin A. Tanner, and Martin T. Wells (2001)*

94 Accelerated Life Models: Modeling and Statistical Analysis
Vilijandas Bagdonavičius and Mikhail Nikulin (2001)

95 Subset Selection in Regression, Second Edition
Alan Miller (2002)

96 Topics in Modelling of Clustered Data
Marc Aerts, Helena Geys, Geert Molenberghs, and Louise M. Ryan (2002)

97 Components of Variance *D.R. Cox and P.J. Solomon* (2002)

Nonlinear Models for Repeated Measurement Data

Marie Davidian
Associate Professor
Department of Biostatistics
Harvard School of Public Health
Boston
USA

David M. Giltinan
Principal Biostatistician
Genentech Inc.
San Francisco
USA

CHAPMAN & HALL/CRC

A CRC Press Company
Boca Raton London New York Washington, D.C.

Library of Congress Cataloging-in-Publication Data

Catalog record is available from the Library of Congress

Direct all inquiries to CRC Press LLC, 2000 N.W. Corporate Blvd., Boca Raton, Florida 33431.

Visit the CRC Press Web site at www.crcpress.com

First edition 1995
First CRC press reprint 1998
Originally published by Chapman & Hall

International Standard Book Number 0-412-98341-9
Printed in the United States of America 5 6 7 8 9 0
Printed on acid-free paper

Contents

TO OUR PARENTS

Preface

Analysis of repeated measurement data is a recurrent challenge to statisticians engaged in biological and biomedical applications. For example, data from clinical trials are often longitudinal in nature, with repeated measures of response taken over time. In pharmacokinetic studies, serial measurements of drug concentrations are taken from each participant. By definition, growth studies involve repeated measurements over time. In other applications, measurements on experimental units may be repeated across some other dimension, e.g. spatially rather than temporally.

Methods for *linear* modeling of repeated measurement data are well developed and well documented in the statistical literature; recent accounts include those by Crowder and Hand (1990), Lindsey (1993), and Diggle, Liang and Zeger (1994). In many biological applications, such as pharmacokinetic analysis and studies of growth and decay, however, *nonlinear* modeling is required for meaningful analysis. For this type of modeling, statistical approaches are less well understood, and discussion of appropriate methodology is scattered across a wide literature. Recent years have seen more attention to nonlinear repeated measurement data in the statistical literature; however, the economy of style imposed by many journals means that the material is sometimes presented in a manner that does not make it readily accessible to practicing statisticians. The result is that, although nonlinear modeling of repeated measurement data represents an area of some practical importance, it is one that still appears to engender a good deal of confusion among data analysts.

Our purpose in writing this monograph is to provide a clear delineation of currently available modeling approaches and inferential methods for nonlinear repeated measures. The goal is to make the material accessible to a wide audience. The book is targeted mainly to practicing biostatisticians in industry and academia, and to graduate students in statistics or biostatistics. We have attempted to keep the exposition at an intermediate level, however, so that the majority of the material should also be accessible to pharmacokineticists and to researchers in the clinical and biological sciences.

The model framework that forms the basis for the inferential methods discussed in the book is that of the hierarchical nonlinear model for continuous response data. This may be viewed as an extension of standard nonlinear modeling techniques to accommodate multiple levels of variability (within and among individuals). Alternatively, it may be regarded as a generalization of the hierarchical linear model framework to include models that are nonlinear in parameters of interest. Hierarchical nonlinear modeling may thus be expected to inherit the computational difficulties intrinsic to both nonlinear regression and to hierarchical linear models. We have certainly found this to be true in practice; computational issues can be formidable at times, so that inference within this framework is not an enterprise to be undertaken lightly. We have included several case studies in later chapters; these represent 'real-life' data sets. We have tried to report analyses in sufficient detail to give the reader a realistic sense, not only of the potential scope and utility of the methods discussed, but also of the potential difficulties. Because computational aspects play a key role in the implementation of all the techniques described in this book, we have tried to include some discussion of available software in each of the relevant chapters. Most of the data sets considered in this book will be available on Statlib, together with code to implement model fits discussed in the text using various software packages.

Several friends and colleagues helped us while writing this book. We are grateful to Sharon Baughman, Doug Bates, Eric Chi, Art DeVault, Jim Frane, Tim Gregoire, Karen Higgins, Debbi Kotlovker, Cynthia Ladd, Nishit Modi, James Reimann, Alan Schumitzky, Anastasios Tsiatis, Jon Wakefield, and Fong Wang-Clow for comments on earlier drafts of the manuscript. Thanks go to researchers at Genentech, Inc., and elsewhere for permission to use their data in the book. Special thanks are due to Alan Hopkins and to Genentech for granting the second author a leave of absence to complete the manuscript and for encouragement throughout the writing process. Moral support was provided by Peter Compton, Carol Deasy, Ellen Gilkerson, Debbi Kotlovker, James Reimann, and Georgia Thompson. Finally, words cannot adequately express our debt of gratitude to Butch Tsiatis, without whose keen statistical insight, ongoing moral and culinary support, and unfailing good humor this manuscript would never have been completed.

This book was typeset using LaTeX; figures were created using Splus.

Boston and San Francisco
March 1995

Marie Davidian
David M. Giltinan

CHAPTER 1

Introduction

Data consisting of repeated measurements taken on each of a number of individuals arise commonly in biological and biomedical applications. For example, in longitudinal clinical studies, measurements are taken on each of a number of subjects over time. Similarly, participants in pharmacokinetic experiments undergo serial blood sampling following administration of a test agent. Pharmacodynamic studies may involve repeated measurement of physiological effect in the same subject in response to differing doses of a drug. By definition, studies of growth and decay involve repeated measurements taken on sample units, which could be human or animal subjects, plants, or cultures.

Modeling data of this kind usually involves characterization of the relationship between the measured response, y, and the repeated measurement factor, or covariate, x. In many applications, the proposed systematic relationship between y and x is nonlinear in unknown parameters of interest. In some cases, the relevant nonlinear model may be derived on physical or mechanistic grounds. In other contexts, a nonlinear relationship may be used simply to provide an empirical description of the data.

The presence of repeated observations on an individual requires particular care in characterizing the random variation in the data. It is important to recognize two sources of variability explicitly: random variation among measurements *within* a given individual and random variation *among* individuals. Inferential procedures accommodate these different variance components within the framework of an appropriate hierarchical statistical model. When the postulated relationship between y and x is linear in the unknown parameters, the relevant framework is that of the classical linear mixed effects model. An alternative is provided by Bayesian inferential methods for a suitable hierarchical linear model. There is an extensive literature on hierarchical linear models; Searle, Casella, and McCulloch (1992) provide a comprehensive overview.

Methods for repeated measurement data where the relationship between y and x is *nonlinear* in the unknown parameters are less well developed. Treatment of existing techniques is scattered through a wide literature. The purpose of this monograph is to provide a unified presentation of methods and issues for nonlinear repeated measurement data. We begin by considering several examples to motivate our subsequent development.

1.1 Motivating examples

1.1.1 Pharmacokinetics of cefamandole

Figure 1.1 shows data from a pilot study to investigate the pharmacokinetics of cefamandole, a cephalosporin antibiotic (Aziz *et al.*, 1978). In this experiment, a dose of 15 mg/kg body weight of cefamandole was administered by ten-minute intravenous infusion to six healthy male volunteers. Blood samples were collected from each subject at each of 14 time points post-dose. Drug concentrations in plasma were determined for each sample by high-performance liquid chromatography (HPLC). The resulting plasma concentration-time profiles for each subject are plotted in the figure.

In characterizing the pharmacokinetics of a drug, it is common to represent the body as a system of compartments and to assume that the rates of transfer between compartments follow first-order or linear kinetics. Solution of the resulting differential equations shows that the relationship between drug concentration and time may be described by a sum of exponential terms. For instance, the biexponential equation

$$C(t) = \beta_1 \exp(-\beta_2 t) + \beta_3 \exp(-\beta_4 t), \quad \beta_1, \ldots, \beta_4 > 0, \qquad (1.1)$$

where $C(t)$ is drug plasma concentration and t is time post-dose, follows from the assumption of a two-compartment model to describe kinetics following intravenous injection (Gibaldi and Perrier, 1982).

The data in Figure 1.1 exhibit similarly shaped profiles for each subject, with possibly different parameter values for different subjects. Variation within each subject about the model in equation (1.1) arises mainly due to the HPLC assay. It is commonly recognized that intra-subject variation of this kind tends to increase with plasma concentration level (Beal and Sheiner, 1988).

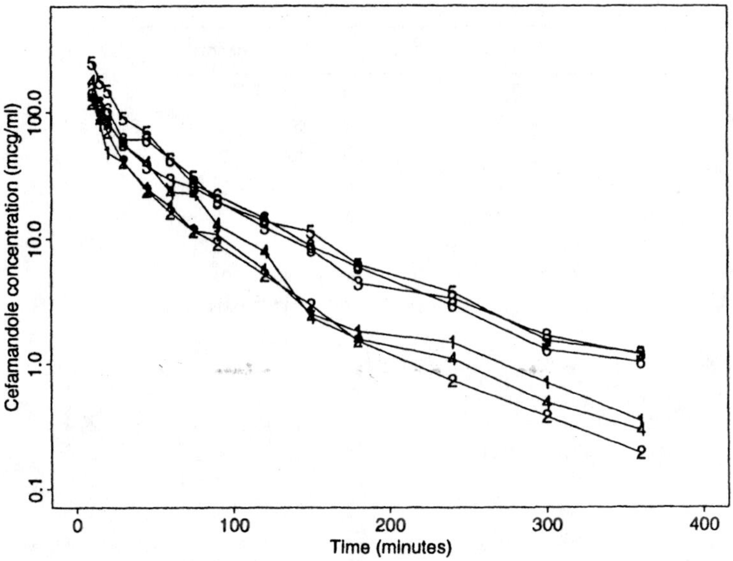

Figure 1.1. *Plasma concentration-time profiles for six subjects, cefamandole data.*

In pilot volunteer studies like this one, primary objectives are to establish an appropriate kinetic model, to obtain preliminary information on values of the model parameters, and to assess the nature of intra-subject variation. Typically, results from the analysis of a pilot study are used as a basis for subsequent investigation of kinetics in a larger, more heterogeneous patient population.

1.1.2 Population pharmacokinetics of quinidine

Data from a clinical study of the pharmacokinetics of the anti-arrhythmic agent quinidine reported by Verme *et al.* (1992) consist of quinidine concentration (mg/L) measurements for 136 hospitalized patients (135 men, 1 woman) treated for either atrial fibrillation or ventricular arrhythmias with oral quinidine therapy. A total of 361 quinidine concentration measurements ranging from one to 11 observations per patient were obtained by enzyme immunoassay during the course of routine clinical treatment.

Measurements were taken within a range of 0.08 hours to 70.5

Table 1.1. *Partial data for two subjects, pharmacokinetic study of quinidine. Units for measurements are given in the text.*

time	*conc.*	*dose*	*SS*[1]	*age*	*wt.*	*creat.*[2]	*glyco.*[3]
			Subject 2				
0.0	–	166	–	58	85	> 50	82
6.0	–	166	–	58	85	> 50	82
12.0	–	166	–	58	85	> 50	82
18.0	–	166	–	58	85	> 50	82
25.0	1.2	–	–	58	85	> 50	82
height = 69, race = Latin, nonsmoker, ethanol abuse, moderate congestive heart failure							
			Subject 10				
0.0	–	201	8	73	79	< 50	254
2.2	3.9	–	8	73	79	< 50	254
288.0	–	201	8	73	79	< 50	176
290.0	5.4	–	8	73	79	< 50	176
504.0	–	201	8	73	79	< 50	150
506.0	2.8	–	8	73	79	< 50	150
816.0	–	201	8	73	79	< 50	127
816.2	–	201	8	73	79	< 50	127
817.0	3.1	–	8	73	79	< 50	127
1241.0	–	201	8	73	79	< 50	98
1249.0	–	201	8	73	79	< 50	98
7897.0	–	201	8	74	82	> 50	158
7897.8	1.6	–	8	74	82	> 50	158
height = 69, race = Caucasian, nonsmoker, no ethanol abuse or congestive heart failure							

[1] if numeric, subject has achieved steady state with given dosing interval; if blank, subject has not achieved steady state

[2] creatinine clearance

[3] α_1-acid glycoprotein concentration

hours after dose. Table 1.1 shows partial data records for two patients selected from the total of 136. Demographic and physiological covariate information was collected for each patient over an observation period ranging from 0.13 hours to 8095.0 hours. The following variables were available for the majority of patients: weight, height, and age, as well as information on race (Latin, Caucasian, Black), smoking status (yes, no), ethanol abuse (yes, no, previously), and status with respect to congestive heart failure (severe, moderate, mild or none). Weight, age, smoking, and cardiac status were recorded periodically during the study. Creatinine clearance (ml/min), a measure of renal function, and α_1-acid glycoprotein concentration (mg/dL), the level of a circulating molecule that binds quinidine, were also measured periodically on all patients, although information on creatinine clearance was recorded in a categorical rather than continuous manner. Baseline albumin concentration (g/dL) measurements were available for some, but not all, patients. Oral quinidine may be administered in two different forms; in this study, it was given as quinidine sulfate to 53 patients, as quinidine gluconate to 57 patients, and in both forms to 26 patients. Doses were adjusted for differences in salt content between the two forms by conversion of both forms to milligrams of quinidine base.

One possible characterization of quinidine disposition is to use a one-compartment open model with first-order absorption (Verme *et al.*, 1992). Let $C(t)$ be the concentration of quinidine and let $C_a(t)$ be the apparent concentration of quinidine in the absorption depot at time t. Written in recursive form, the model is:

For the non-steady state at a dosage time $t = t_\ell$

$$\begin{aligned} C_a(t_\ell) &= C_a(t_{\ell-1})\exp\{-k_a(t_\ell - t_{\ell-1})\} + FD_\ell/V \\ C(t_\ell) &= C(t_{\ell-1})\exp\{-k_e(t_\ell - t_{\ell-1})\} + C_a(t_{\ell-1})\frac{k_a}{k_a - k_e} \\ &\quad \times \Big[\exp\{-k_e(t_\ell - t_{\ell-1})\} - \exp\{-k_a(t_\ell - t_{\ell-1})\}\Big]. \end{aligned} \tag{1.2}$$

For the steady state at a dosage time, $t = t_\ell$

$$\begin{aligned} C_a(t_\ell) &= (FD_\ell/V)(1 - \exp\{-k_a\tau_{ss}\})^{-1} \\ C(t_\ell) &= (FD_\ell/V)\frac{k_a}{k_a - k_e} \\ &\quad \times \Big[(1 - \exp\{-k_e\tau_{ss}\})^{-1} - (1 - \exp\{-k_a\tau_{ss}\})^{-1}\Big]. \end{aligned} \tag{1.3}$$

Between dosage times, $t_\ell < t < t_{\ell+1}$

$$\begin{aligned} C(t) = C(t_\ell)\exp\{-k_e(t - t_\ell)\} + C_a(t_\ell)\frac{k_a}{k_a - k_e} \\ \times \Big[\exp\{-k_e(t - t_\ell)\} - \exp\{-k_a(t - t_\ell)\}\Big]. \end{aligned} \tag{1.4}$$

In (1.2)–(1.4), t_ℓ, $\ell = 0, 1, \ldots$, are the times at which doses D_ℓ

are administered, $C_a(t_0) = FD_0/V$, $C(t_0) = 0$, F is the fraction of dose available, k_a is the absorption rate constant, $k_e = Cl/V$ is the elimination rate constant, Cl is the clearance, V is the apparent volume of distribution, $Cl, V, k_a > 0$, and τ_{ss} is the steady-state dosing interval.

The data in this example share certain characteristics with the cefamandole data: a common nonlinear model form for all subjects, values of the pharmacokinetic parameters that may differ from subject to subject, and probable heterogeneity of assay variation within subject. There are obvious differences as well. In contrast to the experimental setting, where essentially complete concentration-time profiles are collected for each subject, the quinidine data, collected in a clinical setting, are sparse, with relatively few observations per subject. This difference between data collected in a controlled experimental setting and routine clinical data is fairly typical. Clinical data may be available for a much greater number of patients, but data on any one individual patient tend to be sparse.

A second major difference between the quinidine and cefamandole data sets is the availability of demographic and physiological information that may help to explain inter-subject differences in the disposition of quinidine. This difference between experimental and clinical data is also usually the case. Early Phase I data are gathered frequently from a small number of healthy volunteers in a carefully controlled setting. Routine clinical data typically come from a much more extensive and heterogeneous patient population. This allows broader inferences to be drawn, although the task is complicated by the relative paucity of information at the individual subject level.

The major question of interest in the quinidine and similar clinical studies is identification of the demographical and physiological factors affecting drug disposition in a broad patient population. A thorough understanding of this issue may afford important clinical benefit: the dosage regimen for a given patient may be individualized based on relevant physiological and demographic information for the patient if a model is available relating drug disposition to measured patient covariates. Thus, more accurate titration of dosage may be feasible, avoiding possible suboptimal therapeutic benefit resulting from underdosing and minimizing potential toxicity associated with overdosing. Accurate dosing is of particular importance in drugs with a low therapeutic index, where the window of desirable serum concentrations is relatively narrow. For all drugs, however, it is clearly beneficial to maximize understanding of factors affecting drug disposition.

Given a model which accurately predicts a subject's pharmacokinetic parameters as a function of physiological and demographic characteristics, there will still remain a random, or unexplained, component of variation among individuals. Quantifying this inter-subject variation in pharmacokinetic parameters is a secondary objective in population analysis of data such as those from the quinidine study. Achieving the analysis goals is complicated by several issues: (i) the sparsity of information on individual subjects; (ii) the generally nonlinear dependence of response on the relevant pharmacokinetic parameters; and (iii) inter-subject variability. Methods that allow valid inference in the face of these difficulties form the core of this book.

1.1.3 Growth analysis for soybean plants

Figure 1.2 shows data from an experiment to compare growth patterns of two genotypes of soybean: Plant Introduction #416937 (P), an experimental strain, and Forrest (F), a commercial variety.

In this study, data were collected in each of three consecutive years (1988–1990). At the beginning of the growing season in each year, 16 plots were planted with seeds, eight plots with each genotype. To assess growth, each plot was sampled eight to ten times at approximate weekly intervals. At each sampling time, six plants were randomly selected from each plot, leaves from these plants were weighed, and average leaf weight per plant was calculated for the plot. Different plots in different sites were used in different years.

Inspection of Figure 1.2 indicates that the usual logistic function

$$y = \frac{\beta_1}{1 + \beta_2 \exp(\beta_3 x)}, \quad \beta_1, \beta_2 > 0,\ \beta_3 < 0, \tag{1.5}$$

provides a reasonable representation of average leaf weight y and time x. It is evident from the figure that considerable variation in the parameters β_1, β_2 and β_3 that characterize the growth pattern exists among plots for a given genotype. For this kind of growth data, it is reasonable to expect serial correlation among measurements within the same plot. In addition, intra-plot variability may be expected to increase with the average level of response. Thus, these data have several features in common with the previous examples: (i) a nonlinear dependence of response on parameters of

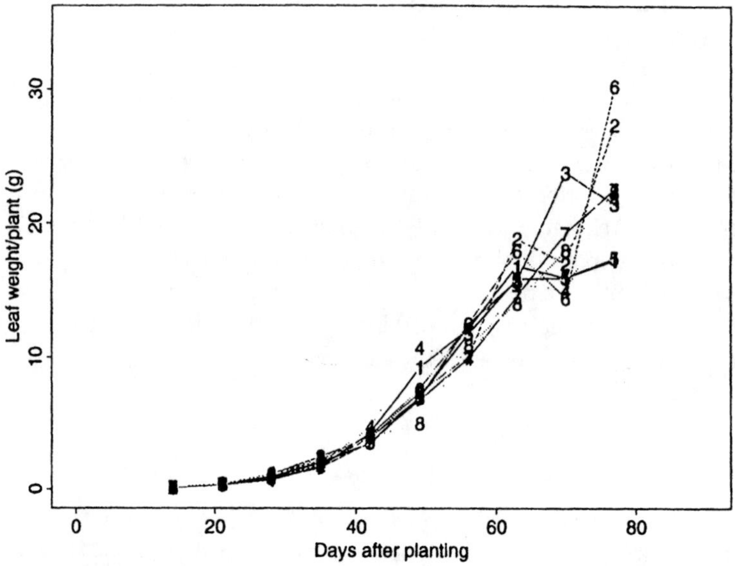

Figure 1.2. *Average leaf weight-time profiles for 8 plots planted with Plant Introduction #416937, 1988.*

interest; (ii) similarly shaped profiles for each plot; and (iii) heterogeneous within-plot variability, which may also include serial correlation. As with the cefamandole example, sufficient data are available for each plot to allow fitting of model parameters based on data from that plot only.

Variation in growth characteristics may depend on several factors. The primary objective of the experiment was comparison of growth characteristics (initial leaf weight, limiting leaf weight, and growth rate) for the two genotypes. Weather patterns differed considerably over the three years: 1988 was unusually dry, 1989 was wet, and conditions in 1990 were normal. Comparison of growth across weather patterns was also of interest.

1.1.4 Bioassay for relaxin by RIA

Determination of the concentration of a particular protein in an unknown sample frequently relies on immunoassay or bioassay techniques. Bioassay methods are generally based on a relevant measure

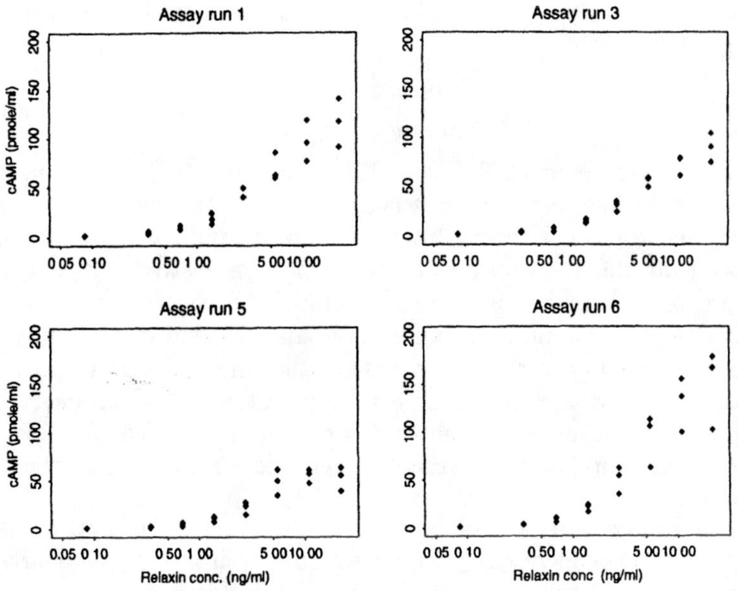

Figure 1.3. *Assay response (cAMP)-concentration data for four runs of the relaxin bioassay.*

of bioactivity of the protein in question and involve measurement of activity at several known (standard) concentrations of the protein (analyte). The resulting concentration-response curve is used to determine the protein concentration in unknown samples by inverse regression (calibration).

Figure 1.3 shows concentration-response data obtained for standard concentrations in four runs of a bioassay for the therapeutic protein relaxin (Fei *et al.*, 1990). For this assay, bioactivity of relaxin is measured by increased generation and release of intracellular adenosine-3′, 5′-cyclic monophosphate (cAMP) by normal human uterine endometrial cells in the presence of relaxin. (cAMP is an enzyme that plays a key role in regulating glycogen metabolism in the cell.) For each a total of nine runs, triplicate cAMP measurements were determined by radioimmunoassay for each of seven known relaxin concentrations. A single measurement at zero standard was also available for each run; by convention, the response at zero concentration has been plotted at two dilutions below the lowest standard in Figure 1.3. A standard choice of dose-response model in describing this kind of assay data is the four-parameter logistic function

$$y = \beta_1 + \frac{\beta_2 - \beta_1}{1 + \exp\{\beta_4(\log x - \beta_3)\}}, \quad \beta_1, \beta_2 > 0. \qquad (1.6)$$

The response y in (1.6) is cAMP level (pmoles/ml), and x represents the known relaxin concentration (ng/ml). The parameters in (1.6) have the following interpretation: β_1 and β_2 represent response at 'infinite' and zero concentration, respectively; β_3 is the $\log EC_{50}$ value, that is, the logarithm of the concentration that gives a response midway between β_1 and β_2; and β_4 is a slope parameter governing the steepness of the concentration-response curve. It is evident from Figure 1.3 that values of the parameters may vary considerably from run to run. It is also clear from the plots that within-assay variability depends strongly on response level.

These data share several features with the previous examples: (i) nonlinear dependence of y on regression parameters; (ii) similarly shaped concentration-response profiles for each run, with possibly different parameter values from run to run; and (iii) within-run variability that increases with the level of the response.

What inferential questions are of interest for assay data of this kind? The primary focus is typically on calibration, or inverse regression; that is, estimation of the concentration of analyte in an unknown sample based on the observed response for that sample. Ancillary questions pertain to assay precision and performance. For example, with what precision are unknown samples calibrated? How may one form accurate confidence intervals about estimated concentrations? What is the 'acceptable range' of the assay, where calibrated estimates are sufficiently precise? What is the lowest limit of reliable assay measurement? Can one exploit the similarity among assay runs to improve calibration? These issues will be discussed in detail in Chapter 10. For now, we remark that the inferential challenge is the same as that in the previous examples, to address the questions of interest within a framework that correctly accommodates both the inter- and intra-assay variation.

1.2 Model specification

In the examples of the previous section, several common features of the data may be identified:

(i) Repeated response measurements taken on a number of different individuals (subjects, plots, or assay runs).

(ii) Nonlinear dependence of the response y on a set of unknown parameters, $\boldsymbol{\beta}$, for each individual.

(iii) Response profiles that are similarly shaped across individuals, but that may have different values of the parameter vector $\boldsymbol{\beta}$ for different individuals.

(iv) A pattern of within-individual variability that is not necessarily homogeneous. Possible deviations from constant variation, or homoscedasticity, include a dependence of the variability on mean response, serial correlation among measurements within an individual, or both.

(v) Inter-individual variability between regression parameters that may be considered to be random, to be systematically related to individual-specific characteristics, or a combination of both.

To incorporate these features in an inferential setting, a useful strategy is to build a *hierarchical*, or staged, model. Full details are presented in Chapter 4; here, we sketch an outline of the approach, using a two-stage model.

The first stage specifies the mean and covariance structure for a given individual. For the ith of m individuals, assume that the $(n_i \times 1)$ response vector $\boldsymbol{y}_i$, satisfies

$$\mathrm{E}(\boldsymbol{y}_i|\boldsymbol{\beta}_i) = \boldsymbol{f}_i(\boldsymbol{\beta}_i) = \begin{bmatrix} f(\boldsymbol{x}_{i1}, \boldsymbol{\beta}_i) \\ \vdots \\ f(\boldsymbol{x}_{in_i}, \boldsymbol{\beta}_i) \end{bmatrix},$$

$$\mathrm{Cov}(\boldsymbol{y}_i|\boldsymbol{\beta}_i) = \boldsymbol{R}_i. \tag{1.7}$$

In (1.7), the function f characterizes the systematic dependence of the response on the repeated measurement conditions for the ith individual, summarized in the covariate vectors $\boldsymbol{x}_{i1}, \ldots, \boldsymbol{x}_{in_i}$. The regression function f depends in a nonlinear fashion on a regression parameter $\boldsymbol{\beta}_i$ specific to the ith individual and has the same basic functional form for all individuals; different individual response patterns are accommodated through the possibility of different $\boldsymbol{\beta}_i$ values for different individuals as well as through the individual-specific covariate vectors $\boldsymbol{x}_{ij}$. The matrix $\boldsymbol{R}_i$ is a covariance matrix summarizing the pattern of random variability associated with the data for the ith individual. Along with an assumption about the distribution of $\boldsymbol{y}_i$, the first stage model (1.7) thus describes both systematic and random features of the data at the *intra-individual* level.

Inter-individual variation is characterized in a second stage, consisting of a *model* for variation in the regression parameters $\boldsymbol{\beta}_i$. This variation can be modeled using a distributional assumption for the $\boldsymbol{\beta}_i$ at various levels of complexity. For instance, one might specify a parametric 'regression' model for the $\boldsymbol{\beta}_i$, e.g.,

$$\boldsymbol{\beta}_i = \boldsymbol{A}_i\boldsymbol{\beta} + \boldsymbol{b}_i. \tag{1.8}$$

In (1.8), $\boldsymbol{\beta}_i$ is assumed to depend linearly and systematically on a vector of parameters $\boldsymbol{\beta}$ and on individual-specific information, such as physiological and demographic characteristics in the quinidine study or genotype and weather condition in the soybean growth study, summarized in a design matrix $\boldsymbol{A}_i$. The 'error' $\boldsymbol{b}_i$ corresponds to the random component of inter-individual variation, which might be taken to have mean zero and covariance matrix $\boldsymbol{D}$. One could add the further restriction that the distribution of the $\boldsymbol{\beta}_i$ belongs to a particular parametric family, for example, the multivariate normal distribution. In the example, this would amount to the assumption that

$$\boldsymbol{\beta}_i \sim \mathcal{N}(\boldsymbol{A}_i\boldsymbol{\beta}, \boldsymbol{D}).$$

We shall refer to the case where the variation in the inter-individual random parameters is specified by a parametric model like (1.8) and the random component is assumed to belong to a particular distributional family as the *fully parametric* model specification. At the other end of the spectrum, one might make no assumptions at all about the form or distribution of the $\boldsymbol{\beta}_i$. We refer to this as the *nonparametric* model specification. An intermediate possibility is to specify a parametric model for the $\boldsymbol{\beta}_i$ as in (1.8), but to avoid the assumption of a particular distributional family for the random component. We refer to this kind of model specification as being *semiparametric*. Finally, the kind of two-stage model that we consider may be arrived at naturally from a Bayesian perspective. In the Bayesian view, individual-specific regression parameters $\boldsymbol{\beta}_i$ are considered to arise from a distribution whose mean and covariance are drawn from an appropriate prior distribution.

Each of the four kinds of model specifications for the random parameters – parametric, nonparametric, semiparametric, or Bayesian – leads to different inferential approaches. One other factor is a major determinant of the inferential technique to be applied: the relative amount of information that is available per individual. We have seen in the context of pharmacokinetics that two quite distinct scenarios are possible: (i) sparse information on each of a large number of individuals and (ii) rich information on each of a small number of individuals. Intermediate scenarios are also a possibility, of course. Not surprisingly, the sampling design plays a role in determining what analysis methods may be employed; certain methods that can be used when sampling on an individual basis is relatively dense are not applicable to the sparse data case.

We shall use the term 'individual' throughout this book to refer to the experimental unit over which repeated measurements are available. In the repeated measurement literature, use of the term 'subject' is common in this context. Other terms include 'block' and 'stratum;' the term 'cluster' has also been proposed (Lindstrom and Bates, 1990). We avoid the latter term due to its more common usage elsewhere in the statistical literature, and use 'individual' in preference to 'subject' because of its greater generality.

1.3 Outline of this book

The goal of this book is to provide a unified presentation of modeling strategies and inferential procedures for the types of continuous repeated measurement data exemplified by the data sets described in section 1.1. This kind of data has recently received considerable attention, but discussion of inferential methods is scattered across a wide variety of sources, both in the statistical and subject-matter literature. We present methods suitable for *continuous* data of this type; techniques for binary or discrete data are not discussed. By providing a clear delineation of models and methods, we hope to make the relevant techniques more accessible both to statisticians and investigators faced with the challenge of analyzing nonlinear repeated measurement data.

For the most part, we have tried to keep exposition at an intermediate level, with emphasis throughout on applications. Familiarity with regression analysis at the level of a text such as Draper and Smith (1981) and with statistical inference at a first-year graduate level should provide adequate background for most of the material in this book. Chapters 7 and 8 are at a somewhat higher mathematical level than the remainder of the book, but can be omitted without significant loss of continuity.

Hierarchical *nonlinear* modeling provides the central framework for everything else in this book. A review of nonlinear regression methods is given in Chapter 2, which, in the context of repeated measurement data, is relevant to data from a single individual only. Many of the techniques applicable to hierarchical nonlinear models are direct extensions of methods for individual data. Chapter 3 provides a comprehensive review of hierarchical *linear* models; a good understanding of this material is helpful in making the transition to the nonlinear case. Chapter 4 is central to the rest of the book; here, we lay out the various approaches to hierarchical nonlinear modeling in some detail. As mentioned above, these may be categorized as (i) fully parametric, (ii) semi- and nonparametric, and (iii) Bayesian. Each of these perspectives leads to different inferential strategies, discussed in Chapters 5–8. In Chapters 5 and 6, inferential procedures for the fully parametric (normal theory) hierarchical nonlinear model are presented. Chapter 5 discusses *two-stage* methods, which are applicable only in the case where sufficient data are available for each individual to allow estimation of individual-specific regression parameters based on data for that individual only. Methods presented in Chapter 6 are based on some type of *linearization* of the model and may also be applied to the 'sparse data' case. Chapter 7 is devoted to semiparametric and nonparametric inference. Bayesian methods are described in Chapter 8. Each of Chapters 9 and 10 treats a particular area of application in detail. Chapter 9 discusses population pharmacokinetic and pharmacodynamic modeling. The analysis of assay data, with particular reference to immunoassays and bioassays, is covered in Chapter 10. Case studies in the areas of crop science, forestry, and seismology, illustrating the general applicability of the methods, are presented in Chapter 11. The book concludes with a discussion of open issues and general comments (Chapter 12).

Schematically, the organization of the material may be represented as follows:

Chapter 1	Introduction
Chapters 2 and 3	Background material
Chapter 4	Model specification
Chapters 5–8	Inferential methods
Chapters 9–11	Applications and case studies
Chapter 12	Conclusion

It is inevitable that different readers will approach this book with different backgrounds and prior exposure to this material. Depending on background and the primary focus of the reader's interest, different reading strategies are possible. For instance, a researcher in pharmacokinetics, whose primary interest is in population pharmacokinetic and pharmacodynamic modeling, might choose to include the following sections on a first reading: Chapter 1, Chapter 2, Chapter 3 (excluding sections 3.2.3, 3.3.3, 3.4, and 3.6), sections 4.1, 4.2, and 4.3.1, Chapters 5 and 6, and Chapter 9. This would provide comprehensive coverage of the fully parametric approach to population modeling. Other approaches, such as the nonparametric or Bayesian paradigms in Chapters 7 and 8, could be covered on subsequent readings.

Similarly, the above sequence might provide a useful first approach to statisticians unfamiliar with this material. For these readers, it would also be useful to include Chapters 10 and 11 to achieve a better understanding of the scope of the methods. Readers with interests in applications in biometry not directly related to pharmacokinetics or pharmacodynamics might cover the following chapters: Chapters 1–3 (excluding sections 3.2.3 and 3.3.3), Chapters 4–6, and Chapter 11, possibly including Chapter 10, depending on interest.

Computational aspects play an important role in the practical implementation of all of the techniques described in this book. For this reason, we have generally included a section in each of the relevant chapters that discusses software implementation. Some appreciation of potential computational difficulties is necessary if the methods discussed in this book are to be implemented sensibly.

CHAPTER 2

Nonlinear regression models for individual data

2.1 Introduction

Individual repeated measurement data often exhibit a relationship between response and measurement factors that is best characterized by a model nonlinear in its parameters. In some settings, an appropriate nonlinear model may be derived on the basis of theoretical considerations. In other situations, a nonlinear relationship may be employed to provide an empirical description of the data. In this chapter, as a prelude to discussion of the hierarchical nonlinear model for data from several individuals, we review techniques for nonlinear modeling and inference for data from a single individual only.

To fix ideas, consider the data in Table 2.1, taken from a study reported by Kwan *et al.* (1976) of the pharmacokinetics of indomethacin following bolus intravenous injection of the same dose in six human volunteers. For each subject, plasma concentrations of indomethacin were measured at 11 time points ranging from 15 minutes to 8 hours post-injection. In this chapter, we focus on the data for the fifth subject only for purposes of illustration; the concentration-time profile for this individual is shown in Figure 2.1. The usual approach to derivation of a suitable model for pharmacokinetic data of this kind is predicated upon the assumption of a compartment model for the human body, as described in section 1.1.1. This approach suggests that, except for random intra-individual variation, the biexponential function

$$y = \beta_1 \exp(-\beta_2 x) + \beta_3 \exp(-\beta_4 x), \quad \beta_1, \ldots, \beta_4 > 0,$$

is a reasonable representation of plasma concentration y as a function of time x.

Table 2.1. *Plasma concentrations ($\mu g/ml$) following intravenous injection of indomethacin for six human subjects.*

	Subject					
time (hrs)	1	2	3	4	5	6
0.25	1.50	2.03	2.72[1]	1.85	2.05	2.31
0.50	0.94	1.63	1.49	1.39	1.04	1.44
0.75	0.78	0.71	1.16	1.02	0.81	1.03
1.00	0.48	0.70	0.80	0.89	0.39	0.84
1.25	0.37	0.64	0.80	0.59	0.30	0.64
2.00	0.19	0.36	0.39	0.40	0.23	0.42
3.00	0.12	0.32	0.22	0.16	0.13	0.24
4.00	0.11	0.20	0.12	0.11	0.11	0.17
5.00	0.08	0.25	0.11	0.10	0.08	0.13
6.00	0.07	0.12	0.08	0.07	0.10	0.10
8.00	0.05	0.08	0.08	0.07	0.06	0.09

[1] outlier; not included in later analyses

Individual data that follow a nonlinear model often exhibit response variation that changes systematically with the level of the response. Heterogeneous variation occurs in nearly all fields of application, including those that are the focus of this book. For example, assay data typically exhibit intra-run variance that is an increasing function of the response level. Variation in pharmacokinetic data and data from growth studies is also widely acknowledged to be related systematically to mean response.

Another complication for repeated measurement data may arise from the tendency for observations on a given individual to be related. When measurement is repeated over time, serial correlation may be evident; in other contexts, correlation patterns may be due to factors such as adjacent positioning of samples on an assay microtiter plate, similarity in genetic composition of litter-mates, or spatial orientation of field samples.

Because of the frequency with which these features arise in practice, the overview of nonlinear regression modeling and inference given in this chapter includes detailed discussion of generalizations of the classical ordinary least squares approach to allow for heterogeneous response variance, often called heteroscedasticity, and for correlation. In section 2.2, the nonlinear regression model framework is introduced, and we discuss inference, including methods for

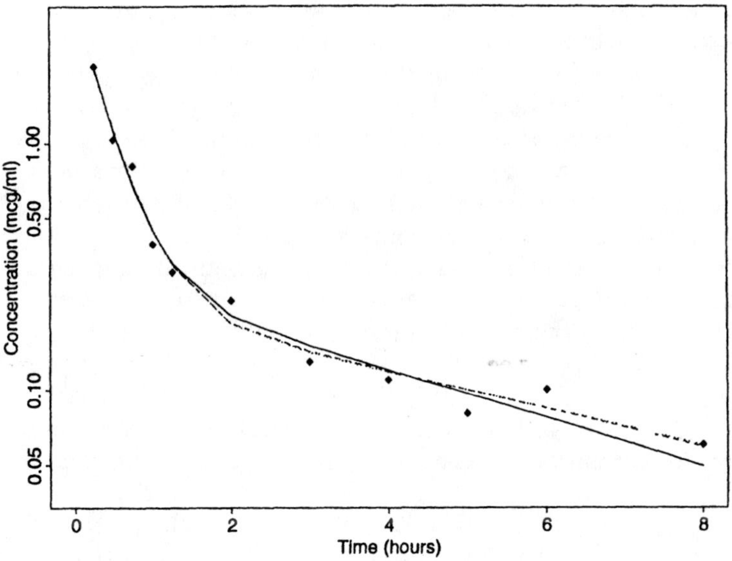

Figure 2.1. *Plasma concentration-time profile, indomethacin data, subject 5. Fits of model (2.43) are superimposed; see section 2.5. The solid line is the OLS fit ($\theta = 0$), the dotted line is the GLS fit with $\theta = 1$, and the dashed line is the GLS fit with θ estimated by PL ($\theta = 0.82$).*

handling heterogeneity of variance and correlation, in section 2.3. Notes on computational methods are given in section 2.4, and two examples are considered in section 2.5. In section 2.6, we comment on related approaches. Section 2.7 contains a brief discussion, and the chapter concludes with bibliographic notes in section 2.8.

2.2 Model specification

The classical nonlinear regression model described in this section is a direct extension of the linear case. We begin our discussion of this model in section 2.2.1 with a statement of the basic statistical model and notation. In section 2.2.2, we set out the classical assumptions and describe how they may be violated in practice. We describe a number of generalizations in section 2.2.3 to account for departures from the assumptions. For simplicity, the index i for individual is suppressed throughout this chapter.

2.2.1 Basic nonlinear regression model

The basic model for a response variable y has two main components: the nonlinear function characterizing mean response and a specification for intra-individual response variance. Let y_j denote the response taken at the jth covariate value $\boldsymbol{x}_j$, $j = 1, \ldots, n$. The $\boldsymbol{x}_j$ are most often viewed as fixed quantities, and, in the case where they are random, the model and assumptions below are understood to be conditional on their values. In the repeated measurement context, the response vector $\boldsymbol{y} = [y_1, \ldots, y_n]'$ summarizes the information available for a single individual taken at values $\boldsymbol{X} = (\boldsymbol{x}_1', \ldots, \boldsymbol{x}_n')'$ of the repeated measurement factor(s).

The model for the jth observation is usually written as

$$y_j = f(\boldsymbol{x}_j, \boldsymbol{\beta}) + e_j. \tag{2.1}$$

In (2.1), the regression function f depends on $\boldsymbol{\beta}$ $(p\times 1)$, the vector of regression parameters, in a nonlinear fashion. For the indomethacin data,

$$f(x, \boldsymbol{\beta}) = \beta_1 \exp(-\beta_2 x) + \beta_3 \exp(-\beta_4 x), \tag{2.2}$$

with $\boldsymbol{\beta} = [\beta_1, \ldots, \beta_4]'$. The random errors e_j reflect uncertainty in the measured response. For repeated measurement data, the vector $\boldsymbol{e} = [e_1, \ldots, e_n]'$ thus summarizes the uncertainty for all observations on a given individual.

2.2.2 Classical assumptions

The classical nonlinear regression framework specifies that data arise according to (2.1) together with the following assumptions:

(i) The errors e_j have mean zero.

(ii) The errors e_j are uncorrelated.

(iii) The errors e_j have common variance σ^2, $\text{Var}(e_j) = \sigma^2$, and are identically distributed for all $\boldsymbol{x}_j$.

(iv) The errors e_j are normally distributed.

The first assumption ensures that the model f for mean response is correctly specified. This assumption is rarely called into question, as it is usually the case that the form of the covariate-response relationship is fairly well understood, especially for nonlinear relationships, where the model may result directly from theoretical considerations.

The remaining three assumptions are fairly restrictive and may not hold in some applications. Assumption (iv) is a reflection of the emphasis of much of statistical methodology on the historical Gaussian paradigm and forms the basis for the standard approach to inference. In applications for which nonlinear models are appropriate, however, this assumption may be particularly unrealistic. For instance, for given $\boldsymbol{x}_j$, the distribution of biological data may be heavily skewed or subject to a higher intensity of outlying observations than would be expected under normality.

If the assumption of normality (iv) does hold, along with that of uncorrelated errors (ii), then the errors are independently distributed. Thus, the assumption of independence is usually made directly, and under the full set of assumptions, the y_j are independently normally distributed with

$$\mathrm{E}(y_j) = f(\boldsymbol{x}_j, \boldsymbol{\beta}), \qquad \mathrm{Var}(y_j) = \sigma^2. \tag{2.3}$$

For repeated measurement data, however, even the less stringent specification of uncorrelated errors may be unrealistic; for example, measurements taken over time on a given individual may be serially related.

Assumption (iii), that of constant intra-individual response variance, is violated frequently in practice. For example, growth data often exhibit constant coefficient of variation rather than constant variance; that is, variance proportional to the square of the mean response. In this case, a more appropriate assumption than (2.3) would be

$$\mathrm{E}(y_j) = f(\boldsymbol{x}_j, \boldsymbol{\beta}), \qquad \mathrm{Var}(y_j) = \sigma^2\{f(\boldsymbol{x}_j, \boldsymbol{\beta})\}^2, \tag{2.4}$$

where the scale parameter σ is the coefficient of variation. A general discussion of extensions of the classical model to accommodate heterogeneity of variance is given in the next section.

2.2.3 Generalizations of the classical framework

The classical nonlinear regression framework may be generalized to accommodate departures from the assumptions in a variety of ways. The following exposition is by no means exhaustive. We adopt the perspective taken in Chapters 2 and 3 of Carroll and Ruppert (1988) and Chapter 6 of Seber and Wild (1989); specifically, that of modeling systematic response variance and correlation patterns explicitly. We focus on this strategy because the hierarchical nonlinear model discussed in the remainder of the book adapts readily to account for these features. Other approaches are noted in section 2.7.

Intra-individual variance heterogeneity

A standard way to account for variance that depends systematically on the level of response or some other factor is to postulate a formal *model* for response variance, much as one models the mean response. For example, in the constant coefficient of variation model (2.4), variance is proportional to a known function (the square) of the mean response. Depending on the application and nature of the data, different functions of the mean, or functions of the covariates or other values, may be appropriate models for variance.

A general statement of this idea entails a modification of the basic nonlinear model framework. Assumption (iii) is relaxed to incorporate possible heteroscedasticity by specification of a *variance function*, g. The function g may depend on the mean response; on constants $\boldsymbol{z}_j$, which may include some or all of the components of $\boldsymbol{x}_j$; and on an additional q-dimensional parameter vector $\boldsymbol{\theta}$. A very general specification for mean and variance is

$$\mathrm{E}(y_j) = f(\boldsymbol{x}_j, \boldsymbol{\beta}), \quad \mathrm{Var}(y_j) = \sigma^2 g^2(\mu_j, \boldsymbol{z}_j, \boldsymbol{\theta}), \quad \mu_j = f(\boldsymbol{x}_j, \boldsymbol{\beta}). \tag{2.5}$$

Note that in (2.5), the variance function g depends on the regression parameter $\boldsymbol{\beta}$ through the mean function f. Variance functions that allow more general dependence on the regression parameter are possible, and the inferential techniques discussed in section 2.3.2 are applicable for more general models; however, in practice, dependence on $\boldsymbol{\beta}$ is almost always directly through the mean response.

This framework offers considerable flexibility in characterizing intra-individual variance. The general form of the relationship is given by the variance function g. The scale parameter σ governs the overall level of precision in the response, while the variance parameter $\boldsymbol{\theta}$ specifies fully the functional form. The choice of g may not require specification of additional parameters, as in the case of (2.4). In other instances, a reasonable function that reflects the nature of the relationship between variance and other factors may be apparent, but it may not be possible to identify fully the entire function. The following examples illustrate this point.

Natural choices for the variance function g are often dictated by the application. For biological data, variance is often thought to depend on the mean response. A common variance function in a number of fields, including assay analysis and pharmacokinetics, is the power model

$$g(\mu_j, z_j, \boldsymbol{\theta}) = \mu_j^{\theta}, \tag{2.6}$$

where, usually, $\theta > 0$. Other proposals include the exponential model

$$g(\mu_j, z_j, \boldsymbol{\theta}) = \exp(\mu_j \theta), \tag{2.7}$$

with θ usually > 0, and a model that allows for two components of variance:

$$g^2(\mu_j, z_j, \boldsymbol{\theta}) = \theta_1 + \mu_j^{2\theta_2}, \quad \theta_1, \theta_2 > 0, \tag{2.8}$$

$\boldsymbol{\theta} = [\theta_1, \theta_2]'$. Equation (2.8) is often used to characterize assay data, which may exhibit constant measurement error at low response values and variance increasing with the mean at higher levels of the response.

In all of these cases, the chosen functional form may accurately reflect the nature of the relationship between variance and level of response, but it may not be possible to specify an appropriate value for $\boldsymbol{\theta}$. For example, in the development phase of an assay, response variance may not yet be well understood. The power model (2.6) may be a reasonable representation of the pattern of variation, but the value of θ may be unknown. Standard choices include $\theta = 1.0$, which gives the constant coefficient of variation model (2.4), and $\theta = 0.5$, which implies a Poisson-like variance structure. Frequently, for assay data, the value of θ that best characterizes the pattern of variation is neither of these common choices (Finney, 1976). Similarly, it may be difficult to specify values *a priori* for θ in the exponential model (2.7) or both θ_1 and θ_2 in the components of variance model (2.8). In such situations, a data-driven approach to determination of an appropriate value for $\boldsymbol{\theta}$ is often adopted. We discuss this general idea in section 2.3.3.

In some instances, variance models depending on other information may be proposed. For example, it is common in econometric applications to model variance as a function of covariate information $\boldsymbol{z}_j$ exogenous to the mean model f. Variance of biological data is sometimes thought to depend on the value of $\boldsymbol{x}_j$; for example, variance may increase rapidly with the value of a scalar covariate x_j, suggesting a model for the logarithm of variance as a linear function of $z_j = x_j$:

$$g(\mu_j, z_j, \boldsymbol{\theta}) = \exp(\theta x_j). \tag{2.9}$$

The modeling framework given in (2.5) accommodates all of these possibilities.

A number of diagnostic procedures have been proposed for detecting and modeling heteroscedasticity. A full discussion of these is beyond the scope of this chapter; Chapters 2 and 3 of Carroll and Ruppert (1988) provide an overview. In the applications that are the focus of this book, heterogeneity of variance is a well-established phenomenon, and standard variance functions are accepted as providing reasonable empirical representations in many cases. Furthermore, it may be difficult to distinguish reliably among proposed variance models, because of the inherent difficulty of characterizing second moment behavior; see section 2.3.4. In general, it is desirable to achieve a parsimonious empirical description g of the prominent aspects of the pattern of heterogeneous variance, with a variance parameter $\boldsymbol{\theta}$ of low dimension (one or two).

Intra-individual correlation

Correlation among observations on a given individual is especially likely in the repeated measurement context. In many cases, a systematic pattern of correlation is evident, which may be characterized accurately by a relatively simple model. To accommodate intra-individual correlation, assumption (ii) is replaced by a description of the assumed correlation pattern among the elements of $\boldsymbol{e}$. Formally, suppose that

$$\text{Corr}(\boldsymbol{e}) = \boldsymbol{\Gamma}(\boldsymbol{\alpha}),$$

where the correlation matrix $\boldsymbol{\Gamma}(\boldsymbol{\alpha})$ is a function of a vector of correlation parameters $\boldsymbol{\alpha}$ $(s \times 1)$. The choice of a suitable correlation matrix depends on the nature of the repeated measurement factor; some common correlation structures are described below. When variance is constant across the response range and equal to some value σ^2, as in assumption (iii),

$$\text{Cov}(\boldsymbol{e}) = \sigma^2 \boldsymbol{\Gamma}(\boldsymbol{\alpha}).$$

The more general case where errors are both heterogeneous and correlated is addressed shortly.

For the important special case where the repeated observations are taken over time, standard models for serial correlation patterns are available, for example, the autoregressive (AR) model of order one. For definiteness, assume that the observations are indexed in the order in which they were collected. In the simplest case where the n repeated measurements are equally spaced in time, if the correlation between two adjacent observations is some value α, then the correlation between any two measurements is given by

$$\mathrm{Corr}(e_{j_1}, e_{j_2}) = \alpha^{|j_1 - j_2|}.$$

This may be expressed by specification of the correlation matrix

$$\boldsymbol{\Gamma}(\alpha) = \begin{bmatrix} 1 & \alpha & \alpha^2 & \cdots & \alpha^{n-1} \\ & 1 & \alpha & \cdots & \alpha^{n-2} \\ & & \ddots & \ddots & \vdots \\ & & & \ddots & \alpha \\ & & & & 1 \end{bmatrix}. \tag{2.10}$$

The AR(1) correlation matrix may be generalized to accommodate situations where the observations are not equally spaced. If the j_1th and the j_2th measurements are taken at times t_{j_1} and t_{j_2}, respectively, $j_1 \neq j_2$, then the (j_1, j_2) and (j_2, j_1) elements of (2.10) would be replaced by $\alpha^{|t_{j_1} - t_{j_2}|}$; see Liang and Zeger (1986) and Chi and Reinsel (1989). An alternative model suggested by Diggle (1988) is given by

$$\mathrm{Corr}(e_{j_1}, e_{j_2}) = \exp(-\alpha_1 |t_{j_1} - t_{j_2}|^{\alpha_2}),\ \alpha_2 = 1 \text{ or } 2.$$

Another example of a serial correlation model is given in section 2.5.

When the repeated measurement factor is something other than time, different correlation models may provide an appropriate representation. A detailed discussion of correlation models is beyond the scope of this book; Chapter 6 of Seber and Wild (1989) gives an account in the nonlinear regression context.

A word of caution is in order. Unless a large number of observations covering a broad range of the response are available, the paucity of information on associations may preclude reliable inference on within-individual correlation patterns. We return to this issue in subsequent chapters; for now, we remark that, although it may be tempting to specify intra-individual correlation patterns, practical implementation may be difficult and may induce complicated interplay among components of the overall response model.

General intra-individual covariance structures

In some situations, both correlation among measurements and heterogeneity of intra-individual variance may be evident. This might arise when measurements are taken over time and variance tends to increase with the level of the response, as in the case of data from pharmacokinetic experiments and growth studies. It is straightforward to relax both assumptions (ii) and (iii) to accommodate this possibility by simultaneous specification of a variance function and a model for the correlation structure.

For variance function g, define the diagonal variance matrix

$$G(\boldsymbol{\beta}, \boldsymbol{\theta}) = \text{diag}[g^2(\mu_1, z_1, \boldsymbol{\theta}), \ldots, g^2(\mu_n, z_n, \boldsymbol{\theta})], \tag{2.11}$$

with $\boldsymbol{G}^{1/2}(\boldsymbol{\beta}, \boldsymbol{\theta})$ the diagonal matrix with elements the square root of those of $\boldsymbol{G}(\boldsymbol{\beta}, \boldsymbol{\theta})$. Here, $\boldsymbol{\beta}$ appears as an explicit argument to emphasize the possible dependence of intra-individual variance on the regression parameter (through the mean responses $\mu_j = f(\boldsymbol{x}_j, \boldsymbol{\beta})$). If the model for the correlation pattern is given by a matrix $\boldsymbol{\Gamma}(\boldsymbol{\alpha})$, then the specification

$$\begin{aligned} \text{Cov}(\boldsymbol{e}) &= \sigma^2 \boldsymbol{G}^{1/2}(\boldsymbol{\beta}, \boldsymbol{\theta}) \boldsymbol{\Gamma}(\boldsymbol{\alpha}) \boldsymbol{G}^{1/2}(\boldsymbol{\beta}, \boldsymbol{\theta}) \\ &= \boldsymbol{R}(\boldsymbol{\beta}, \boldsymbol{\xi}), \quad \boldsymbol{\xi} = [\sigma, \boldsymbol{\theta}', \boldsymbol{\alpha}']', \end{aligned} \tag{2.12}$$

implies that

$$\text{Var}(y_j) = \sigma^2 g^2(\mu_j, z_j, \boldsymbol{\theta}), \quad \text{Corr}(y_{j_1}, y_{j_2}) = \boldsymbol{\Gamma}_{(j_1, j_2)}(\boldsymbol{\alpha}).$$

In (2.12), $\boldsymbol{\xi}$ is the $\{(q + s + 1) \times 1\}$ combined vector of all intra-individual covariance parameters. The general covariance structure (2.12) provides a great deal of versatility in characterizing uncertainty in intra-individual responses, as it allows separate consideration of heterogeneity of variance and the pattern of correlation.

2.3 Inference

Standard inferential methods for the nonlinear regression model are based on the familiar principle of least squares. In this section, we review procedures for both estimation and testing under the classical assumptions and then show how these may be expanded to accommodate more general covariance structures. The principal focus is usually inference on the regression parameter $\boldsymbol{\beta}$. In situations where variance or correlation models are involved, however, inference on the parameters $\boldsymbol{\theta}$ and $\boldsymbol{\alpha}$ may also be of interest. Methods described in this section are illustrated in section 2.5.

2.3.1 Ordinary least squares

The ordinary least squares (OLS) estimator for $\boldsymbol{\beta}$, $\hat{\boldsymbol{\beta}}_{OLS}$, minimizes

$$\sum_{j=1}^{n}\{y_j - f(\boldsymbol{x}_j, \boldsymbol{\beta})\}^2; \tag{2.13}$$

equivalently, $\hat{\boldsymbol{\beta}}_{OLS}$ solves the p estimating equations

$$\sum_{j=1}^{n}\{y_j - f(\boldsymbol{x}_j, \boldsymbol{\beta})\}\boldsymbol{f}_{\beta}(\boldsymbol{x}_j, \boldsymbol{\beta}) = \boldsymbol{0}, \tag{2.14}$$

where $\boldsymbol{f}_{\beta}(\boldsymbol{x}_j, \boldsymbol{\beta})$ is the $(p \times 1)$ vector of derivatives of f with respect to the elements of $\boldsymbol{\beta}$; that is, the vector with kth element $\partial/\partial\beta_k\{f(\boldsymbol{x}_j, \boldsymbol{\beta})\}, k = 1, \ldots, p$.

Under the classical assumptions of independence, constant variance, and normality, the OLS estimator is also the maximum likelihood estimator for $\boldsymbol{\beta}$. Thus, the OLS estimator may be motivated in a number of ways. From the standpoint of classical statistical theory, $\hat{\boldsymbol{\beta}}_{OLS}$ is the estimator obtained by application of the usual likelihood principle to the nonlinear regression model, under full distributional assumptions. As long as the assumptions hold, the estimator enjoys standard optimality properties. A second view of least squares estimation is as an omnibus method, regardless of the distribution of the data. Under the assumptions of independence and the mean-variance specification (2.3), a sensible criterion, the sum of squared deviations between observed responses and their assumed mean, is minimized. All deviations receive equal weight in the minimization in accordance with the constant variance assumption, which implies that all responses are subject to the same degree of uncertainty. A third perspective is that of the so-called 'estimating equation' approach. Rather than consider an objective function to be minimized, focus on the form of an estimating equation to be solved in terms of $\boldsymbol{\beta}$ that is 'optimal' in some sense. The equations (2.14), which are linear in y_j, have the desirable property of unbiasedness; that is, they have expectation zero at the true value of $\boldsymbol{\beta}$. Moreover, (2.14) is in fact the 'best' such equation, in terms of quality of the estimate, if one is willing to make no assumption about the distribution of y_j other than that about the first two moments given in (2.3); see, for example, McCullagh and Nelder (1989, section 9.5) for general discussion. Under the latter two interpretations, then, the only requirement for application of the method is knowledge of the form of the first two moments of the response. In the case of the classical assumptions, all approaches lead to the same estimation method; in more complicated settings, this may not be the case, as we indicate in section 2.3.2.

The maximum likelihood estimator for σ^2 is given by

$$\tilde{\sigma}^2 = \frac{1}{n}\sum_{j=1}^{n}\{y_j - f(\boldsymbol{x}_j, \hat{\boldsymbol{\beta}}_{OLS})\}^2.$$

This estimate is generally biased downward. By analogy to the linear case, $\tilde{\sigma}^2$ is usually replaced by

$$\hat{\sigma}^2_{OLS} = \frac{1}{n-p}\sum_{j=1}^{n}\{y_j - f(\boldsymbol{x}_j, \hat{\boldsymbol{\beta}}_{OLS})\}^2. \qquad (2.15)$$

The estimator $\hat{\sigma}^2_{OLS}$ may be viewed as incorporating a correction for loss of degrees of freedom in estimating $\boldsymbol{\beta}$. We discuss generalizations of this idea, known as restricted maximum likelihood estimation, in section 2.3.3 and Chapter 3.

Estimation for the nonlinear regression model under the classical assumptions is thus, in principle, entirely analogous to that for the linear case. The fundamental difference is that, in general, it is not possible to solve (2.14) explicitly for $\hat{\boldsymbol{\beta}}_{OLS}$. Rather, numerical methods must be employed to obtain the solution. We discuss this issue in section 2.4.

2.3.2 Generalized least squares

When heterogeneity of variance is evident, ordinary least squares estimation of $\boldsymbol{\beta}$ may be inefficient relative to methods that take heteroscedasticity into account. To motivate our discussion of this issue and of how the classical least squares procedures may be modified, consider first the situation where variance is nonconstant across the response range such that the variances of the y_j are *known* up to a constant of proportionality; that is,

$$\mathrm{E}(y_j) = f(\boldsymbol{x}_j, \boldsymbol{\beta}), \qquad \mathrm{Var}(y_j) = \sigma^2/w_j \qquad (2.16)$$

for some known constants $w_j, j = 1, \ldots, n$. A setting in which (2.16) might arise is the case where the responses y_j are themselves averages of w_j uncorrelated replicate measurements, with all such measurements having common variance σ^2. Another situation in which (2.16) is applicable is where variance is a known function of constants z_j, as in the loglinear model (2.9) with $\boldsymbol{\theta}$ *known*. This might be the case, for example, if intra-individual variance has been studied extensively in preliminary studies so that a full functional form may be specified.

Under (2.16), with the assumption of independent, normal errors, it is straightforward to show that the maximum likelihood estimator of $\boldsymbol{\beta}$ is the value $\hat{\boldsymbol{\beta}}_{WLS}$ that minimizes

$$\sum_{j=1}^{n} w_j\{y_j - f(\boldsymbol{x}_j, \boldsymbol{\beta})\}^2; \tag{2.17}$$

equivalently, $\hat{\boldsymbol{\beta}}_{WLS}$ solves the estimating equations

$$\sum_{j=1}^{n} w_j\{y_j - f(\boldsymbol{x}_j, \boldsymbol{\beta})\}\boldsymbol{f}_{\beta}(\boldsymbol{x}_j, \boldsymbol{\beta}) = \boldsymbol{0}. \tag{2.18}$$

This method is referred to as *weighted least squares* (WLS) estimation.

As with ordinary least squares, the maximum likelihood estimator for σ^2 is usually replaced by

$$\hat{\sigma}^2_{WLS} = \frac{1}{n-p}\sum_{j=1}^{n} w_j\{y_j - f(\boldsymbol{x}_j, \hat{\boldsymbol{\beta}}_{WLS})\}^2. \tag{2.19}$$

From (2.17) and (2.18), $\hat{\boldsymbol{\beta}}_{WLS}$ has a natural interpretation: a reasonable criterion, the sum of squared deviations, is minimized, but each deviation is weighted in inverse proportion to the magnitude of uncertainty in the associated response. Intuitively, this should be preferred to OLS when (2.16) holds: if all responses were weighted equally, measurements of lower quality would be given too much importance in determining the fit, and conversely. Moreover, (2.18) may be shown to be optimal among all linear equations if one makes no other assumption than (2.16). Again, from these points of view, the method has global appeal: all that is required for implementation is willingness to specify the first two moments of the response as in (2.16).

In many cases, it is unlikely that such complete knowledge of the variances would be available; in a pharmacokinetic experiment, for example, replicate measurements at a single time on a given subject are not taken. More generally, heterogeneity of variance is likely to be a systematic function of the mean or other factors, as described in section 2.2.3, where some components of the function, such as the mean response itself, are *unknown*. Although not directly applicable in this situation, the WLS method suggests a natural approach to estimation to take account of heterogeneous variance.

For definiteness, consider the constant coefficient of variation model (2.4). In this model, except for the multiplicative constant σ^2, variance is known up to the value of the regression parameter $\boldsymbol{\beta}$, which appears through the mean response. An obvious approach is thus to take advantage of the functional form for variance to construct *estimated* weights, replacing $\boldsymbol{\beta}$ by a suitable estimate, and to apply the weighted least squares idea. The OLS estimator $\hat{\boldsymbol{\beta}}_{OLS}$ is a natural choice to use for construction of estimated weights. Formally, an estimator for $\boldsymbol{\beta}$ that takes into account the assumed mean-variance relationship may be obtained by forming estimated weights

$$\hat{w}_j = 1/f^2(\boldsymbol{x}_j, \hat{\boldsymbol{\beta}}_{OLS})$$

and then solving (2.18), using the $\hat{w}_j$ in place of w_j. Intuition suggests that the resulting estimator for $\boldsymbol{\beta}$ will be preferred to that obtained by ordinary least squares.

This example is a special case of a very general class of methods for estimation of $\boldsymbol{\beta}$ known in the context of nonlinear regression as *generalized least squares* (GLS).

For the general mean-variance specification (2.5) with $\boldsymbol{\theta}$ known, GLS methods can be characterized by the following scheme:

1. Estimate $\boldsymbol{\beta}$ by a preliminary estimator $\hat{\boldsymbol{\beta}}^{(p)}$, e.g. the OLS estimator $\hat{\boldsymbol{\beta}}_{OLS}$.
2. Form estimated weights
$$\hat{w}_j = 1/g^2(\hat{\mu}_j, \boldsymbol{z}_j, \boldsymbol{\theta}); \qquad \hat{\mu}_j = f(\boldsymbol{x}_j, \hat{\boldsymbol{\beta}}^{(p)}).$$
3. Using the weights from step 2, reestimate $\boldsymbol{\beta}$ by weighted least squares. Treating the resulting estimator as a new preliminary estimator, return to step 2.

Denote the final estimate by $\hat{\boldsymbol{\beta}}_{GLS}$.

Step 3 provides for the possibility of iteration. It is natural to expect that using a weighted rather than unweighted estimator to form weights in step 2 should lead to improved final estimation of $\boldsymbol{\beta}$. No general consensus exists on a recommended number of iterations, however. Full iteration of the algorithm may be undertaken, this is commonly referred to as *iteratively reweighted least squares*, and is usually performed in practice using alternative computational methods; see, for example, McCullagh and Nelder (1989, section 2.5). A complete discussion of these issues is given in Chapter 2 of Carroll and Ruppert (1988). Practical experience indicates that the algorithm will often converge after a few iterations, although there is no guarantee in theory that the algorithm should converge. Theory (e.g. Carroll, Wu and Ruppert, 1988) suggests that at least two iterations of the algorithm are desirable to 'wash out' the effect of the inefficient preliminary estimator $\hat{\boldsymbol{\beta}}^{(p)}$.

In the GLS procedure outlined above, weights are formed by estimating the mean at each $\boldsymbol{x}_j$ by the regression function at a preliminary estimator for $\boldsymbol{\beta}$ and then evaluating the variance function g at these estimated means $\hat{\mu}_j$. A common alternative strategy for constructing weights is to replace these estimated means by the actual data values y_j in the evaluation of g. This latter approach can be highly unreliable, as it is based on how well the y_j approximate the true means μ_j, and is to be avoided. The method for constructing weights based on predicted values from a preliminary fit is preferred, as the information on the mean response from all the data is used to better represent the mean, and consequently the variance, at each $\boldsymbol{x}_j$.

Estimation of the scale parameter σ^2 is also conducted by analogy to the WLS method. Specifically, the estimate is

$$\hat{\sigma}^2_{GLS} = \frac{1}{n-p} \sum_{j=1}^{n} \hat{w}_j \{y_j - f(\boldsymbol{x}_j, \hat{\boldsymbol{\beta}}_{GLS})\}^2,$$

$$\hat{w}_j = 1/g^2(\hat{\mu}_j, \boldsymbol{z}_j, \boldsymbol{\theta}), \qquad \hat{\mu}_j = f(\boldsymbol{x}_j, \hat{\boldsymbol{\beta}}_{GLS}), \tag{2.20}$$

so that the weights used are based on the final estimate of $\boldsymbol{\beta}$ obtained from the GLS algorithm.

The GLS method may be used even without knowledge of the distribution of the data, and this gives it broad appeal. All that is required is the ability to specify the first two moments of the response.

2.3.3 Variance function estimation

As noted in section 2.2.3, although a suitable model for intra-individual response variance may be apparent, it may not be possible to specify ***a priori*** the value of the variance parameter $\boldsymbol{\theta}$. In such cases, a reasonable strategy is to estimate $\boldsymbol{\theta}$ from the data. Estimation of variance parameters is commonly referred to as variance function estimation, as knowledge of the parameter value provides a complete specification of the variance model.

The GLS algorithm may be modified to incorporate variance function estimation by replacing step 2 by the following:

2. Estimate $\boldsymbol{\theta}$ by $\hat{\boldsymbol{\theta}}$, and form estimated weights based on the preliminary estimate $\hat{\boldsymbol{\beta}}^{(p)}$:

$$\hat{w}_j = 1/g^2(\hat{\mu}_j, \boldsymbol{z}_j, \hat{\boldsymbol{\theta}}); \qquad \hat{\mu}_j = f(\boldsymbol{x}_j, \hat{\boldsymbol{\beta}}^{(p)}).$$

A number of approaches to estimation of variance parameters have been proposed. We focus on methods based on transformations of absolute residuals from a preliminary fit. A detailed discussion of this class of methods is given in Chapter 3 of Carroll and Ruppert (1988). Other techniques are noted in section 2.5.

Given a preliminary estimator $\hat{\boldsymbol{\beta}}^{(p)}$ for $\boldsymbol{\beta}$, the method of *pseudolikelihood* (PL) minimizes in σ and $\boldsymbol{\theta}$

$$PL(\hat{\boldsymbol{\beta}}^{(p)}, \sigma, \boldsymbol{\theta}) = \sum_{j=1}^{n} \left(\frac{\{y_j - f(\boldsymbol{x}_j, \hat{\boldsymbol{\beta}}^{(p)})\}^2}{\sigma^2 g^2\{f(\boldsymbol{x}_j, \hat{\boldsymbol{\beta}}^{(p)}), \boldsymbol{z}_j, \boldsymbol{\theta}\}} + \log[\sigma^2 g^2\{f(\boldsymbol{x}_j, \hat{\boldsymbol{\beta}}^{(p)}), \boldsymbol{z}_j, \boldsymbol{\theta}\}] \right). \tag{2.21}$$

Minimizing (2.21) corresponds to maximizing the normal loglikelihood evaluated at $\hat{\boldsymbol{\beta}}^{(p)}$, which lends the procedure its name; however, despite the dependence on the normal distribution, PL may be regarded as an omnibus method, as we now illustrate. Following Carroll and Ruppert (1988, section 2.2) and Davidian and Haaland (1990), note that the objective function (2.21) depends on the data through the residual $r_j = \{y_j - f(\boldsymbol{x}_j, \hat{\boldsymbol{\beta}}^{(p)})\}$ from the preliminary fit. Differentiation of (2.21) shows that the resulting PL estimating equations for σ and $\boldsymbol{\theta}$ have form similar to the weighted least squares estimating equations (2.17) for $\boldsymbol{\beta}$; specifically, writing $\hat{\mu}_j^{(p)} = f(\boldsymbol{x}_j, \hat{\boldsymbol{\beta}}^{(p)})$, the estimating equations are given by

$$\sum_{j=1}^{n} g^{-4}(\hat{\mu}_j^{(p)}, \boldsymbol{z}_j, \boldsymbol{\theta})\{r_j^2 - \sigma^2 g^2(\hat{\mu}_j^{(p)}, \boldsymbol{z}_j, \boldsymbol{\theta})\}\boldsymbol{\tau}_\theta(\hat{\mu}_j^{(p)}, \sigma, \boldsymbol{\theta}) = \mathbf{0}, \tag{2.22}$$

where $\boldsymbol{\nu}_\theta(\mu_j, \boldsymbol{z}_j, \boldsymbol{\theta})$ is the q-vector of derivatives of $\log g(\mu_j, \boldsymbol{z}_j, \boldsymbol{\theta})$ with respect to $\boldsymbol{\theta}$ and

$$\boldsymbol{\tau}_\theta(\mu_j, \sigma, \boldsymbol{\theta}) = g^2(\hat{\mu}_j^{(p)}, \boldsymbol{z}_j, \boldsymbol{\theta})[1/\sigma, \boldsymbol{\nu}_\theta'(\mu_j, \boldsymbol{z}_j, \boldsymbol{\theta})]'.$$

Inspection of (2.22) shows that PL estimation of σ and $\boldsymbol{\theta}$ may be interpreted as a 'weighted least squares regression' of 'responses' r_j^2 on the 'regression function' $\sigma^2 g^2(\hat{\mu}_j^{(p)}, \boldsymbol{z}_j, \boldsymbol{\theta})$ with $(q + 1 \times 1)$

'regression parameter' $[\sigma, \boldsymbol{\theta}']'$, 'weights' $g^{-4}(\hat{\mu}_j^{(p)}, \boldsymbol{z}_j, \boldsymbol{\theta})$, and 'gradient' $\boldsymbol{\tau}_\theta(\hat{\mu}_j^{(p)}, \sigma, \boldsymbol{\theta})$. Thus, PL estimation of variance parameters enjoys the same broad appeal as GLS methods for regression parameters: it may be used without knowledge of the distribution of the data, requiring only the ability to specify moments. This analogy applies to all of the variance function estimation techniques we discuss; these procedures exploit the smooth nature of the mean-variance relationship in the same way that regression methods for estimation of $\boldsymbol{\beta}$ take advantage of the smoothness of the mean response function.

Appreciation of the form of the PL estimating equations (2.22) is also important in understanding the connection between generalized least squares for nonlinear regression and generalized estimating equation methods. We discuss this connection in section 2.6.2.

One criticism of PL estimation is that it takes no account of the loss of degrees of freedom due to preliminary estimation of $\boldsymbol{\beta}$ An alternative is to replace (2.21) by a suitably modified objective function, as may be derived using the ideas of restricted maximum likelihood estimation (REML). A detailed discussion of restricted maximum likelihood is given in section 3.3.2. In the context of the mean-variance model (2.5), the REML objective function is

$$\begin{aligned} REML(\hat{\boldsymbol{\beta}}^{(p)}, \sigma, \boldsymbol{\theta}) &= PL(\hat{\boldsymbol{\beta}}^{(p)}, \sigma, \boldsymbol{\theta}) \\ &\quad -p \log \sigma^2 + \log |\boldsymbol{X}'(\hat{\boldsymbol{\beta}}^{(p)}) \boldsymbol{G}^{-1}(\hat{\boldsymbol{\beta}}^{(p)}, \boldsymbol{\theta}) \boldsymbol{X}(\hat{\boldsymbol{\beta}}^{(p)})|, \end{aligned} \tag{2.23}$$

where $\boldsymbol{G}(\boldsymbol{\beta}, \boldsymbol{\theta})$ is the diagonal matrix defined in (2.11), and $\boldsymbol{X}(\boldsymbol{\beta})$ is the the $(n \times p)$ matrix with jth row equal to $\boldsymbol{f}'_\beta(\boldsymbol{x}_j, \boldsymbol{\beta})$. The REML method in the context of variance function estimation is discussed in detail in section 3.2 of Carroll and Ruppert (1988); see also Beal and Sheiner (1988).

Both PL (2.21) and REML (2.23) base estimation of the variance parameters on squared residuals, which follows from dependence on the normal theory maximum likelihood objective function. These methods may thus be expected to give good results for normal and nearly normal e_j, but may be adversely affected by outlying observations. To allow for the possibility of a heavier-tailed distribution of the e_j, an attractive proposal is to replace the role of squared residuals by absolute residuals. Analogous to (2.21), the absolute residual (AR) method estimates $\boldsymbol{\theta}$ by minimizing in η and $\boldsymbol{\theta}$

$$AR(\hat{\beta}^{(p)}, \eta, \theta) =$$
$$\sum_{j=1}^{n} \left(\frac{|y_j - f(x_j, \hat{\beta}^{(p)})|}{\eta g\{f(x_j, \hat{\beta}^{(p)}), \theta\}} + \log[\eta g\{f(x_j, \hat{\beta}^{(p)}), z_j, \theta\}] \right). \tag{2.24}$$

Here, the role of σ^2 is replaced by $\eta = \mathrm{E}|e_j|/g\{f(x_j, \beta), z_j, \theta\}$, assuming that $\mathrm{E}|e_j|/g\{f(x_j, \beta) z_j, \theta\}$ is the same for all j. The objective function (2.24) may also be motivated by considering a suitable double exponential distribution. As with PL, AR estimation may be viewed as a 'weighted least squares regression' (Carroll and Ruppert, 1988, section 3.3.3; Davidian and Haaland, 1990). Use of (2.24) can be quite competitive relative to (2.21) or (2.23) for estimation of θ for normal data and significantly better if only a few outliers are present (Davidian and Carroll, 1987).

Whatever method one chooses for estimation of θ in step 2, it is essential to iterate the GLS scheme at least once so that the residuals used are obtained from a weighted fit. The final estimate of σ^2 is obtained by evaluation of (2.20) at the final estimated values $\hat{\beta}_{GLS}$ and $\hat{\theta}$.

Other approaches to variance function estimation that may be used with the GLS procedure have been proposed. Section 2.7 contains a brief review of some alternative methods.

Whether θ is known or must be estimated, other approaches are possible, such as joint maximum likelihood (ML) estimation of all unknown parameters under the assumption of normality. In contrast to the cases of constant variance (2.3) or known weights (2.16), in the model (2.5) it is not true in general that the GLS and maximum likelihood estimators for β coincide. We prefer GLS to ML estimation as an omnibus method, for reasons discussed in section 2.7.

2.3.4 General covariance structures

For general mean-covariance models with assumed covariance structure

$$\begin{aligned} \mathrm{Cov}(y) &= \sigma^2 G^{1/2}(\beta, \theta) \Gamma(\alpha) G^{1/2}(\beta, \theta) \\ &= R(\beta, \xi), \quad \xi = [\sigma, \theta', \alpha']', \end{aligned} \tag{2.25}$$

it is possible to extend the generalized least squares principle to accommodate simultaneous estimation of $\boldsymbol{\beta}$ and $\boldsymbol{\xi}$. For convenience, write

$$\boldsymbol{R}(\boldsymbol{\beta}, \boldsymbol{\xi}) = \sigma^2 \boldsymbol{S}(\boldsymbol{\beta}, \boldsymbol{\gamma}), \qquad \boldsymbol{\gamma} = [\boldsymbol{\theta}', \boldsymbol{\alpha}']',$$

so that γ is the v-dimensional vector of parameters in the functional part of the covariance model, $v = q + s$. The development is analogous to that for weighted least squares in the simpler model (2.5), so we use the same subscript conventions to identify the estimators. If $\text{Cov}(\boldsymbol{y}) = \sigma^2 \boldsymbol{W}^{-1}$ for some known matrix $\boldsymbol{W}$, then, under normality, the maximum likelihood estimator for $\boldsymbol{\beta}$ is the value $\hat{\boldsymbol{\beta}}_{WLS}$ minimizing

$$\{\boldsymbol{y} - \boldsymbol{f}(\boldsymbol{\beta})\}' \boldsymbol{W} \{\boldsymbol{y} - \boldsymbol{f}(\boldsymbol{\beta})\}; \tag{2.26}$$

that is, the estimator for $\boldsymbol{\beta}$ solving the p-dimensional set of estimating equations

$$\boldsymbol{X}'(\boldsymbol{\beta}) \boldsymbol{W} \{\boldsymbol{y} - \boldsymbol{f}(\boldsymbol{\beta})\} = \boldsymbol{0}. \tag{2.27}$$

The scale parameter σ^2 may be estimated by

$$\hat{\sigma}^2_{WLS} = \frac{1}{n-p} \{\boldsymbol{y} - \boldsymbol{f}(\hat{\boldsymbol{\beta}}_{WLS})\}' \boldsymbol{W} \{\boldsymbol{y} - \boldsymbol{f}(\hat{\boldsymbol{\beta}}_{WLS})\}, \tag{2.28}$$

where $\boldsymbol{f}(\boldsymbol{\beta}) = [f(\boldsymbol{x}_1, \boldsymbol{\beta}), \ldots, f(\boldsymbol{x}_n, \boldsymbol{\beta})]'$. This suggests the following generalization of the GLS algorithm:

1. Estimate $\boldsymbol{\beta}$ by a preliminary estimator $\hat{\boldsymbol{\beta}}^{(p)}$, e.g. the OLS estimator $\hat{\boldsymbol{\beta}}_{OLS}$.
2. Obtain an estimator $\hat{\gamma}$ and form the estimated weight matrix based on $\hat{\boldsymbol{\beta}}^{(p)}$ and $\hat{\gamma}$

$$\hat{\boldsymbol{W}} = \boldsymbol{S}^{-1}(\hat{\boldsymbol{\beta}}^{(p)}, \hat{\gamma}).$$

3. Using the weight matrix from step 2, reestimate $\boldsymbol{\beta}$ by minimizing (2.26). Treating the resulting estimator as a new preliminary estimator, return to step 2.

As in the simpler case, we denote the final estimator for $\boldsymbol{\beta}$ as $\hat{\boldsymbol{\beta}}_{GLS}$. This scheme may be iterated a fixed number of times or to convergence, with at least one iteration recommended.

Estimation of γ can be accomplished by extension of variance function methods described in section 2.3.2 to the general covariance structure (2.25). This is straightforward for PL and REML, which are based on the normal distribution. Specifically, the pseudolikelihood method minimizes in σ and γ

$$PL(\hat{\boldsymbol{\beta}}^{(p)}, \sigma, \gamma) = \log |\sigma^2 \boldsymbol{S}(\hat{\boldsymbol{\beta}}^{(p)}, \gamma)| \\ + \{\boldsymbol{y} - \boldsymbol{f}(\hat{\boldsymbol{\beta}}^{(p)})\}' \boldsymbol{S}^{-1}(\hat{\boldsymbol{\beta}}^{(p)}, \gamma)\{\boldsymbol{y} - \boldsymbol{f}(\hat{\boldsymbol{\beta}}^{(p)})\}/\sigma^2. \quad (2.29)$$

Differentiation of (2.29) shows that PL estimation of the covariance parameters may be viewed as a 'weighted least squares' approach, analogous to the form of (2.27). Alternatively, to account for preliminary estimation of $\boldsymbol{\beta}$, a restricted maximum likelihood approach would use

$$REML(\hat{\boldsymbol{\beta}}^{(p)}, \sigma, \gamma) = PL(\hat{\boldsymbol{\beta}}^{(p)}, \sigma, \gamma) \\ -p \log \sigma^2 + \log |\boldsymbol{X}'(\hat{\boldsymbol{\beta}}^{(p)})\boldsymbol{S}^{-1}(\hat{\boldsymbol{\beta}}^{(p)}, \gamma)\boldsymbol{X}(\hat{\boldsymbol{\beta}}^{(p)})|. \quad (2.30)$$

In either case, the final estimate of σ^2 is calculated based on the final estimates for $\boldsymbol{\beta}$ and γ according to

$$\hat{\sigma}^2_{GLS} = \frac{1}{n-p}\{\boldsymbol{y} - \boldsymbol{f}(\hat{\boldsymbol{\beta}}_{GLS})\}' \boldsymbol{S}^{-1}(\hat{\boldsymbol{\beta}}_{GLS}, \hat{\gamma})\{\boldsymbol{y} - \boldsymbol{f}(\hat{\boldsymbol{\beta}}_{GLS})\}. \quad (2.31)$$

Full maximum likelihood estimation assuming normality is possible, but suffers from the same drawbacks as in the simpler case.

It is important to recognize that estimation of covariance parameters is considerably more difficult than estimation of regression parameters. Information on higher moment properties is typically not as good as that on the mean response relationship. Consequently, although it may be possible to obtain reliable estimates of $\boldsymbol{\beta}$ with a limited sample size, it is unrealistic to expect similar precision from estimates of variance and correlation parameters with sparse information. In the context of repeated measurement data, it may be unwise to attempt estimation of intra-individual covariance parameters at the individual level. Rather, a more sensible approach may be to combine information across individuals. We discuss this idea in Chapter 5.

2.3.5 Confidence intervals and hypothesis testing

In the case of *linear* least squares, under the full set of classical assumptions, exact distributional results for the OLS estimator for $\boldsymbol{\beta}$ and the unbiased estimator (2.15) for the common variance σ^2 may be derived; see, for example, Draper and Smith (1981). This theory may be used to construct exact confidence intervals and hypothesis tests regarding $\boldsymbol{\beta}$ and σ^2, as long as the assumption of normality holds. Even if the response is not normally distributed, the estimators remain unbiased, and the expression for the covariance matrix of $\hat{\boldsymbol{\beta}}_{OLS}$ remains unchanged, so that standard error estimates may still be obtained.

In the case of a *nonlinear* model, it is no longer possible to obtain exact, fixed sample size results in general, even under the normal distribution with constant variance. This stems from the inability to obtain an explicit expression for the estimator for $\boldsymbol{\beta}$ solving the estimating equations.

It is possible, however, to develop approximate procedures using large sample theory, where, by 'large sample' and 'asymptotic,' we mean that the sample size n grows large while the model parameters remain of fixed dimension. Such asymptotic theory requires certain assumptions about the behavior of the regression function f, the variance function g, and their derivatives, as well as additional technical conditions. A discussion of the theoretical arguments is beyond the scope of this book; see Jennrich (1969), Gallant, (1987, Chapters 3, 4, and 6), and Seber and Wild (1989, Chapters 2 and 12) for rigorous treatment. The monograph of Carroll and Ruppert (1988, Chapters 2, 3, and 7) provides practical discussion of the results and their implications.

In this section, we state the standard large sample theory distributional results for least squares and generalized least squares estimators and show how these may be used to construct approximate confidence intervals and hypothesis tests. These asymptotic approximations hold even when the assumption of normality (iv) does not.

Asymptotic distribution theory

In the case where the intra-individual variance is constant, under assumptions (i)–(iii), with the additional condition that the errors e_j are independently distributed, the OLS estimator $\hat{\boldsymbol{\beta}}_{OLS}$ is approximately (in large samples) normally distributed with mean $\boldsymbol{\beta}$ and covariance matrix $\sigma^2 \boldsymbol{\Sigma}_{OLS}$, where

$$\boldsymbol{\Sigma}_{OLS}^{-1} = \boldsymbol{X}'(\boldsymbol{\beta})\boldsymbol{X}(\boldsymbol{\beta}),$$

and, again, $\boldsymbol{X}(\boldsymbol{\beta})$ is the the $(n \times p)$ matrix with jth row equal to $\boldsymbol{f}'_{\beta}(\boldsymbol{x}_j, \boldsymbol{\beta})$. That is, asymptotically,

$$\hat{\boldsymbol{\beta}}_{OLS} \sim \mathcal{N}(\boldsymbol{\beta}, \sigma^2 \boldsymbol{\Sigma}_{OLS}). \tag{2.32}$$

Note the analogy with linear regression: if $f(\boldsymbol{x}_j, \boldsymbol{\beta}) = \boldsymbol{x}_j{}'\boldsymbol{\beta}$ for a $(p \times 1)$ vector $\boldsymbol{x}_j$, then the matrix $\boldsymbol{X}(\boldsymbol{\beta})$ is the usual design matrix. In the nonlinear case, $\boldsymbol{\Sigma}_{OLS}$ depends on the regression parameter $\boldsymbol{\beta}$, so in practice, it is estimated by the matrix $\hat{\boldsymbol{\Sigma}}_{OLS}$ obtained by replacing $\boldsymbol{\beta}$ by $\hat{\boldsymbol{\beta}}_{OLS}$. The estimate of the approximate covariance matrix of $\hat{\boldsymbol{\beta}}_{OLS}$ is then given by $\hat{\sigma}^2_{OLS}\hat{\boldsymbol{\Sigma}}^{-1}_{OLS}$. Estimated approximate standard errors for the elements of $\hat{\boldsymbol{\beta}}_{OLS}$ may be obtained as the square root of the diagonal elements of this matrix.

When intra-individual variance is not constant, but has been modeled correctly, a similar result holds. Assume that the errors e_j are independent, and the responses y_j follow the mean-variance specification (2.5). Let $\boldsymbol{G}(\boldsymbol{\beta}, \boldsymbol{\theta})$ be the diagonal matrix defined in section 2.2.3. Under the conditions given above, whether $\boldsymbol{\theta}$ is known (section 2.3.2) or has been estimated by one of the methods PL, REML, or AR (section 2.3.3), the GLS estimator $\hat{\boldsymbol{\beta}}_{GLS}$ has asymptotic normal distribution

$$\hat{\boldsymbol{\beta}}_{GLS} \sim \mathcal{N}(\boldsymbol{\beta}, \sigma^2 \boldsymbol{\Sigma}_{GLS}), \tag{2.33}$$

where

$$\boldsymbol{\Sigma}^{-1}_{GLS} = \boldsymbol{X}'(\boldsymbol{\beta})\boldsymbol{G}^{-1}(\boldsymbol{\beta}, \boldsymbol{\theta})\boldsymbol{X}(\boldsymbol{\beta}).$$

The estimate of the approximate covariance matrix of $\hat{\boldsymbol{\beta}}_{GLS}$ is given by $\hat{\sigma}^2_{GLS}\hat{\boldsymbol{\Sigma}}_{GLS}$, where $\hat{\boldsymbol{\Sigma}}_{GLS}$ is the matrix obtained by replacing $\boldsymbol{\beta}$ by the GLS estimate $\hat{\boldsymbol{\beta}}_{GLS}$ and, if necessary, $\boldsymbol{\theta}$ by its estimate, in the components of $\boldsymbol{\Sigma}_{GLS}$.

This result has several interesting features that deserve comment:

(i) The asymptotic distribution given in (2.33) for $\hat{\boldsymbol{\beta}}_{GLS}$ is identical to that for the WLS estimator $\hat{\boldsymbol{\beta}}_{WLS}$ that could be calculated if the weights were known rather than estimated. The implication is that, in principle, the need to estimate weights should not degrade performance of the GLS estimator relative to that which could be attained if the true weights were known exactly.

(ii) The large sample normal distribution of the GLS estimator is the same regardless of the number of iterations of the GLS algorithm used to obtain $\hat{\boldsymbol{\beta}}_{GLS}$. This explains the lack of consensus regarding iteration; the usual large sample theory does not provide insight into an 'optimal' number of iterations. See Chapter 2 of Carroll and Ruppert (1988); we discuss this issue further below.

(iii) Equation (2.33) may be used to gain insight into the improvement that may be expected by using the GLS estimator rather than OLS under heterogeneity of variance. If the mean-variance specification (2.5) holds, but the OLS estimator is used, effectively ignoring heteroscedasticity, it may be shown that $\hat{\boldsymbol{\beta}}_{OLS}$ still has an asymptotic normal distribution with mean $\boldsymbol{\beta}$ but covariance matrix

$$\boldsymbol{\Sigma}^*_{OLS} = \boldsymbol{\Sigma}_{OLS}\boldsymbol{\Lambda}\boldsymbol{\Sigma}_{OLS}, \quad \boldsymbol{\Lambda} = \boldsymbol{X}'(\boldsymbol{\beta})\boldsymbol{G}(\boldsymbol{\beta},\boldsymbol{\theta})\boldsymbol{X}(\boldsymbol{\beta}).$$

The comparison of the covariance matrices $\boldsymbol{\Sigma}^*_{OLS}$ and $\boldsymbol{\Sigma}_{GLS}$ is the same as that underlying Gauss–Markov theory: $\boldsymbol{\Sigma}^*_{OLS}$ is 'larger' than $\boldsymbol{\Sigma}_{GLS}$ in the sense of nonnegative definiteness. Consequently, when variance heterogeneity can be correctly characterized, use of ordinary least squares methods will entail a loss in efficiency of estimation relative to GLS. Moreover, confidence intervals and hypothesis tests based on OLS will be inaccurate.

Methods exist for construction of 'corrected'covariance estimates that are robust to misspecifications of variance; see Liang and Zeger (1986), Gallant (1987, section 2.1), and Moore and Tsiatis (1991). These so-called 'sandwich' covariance estimators are used in place of $\hat{\boldsymbol{\Sigma}}_{OLS}$ and $\hat{\boldsymbol{\Sigma}}_{GLS}$ to provide protection against disregard or improper modeling of heteroscedasticity. These estimators should be used with caution, however, as they may be highly sensitive to outlying observations.

In applications, it is often preferable to use GLS with a variance model that may not be exactly correct but captures most of the qualitative features of variation rather than to ignore variance heterogeneity altogether by using OLS methods. This is especially true in applications where assessment of precision is a primary concern. This issue is considered further in Chapter 10.

(iv) The large sample distribution of $\hat{\boldsymbol{\beta}}_{GLS}$ when $\boldsymbol{\theta}$ has been estimated is identical to that when $\boldsymbol{\theta}$ is known. This has led to some confusion regarding the importance of estimation of $\boldsymbol{\theta}$. In principle, estimation of $\boldsymbol{\theta}$ should have no effect on the properties of $\hat{\boldsymbol{\beta}}_{GLS}$. In practice, however, this result can be optimistic, resulting in estimated standard errors that are too small. Similar caution must be exercised in interpreting confidence intervals and the results of hypothesis tests. Carroll, Wu and Ruppert (1988) develop a higher order theory to show how estimation of $\boldsymbol{\theta}$ can have an effect on estimation of the regression parameter. An implication of their work is that it is generally desirable to obtain the most reliable estimate of $\boldsymbol{\theta}$ available. If several outlying observations are evident, for example, use of the AR method rather than those based on the normal distribution is prudent.

Correct estimation of $\boldsymbol{\theta}$ may be of interest in its own right. For instance, precise estimation of $\boldsymbol{\theta}$ is essential for accurate calibration inference based on the fit (Belanger, Davidian and Giltinan, 1995); we pay particular attention to this issue in Chapter 10.

Theoretical results for large sample properties of variance function estimation techniques have been derived (Davidian and Carroll, 1987; Carroll and Ruppert, 1988, Chapter 3) and may be used to construct formal inferential procedures for $\boldsymbol{\theta}$. These methods are highly sensitive to assumptions about the distribution of the data and should be used with caution. Coverage of these results is beyond the scope of this review.

For models with general covariance structures (section 2.3.4), the large sample distributional result is entirely similar to (2.33), and the same issues arise. Whether $\boldsymbol{\gamma}$ has been estimated or is known, as long as the specified mean-covariance relationship holds, the estimator $\hat{\boldsymbol{\beta}}_{GLS}$ has asymptotic normal distribution

$$\hat{\boldsymbol{\beta}}_{GLS} \sim \mathcal{N}(\boldsymbol{\beta}, \sigma^2 \boldsymbol{\Sigma}_{GLS}). \tag{2.34}$$

Here,

$$\boldsymbol{\Sigma}_{GLS}^{-1} = \boldsymbol{X}'(\boldsymbol{\beta})\boldsymbol{S}^{-1}(\boldsymbol{\beta}, \boldsymbol{\gamma})\boldsymbol{X}(\boldsymbol{\beta}).$$

In applications, the covariance matrix in (2.34) is estimated as before, by replacing σ^2, $\boldsymbol{\beta}$, and $\boldsymbol{\gamma}$ by the final GLS estimates.

Inferential procedures

The large sample results given above may be used to construct approximate confidence intervals and hypothesis testing procedures regarding functions of the regression parameter in a number of ways. Here, we describe only a few methods. Chapter 5 of Seber and Wild (1989) contains a comprehensive discussion in the specific context of ordinary least squares, which may be extended easily to apply to more general covariance models.

In the following, let $(\hat{\boldsymbol{\beta}}, \hat{\sigma}^2)$ represent any of the OLS or GLS estimators for $(\boldsymbol{\beta}, \sigma^2)$, and let $\sigma^2\boldsymbol{\Sigma}$ represent the corresponding large sample covariance matrix given in (2.32), (2.33), or (2.34). For a linear combination $\boldsymbol{a}'\boldsymbol{\beta}$ of the elements of $\boldsymbol{\beta}$, the asymptotic distributional results imply immediately that

$$\boldsymbol{a}'\hat{\boldsymbol{\beta}} \sim \mathcal{N}(\boldsymbol{a}'\boldsymbol{\beta}, \sigma^2\boldsymbol{a}'\boldsymbol{\Sigma}\boldsymbol{a}).$$

In the nonlinear model setting, it is not uncommon for interest to focus on *nonlinear* functions of the elements of the regression parameter. For instance, in pharmacokinetic analysis based on exponential mean response models such as (2.2), meaningful quantities such as the half-life of a drug or the area under the curve are calculable as nonlinear functions of model parameters. We consider an example in section 2.5. It may be shown using standard theory and the results given in (2.32)–(2.34) that, for a one-dimensional nonlinear function $a(\boldsymbol{\beta})$ of $\boldsymbol{\beta}$,

$$a(\hat{\boldsymbol{\beta}}) \sim \mathcal{N}\{a(\boldsymbol{\beta}), \sigma^2\boldsymbol{a}'\boldsymbol{\Sigma}\boldsymbol{a}\},$$

where $\boldsymbol{a}$ is now the $(p \times 1)$ vector of partial derivatives of a with respect to the elements of $\boldsymbol{\beta}$, holds approximately for large n. Note that this expression subsumes that above in the case that $a(\boldsymbol{\beta}) = \boldsymbol{a}'\boldsymbol{\beta}$. Thus, for either linear or nonlinear functions of $\boldsymbol{\beta}$, an approximate $100(1-\alpha)\%$ confidence interval for $a(\boldsymbol{\beta})$ is

$$a(\hat{\boldsymbol{\beta}}) \pm t_{\alpha/2,n-p}\hat{\sigma}(\boldsymbol{a}'\hat{\boldsymbol{\Sigma}}\boldsymbol{a})^{1/2},$$

where $t_{\alpha/2,n-p}$ is the $100(1-\alpha/2)$ percentage point of the t distribution with $n-p$ degrees of freedom. In practice, $\boldsymbol{\Sigma}$ would be replaced by an appropriate estimate $\hat{\boldsymbol{\Sigma}}$. A test of

$$H_0 : a(\boldsymbol{\beta}) = 0 \text{ vs. } H_1 : a(\boldsymbol{\beta}) \neq 0$$

may be based on the statistic

$$T = \frac{a(\hat{\boldsymbol{\beta}})}{\hat{\sigma}(\boldsymbol{a}'\hat{\boldsymbol{\Sigma}}\boldsymbol{a})^{1/2}}.$$

Note that linear hypotheses of the form $H_0 : \boldsymbol{a}'\boldsymbol{\beta} = \tau_0$ vs. $H_1 : \boldsymbol{a}'\boldsymbol{\beta} \neq \tau_0$ may be put in this form by letting $a(\boldsymbol{\beta})$ be the linear function of $\boldsymbol{\beta}$ minus τ_0. The test is conducted by comparing $|T|$ to the appropriate t_{n-p} critical value; one-sided tests are conducted in the obvious way. For both confidence intervals and hypothesis tests, the normal distribution is often used (when n is large) in place of the t.

The vector $\boldsymbol{a}$ of partial derivatives will in general be a function of the regression parameter $\boldsymbol{\beta}$. Thus, in practical application, $\boldsymbol{a}$ is replaced by the vector $\hat{\boldsymbol{a}}$ obtained by substituting the estimate $\hat{\boldsymbol{\beta}}$ for $\boldsymbol{\beta}$, and, of course, $\boldsymbol{\Sigma}$ is replaced by its estimate.

The one-dimensional results above are special cases of the general framework of Wald inference for d-dimensional functions of the regression parameter. We give results for hypothesis testing; see Seber and Wild (1989, section 5.2) for a discussion of confidence regions for the regression parameter. For a d-dimensional linear hypothesis

$$H_0 : \boldsymbol{A\beta} = \boldsymbol{\tau}_o \;\; \text{vs.} \;\; H_1 : \boldsymbol{A\beta} \neq \boldsymbol{\tau}_o,$$

where $\boldsymbol{A}$ is a $(d \times p)$ matrix of linear constraints and $\boldsymbol{\tau}_o$ is a $(d \times 1)$ vector of constants, the test statistic

$$T^2 = (\boldsymbol{A}\hat{\boldsymbol{\beta}} - \boldsymbol{\tau}_o)'(\hat{\sigma}^2 \boldsymbol{A}\hat{\boldsymbol{\Sigma}}\boldsymbol{A}')^{-1}(\boldsymbol{A}\hat{\boldsymbol{\beta}} - \boldsymbol{\tau}_o)$$

has an approximate (asymptotic) chi-square distribution with d degrees of freedom. A test may be conducted by comparing the value of T^2 to the appropriate χ^2_d critical value. The statistic may be modified to improve accuracy; see Gallant (1975).

When the functions of interest are nonlinear, an analogous statistic may be constructed. Let $\boldsymbol{a}(\boldsymbol{\beta})$ be a d-dimensional vector-valued function of $\boldsymbol{\beta}$; that is, $\boldsymbol{a}(\boldsymbol{\beta}) = [a_1(\boldsymbol{\beta}), \ldots, a_d(\boldsymbol{\beta})]'$ for d nonlinear constraints $a_\ell, \ell = 1, \ldots, d$. Let $\boldsymbol{A}$ be the $(d \times p)$ matrix whose ℓth row is the $(1 \times p)$ vector of derivatives of $a_\ell(\boldsymbol{\beta})$ with respect to the elements of $\boldsymbol{\beta}$, $\ell = 1, \ldots, d$; thus, $\boldsymbol{A}$ is a function of $\boldsymbol{\beta}$. For the d-dimensional nonlinear hypothesis

$$H_0 : \boldsymbol{a}(\boldsymbol{\beta}) = \boldsymbol{0} \;\text{vs.}\; H_1 : \boldsymbol{a}(\boldsymbol{\beta}) \neq \boldsymbol{0},$$

the test statistic is

$$T^2 = \boldsymbol{a}'(\hat{\boldsymbol{\beta}})(\hat{\sigma}^2 \hat{\boldsymbol{A}}\hat{\boldsymbol{\Sigma}}\hat{\boldsymbol{A}}')^{-1}\boldsymbol{a}(\hat{\boldsymbol{\beta}}),$$

where $\hat{\boldsymbol{A}}$ is the matrix $\boldsymbol{A}$ with $\boldsymbol{\beta}$ replaced by its estimator. Again, the test is conducted by comparing the value of T^2 to the appropriate χ^2_d critical value.

An advantage of Wald inference is ease of implementation. However, as pointed out by Bates and Watts (1988) and Carroll and Ruppert (1988), Wald inference may be inaccurate relative to methods based on approximate likelihood ideas. For the case of uncorrelated data, Carroll and Ruppert (1988, section 2.5) suggest use of the likelihood ratio test one would ordinarily construct under the assumptions of normality and the model (2.16), replacing the weights w_j by final estimates from the GLS fit; this idea may be extended to more complicated error structures. Confidence regions for components of $\boldsymbol{\beta}$ would then be constructed as the set of all values of the components for which the null hypothesis is not rejected. While Wald inference may be based on a single fit, this form of likelihood inference requires an additional fit under the null hypothesis.

It is worth remarking that accuracy of asymptotic theory approximations is not guaranteed; this holds true whether one considers Wald or approximate likelihood inference. Whether or not the theory is valid depends on the specific problem. Estimated standard errors may be optimistic, and tests and confidence intervals may be unreliable. This limitation should be taken into account by the analyst in the interpretation of results. Some investigation of this issue has been undertaken; see, for example, Donaldson and Schnabel (1987) for an empirical study and discussion of this issue in the case of ordinary least squares. Bates and Watts (1988, Chapter 6) discuss graphical procedures for inference following a nonlinear model fit.

2.4 Computational aspects

2.4.1 Least squares estimation

Because closed form solutions of the ordinary or generalized least squares estimating equations are rarely available, computation of nonlinear least squares estimates requires the use of iterative numerical algorithms. Detailed discussion of the various numerical techniques used to compute nonlinear least squares estimates is beyond the scope of this book. It is worthwhile, however, to note briefly the simplest computational methods.

We recall the basic idea of the Newton–Raphson technique. Suppose we are interested in maximizing a scalar-valued objective function $O(\boldsymbol{\tau})$ of its vector-valued argument $\boldsymbol{\tau}$ $(t \times 1)$. Given an approximation $\boldsymbol{\tau}^*$ to the value minimizing $O(\boldsymbol{\tau})$, an approximation to $O(\boldsymbol{\tau})$ may be derived by a quadratic Taylor expansion:

$$O(\boldsymbol{\tau}) \approx O(\boldsymbol{\tau}^*) + \boldsymbol{s}'(\boldsymbol{\tau}^*)(\boldsymbol{\tau}^* - \boldsymbol{\tau}) + \frac{1}{2}(\boldsymbol{\tau}^* - \boldsymbol{\tau})'\boldsymbol{J}(\boldsymbol{\tau})(\boldsymbol{\tau}^* - \boldsymbol{\tau}), \tag{2.35}$$

where $\boldsymbol{s}(\boldsymbol{\tau})$ is the $(t \times 1)$ vector of partial derivatives of O with respect to the components of $\boldsymbol{\tau}$ (the gradient vector) and $\boldsymbol{J}(\boldsymbol{\tau})$ is the $(t \times t)$ matrix whose (k_1, k_2) entry is the second partial derivative $\partial^2/\partial\tau_{k_1}\partial\tau_{k_2}O(\boldsymbol{\tau})$ (the Hessian matrix). Maximization of (2.35) yields

$$\boldsymbol{\tau} = \boldsymbol{\tau}^* - \boldsymbol{J}^{-1}(\boldsymbol{\tau}^*)\boldsymbol{s}(\boldsymbol{\tau}^*).$$

This provides a natural basis for an iterative method of computing the value that maximizes O. To move from one iteration to the next, given the hth iterate $\tau^{(h)}$, the $(h+1)$th approximation is

$$\tau^{(h+1)} = \tau^{(h)} - J^{-1}(\tau^{(h)})s(\tau^{(h)}). \tag{2.36}$$

A standard modification of this algorithm suggested by Marquardt (1963) to handle the case of a non-positive definite Hessian matrix is to inflate the diagonal elements of J by a suitably chosen adjustment. A number of other variants on the Newton–Raphson scheme have been proposed; see, for example, Seber and Wild (1989, Chapter 14).

Computation of the Hessian matrix may be quite burdensome. It may be circumvented by using the method of Fisher scoring, wherein $J(\tau)$ is replaced in (2.36) by its expectation. In general, the expectation matrix will be easier to calculate than $J(\tau)$.

The Newton–Raphson algorithm may be used to compute least squares estimates, as one may compute a minimum by maximizing the negative of the criterion to be minimized. For simplicity, consider ordinary least squares estimation. In the particular case of the nonlinear regression model, $\tau = \beta$ and the objective function $O(\tau)$ is given by

$$O(\beta) = \{y - f(\beta)\}'\{y - f(\beta)\}, \tag{2.37}$$

writing (2.13) in vector notation. For this objective function, the gradient vector is $s(\beta) = -2X'(\beta)\{y - f(\beta)\}$, and the Hessian $J(\beta)$ is the symmetric $(p \times p)$ matrix with (k_1, k_2) element given by

$$2\sum_{j=1}^{n}\Big(\partial/\partial\beta_{k_1}\{f/(x_j,\beta)\}\partial/\partial\beta_{k_2}\{f(x_j,\beta)\} -$$
$$\{y_j - f(x_j,\beta)\}\partial^2/\partial\beta_{k_1}\partial\beta_{k_2}\{f(x_j,\beta)\}\Big). \tag{2.38}$$

From (2.38), the expectation of $J(\beta)$ is $2X'(\beta)X(\beta)$. Thus, the method of Fisher scoring leads to a simpler update, as expected. The algorithm obtained by the scoring method may be motivated in other ways and is referred to as the Gauss–Newton method in the context of nonlinear regression. See section 2.2 of Seber and Wild (1989) for an alternative derivation and notes on the relative performance of the two methods in the context of nonlinear regression.

Extension to minimization of the weighted criteria (2.17) and (2.26), in the case where the weights or weight matrix, respectively, are known, is straightforward; see section 2.2 of Seber and Wild (1989).

The Gauss–Newton and Newton–Raphson schemes should converge to the value of the least squares estimate minimizing (2.37). In practical implementation, updating continues until the difference between two successive iterates or values of the objective function (2.37) is sufficiently small. The basic algorithms given here may experience problems when carried out in practice, so a number of modifications have been proposed. See sections 2.1 and 3.2 and Chapter 14 of Seber and Wild (1989) and sections 2.2 and 3.5 of Bates and Watts (1988) for coverage of modifications, convergence issues, and alternative methods. These authors also note that when some of the parameters of the model enter in a linear fashion, maximization and inference can be simplified; see Bates and Watts (1988, section 3.3.5).

An additional practical consideration for fitting nonlinear models is the importance of the parameterization adopted. The choice of parameterization can speed convergence and improve the validity of the linear approximation used to form tests and confidence intervals. Moreover, an appropriate parameterization may be used to enforce constraints on the values of the parameters, which are critical for the model to make physical sense. This technique is used in the analysis of the indomethacin data in section 2.5. It is also worthwhile to try to find a representation of the model where the parameter estimates are approximately independent in order to improve stability and accuracy of estimation. For example, Wakefield (discussion of Smith and Roberts, 1993) has noted that for a monoexponential pharmacokinetic model of the form $f(x, \boldsymbol{\beta}) = D\beta_1 \exp(-\beta_2 x)$, where x is time and D is a known fixed dose, an alternative parameterization in terms of log volume (β_1) and log clearance (β_2), written as $(D/e^{\beta_1})\exp\{-(e^{\beta_2}/e^{\beta_1})x\}$, is better in this regard than other choices; this form also enforces positivity. A full treatment of these issues is beyond the scope of this book; an excellent discussion may be found in Chapters 3, 6, and 7 of Bates and Watts (1988).

A number of standard software packages are available to perform ordinary least squares estimation for the nonlinear regression model. Most of these programs also perform weighted least squares estimation as in section 2.3.2 with known weights specified by the user, and offer the user a choice among several computational strategies. SAS offers an extensive selection of numerical methods in PROC NLIN. Other programs include PCNONLIN

(Statistical Consultants, Inc., 1986), ADAPT II (D'Argenio and Schumitzky, 1992), which is tailored to pharmacokinetic modeling, and Module 3R of BMDP. S-plus routines to fit nonlinear models are also available; see Chambers and Hastie (1992). The reader is referred to the documentation for these software packages for details on implementation.

2.4.2 Variance function estimation

The variance parameter estimators based on the PL, REML, and AR criteria (2.21)–(2.24) may be computed in a number of ways. In all of these cases, the Newton–Raphson technique or a modification may be applied to perform the minimization. This approach requires software that allows minimization of a general, user-specified objective function. In some cases, computation of these estimates may be accomplished using methods for nonlinear regression, as the following development shows.

Consider the pseudolikelihood criterion (2.21). It is possible to solve explicitly for the estimator of σ^2 satisfying (2.21). Substitution of the expression for this estimate into (2.21) yields a function of $\boldsymbol{\theta}$ only, which is minimized by the value of $\boldsymbol{\theta}$ minimizing (2.21). It is easily seen that minimization of this criterion is identical to minimization in $\boldsymbol{\theta}$ of the sum of squares

$$\sum_{j=1}^{n} \left(\frac{\dot{g}(\boldsymbol{\theta})(y_j - \hat{\mu}_j)}{g(\hat{\mu}_j, \boldsymbol{z}_j, \boldsymbol{\theta})} \right)^2, \tag{2.39}$$

where, as before, $\hat{\mu}_j = f(\boldsymbol{x}_j, \hat{\boldsymbol{\beta}}^{(p)})$, and $\dot{g}(\boldsymbol{\theta})$ is the geometric mean of the n $g(\hat{\mu}_j, \boldsymbol{z}_j, \boldsymbol{\theta})$ values, that is

$$\left(\prod_{j=1}^{n} g(\hat{\mu}_j, \boldsymbol{z}_j, \boldsymbol{\theta}) \right)^{1/n}.$$

From (2.39), estimation of $\boldsymbol{\theta}$ may thus be accomplished by regressing a dummy variable d_j, which is identically zero for all j, on the 'regression function'

$$\frac{\dot{g}(\boldsymbol{\theta})(y_j - \hat{\mu}_j)}{g(\hat{\mu}_j, \boldsymbol{z}_j, \boldsymbol{\theta})}$$

(Carroll and Ruppert, 1988, section 3.2).

By a similar argument, it may be shown that the objective function (2.23) for REML is minimized by regressing d_j on

$$\frac{\{\dot{g}(\boldsymbol{\theta})\}^{n/(n-p)}(y_j - \hat{\mu}_j)|\boldsymbol{X}'(\hat{\boldsymbol{\beta}}^{(p)})\boldsymbol{G}^{-1}(\hat{\boldsymbol{\beta}}^{(p)},\boldsymbol{\theta})\boldsymbol{X}(\hat{\boldsymbol{\beta}}^{(p)})|^{1/[2(n-p)]}}{g(\hat{\mu}_j, \boldsymbol{z}_j, \boldsymbol{\theta})}. \tag{2.40}$$

To minimize the absolute residuals objective function (2.24), regress d_j on

$$\left(\frac{\dot{g}(\boldsymbol{\theta})|y_j - \hat{\mu}_j|}{g(\hat{\mu}_j, \boldsymbol{z}_j, \boldsymbol{\theta})}\right)^{1/2}. \tag{2.41}$$

In two common cases, the power and exponential variance models (2.6) and (2.7), the relevant dummy variable regressions for PL and AR can be carried out using any standard nonlinear least squares package. Giltinan and Ruppert (1989) give details of implementation using PROC NLIN of SAS. These functions share the common feature that the geometric mean may be written in the form

$$\left(\prod_{j=1}^{n} v(\hat{\mu}_j, \boldsymbol{z}_j)\right)^{\theta/n}, \tag{2.42}$$

where v is a known function of μ and $\boldsymbol{z}$, so that the product in large braces need only be computed once during the minimization process. For more general g, evaluation of the geometric mean at each internal iteration of the nonlinear minimization algorithm requires a pass through the data. Similarly, for REML estimation, a pass through the data is required for each evaluation of the determinant in (2.40). The limitation to models that satisfy (2.42) and the difficulty implementing the REML scheme with routine software thus arise because most common nonlinear regression programs are not flexible enough to allow multiple passes through the data at each internal iteration.

2.4.3 General covariance structures

The computational 'trick' that allows use of standard nonlinear regression software to perform variance function estimation may be extended to minimization of the objective functions (2.29) and (2.30) for general covariance models; however, the resulting form is complex, and the problem of the need for multiple passes through the data at internal iterations arises again. Alternatively, these criteria can be minimized by the Newton–Raphson method or the method of scoring. See Davidian and Giltinan (1993b) for explicit calculation of the expected Hessian matrix and the score vector $\boldsymbol{s}$ for the case of estimation of the covariance parameters σ and γ by minimization of the PL (2.29) and REML (2.30) criteria.

2.5 Examples

Pharmacokinetics of indomethacin

The parameters of the biexponential regression function as written in (2.2) must be positive for the model to make physical sense. In order to ensure positivity, it is useful to adopt a different parameterization of the model:

$$f(x, \boldsymbol{\beta}) = e^{\beta_1} \exp(-e^{\beta_2} x) + e^{\beta_3} \exp(-e^{\beta_4} x). \qquad (2.43)$$

In (2.43), e^{β_2} and e^{β_4} are the rate constants corresponding to the two apparent exponential phases of drug disposition. A number of quantities of interest may be derived as nonlinear functions of the components of $\boldsymbol{\beta}$; for example, the half-life of the terminal phase of drug disposition is given by $t_{1/2,term} = a(\boldsymbol{\beta}) = \log 2/e^{\beta_4}$, where $\beta_4 < \beta_2$.

The results of OLS estimation of the parameters in (2.43) based on the 11 responses for the fifth subject are given in Table 2.2. The estimated mean response curve based on these estimates is shown in Figure 2.1. From the table, the OLS estimate of the terminal half-life is $\hat{t}_{1/2,term} = 3.13$. Using the Wald methods of section 2.3.5, an approximate 95% confidence interval for half-life may be calculated as $(1.89, 4.37)$.

Figure 2.2 shows a plot of residuals versus the logarithm of predicted values from the OLS fit. The residuals are Studentized as described in Carroll and Ruppert (1988, p. 31–3) to minimize the effect of the design on the plot, and the logarithm of predicted values is used so that the residual pattern may be seen more easily. The data indicate an increasing relationship between variance and mean response. Variance heterogeneity is not unexpected with pharmacokinetic data, and a standard model for variance is the power function (2.6) with $\theta = 1$. Assuming this model, the GLS algorithm was used to estimate $\boldsymbol{\beta}$. The procedure converged, in the sense that the relative change in the estimates of all parameters was small ($< 10^{-4}$), after eight iterations to the values given in Table 2.2. The Wald 95% confidence interval for terminal half-life is $(3.13, 5.01)$; comparison with that obtained by OLS reveals substantial disagreement.

Table 2.2. *Parameter estimates for fits of (2.43) for the data from subject 5 of the indomethacin study. The four entries for $\hat{\beta}$ are the estimates of the components of β.*

	$\hat{\beta}$ (SE)	$\hat{\sigma},\hat{\theta}$	$\hat{t}_{1/2,term}$ (SE)
OLS	$1.27, 1.04, -1.23, -1.51$	0.07,–	3.13
	$(0.08, 0.15, 0.49, 0.64)$		(0.52)
GLS, $\theta = 1$	$1.21, 0.95, -1.45, -1.77$	0.17,1.00[1]	4.07
	$(0.24, 0.16, 0.21, 0.24)$		(.40)
GLS-PL	$1.24, 0.97, -1.43, -1.74$	0.13,0.82	3.96
	$(0.18, 0.14, 0.23, 0.26)$		(0.38)
GLS-REML	$1.23, 0.96, -1.44, -1.75$	0.14,0.88	3.99
	$(0.19, 0.14, 0.22, 0.25)$		(0.38)
GLS-AR	$1.24, 0.97, -1.43, -1.74$	0.13,0.82	3.96
	$(0.18, 0.14, 0.23, 0.26)$		(0.38)

[1] known value

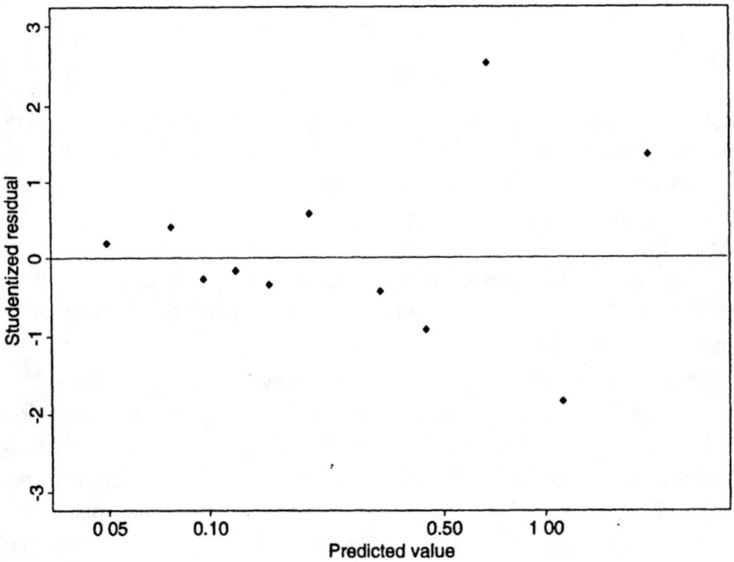

Figure 2.2. *Plot of Studentized OLS residuals versus predicted values for the data of subject 5, indomethacin data.*

Alternatively, one might assume the power variance function, but leave θ unspecified and estimate θ from the data. With only 11 measurements on this individual, estimates of θ are unlikely to be very good; however, for purposes of illustration, we summarize the results of using each of the variance function estimation techniques PL, REML, and AR applied to these data in Table 2.2. In each case, the procedure converged within ten iterations. With AR, the estimates to this level of accuracy are identical to the PL estimates. The REML estimate of θ is larger than those obtained from the other two methods; a possible explanation is that REML corrects for downward bias due to the failure of PL and AR to adjust for the loss of degrees of freedom associated with estimation of $\boldsymbol{\beta}$, although the difference is slight. Wald 95% confidence intervals for terminal half-life are $(3.07, 4.85)$ for PL, $(3.09, 4.90)$ for REML, and $(3.07, 4.85)$ for AR, which are similar across methods, differing somewhat from that obtained assuming $\theta = 1$; again, the OLS interval is in poor agreement.

Figure 2.3 shows a plot of weighted residuals

$$\{y_j - f(\boldsymbol{x}_j, \hat{\boldsymbol{\beta}}_{GLS})\}/g\{f(\boldsymbol{x}_j, \hat{\boldsymbol{\beta}}_{GLS}), \hat{\theta}\},$$

Studentized as in section 2.7 of Carroll and Ruppert (1988), versus predicted values for the fit using PL. The haphazard scatter of residuals in this plot suggests that variance heterogeneity has been adequately taken into account. Comparison of this plot with that in Figure 2.2 indicates that the OLS analysis is inappropriate. This example illustrates that failure to account adequately for heterogeneity of variance may lead to erroneous inference regarding quantities of interest.

Because the responses in the indomethacin study are repeated measurements over time, serial correlation within a given individual is likely. One might be tempted to consider a general covariance model to take into account both heterogeneity of variance and intra-individual correlation. However, with only 11 response measurements for this individual, estimation of the relevant covariance parameters would be difficult and the results highly suspect; thus, we do not attempt such an analysis here. We revisit this example in Chapter 5, where we apply methods for estimation of covariance parameters that make use of the data from all six individuals.

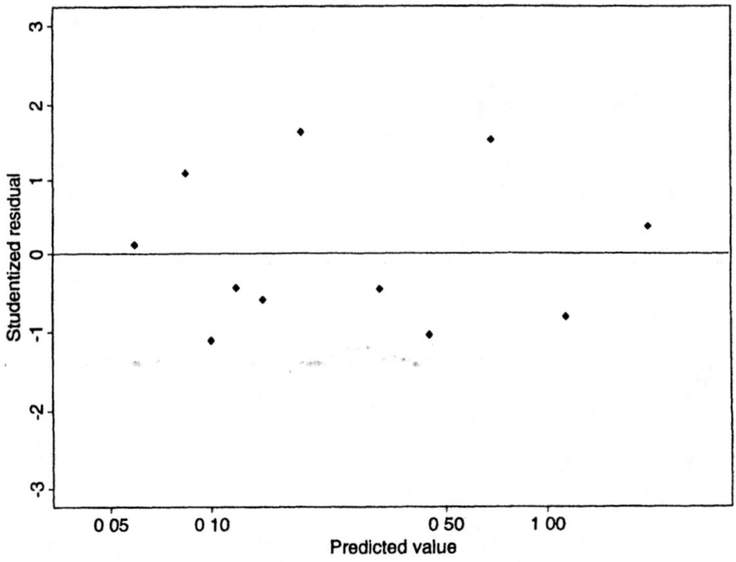

Figure 2.3. *Plot of weighted Studentized GLS residuals based on PL versus predicted values for the data of subject 5, indomethacin data.*

Sweetgum tree data

We apply the methods in this chapter to estimation of intra-individual correlation parameters by considering data on bole volumes of sweetgum trees, discussed in detail in section 11.3. It is of interest to develop a predictive equation for the cumulative volume of the bole of a sweetgum tree, y, from breast height (4.5 ft above ground) up to a point along the bole of specific diameter, d, given D, the diameter at breast height. For each of a number of such trees, y was measured at a number of values of d at equally spaced intervals along the length of the bole. Figure 2.4 shows data for tree #10, where the quantity $x = D - d$ is plotted along the horizontal axis; there were $n = 33$ (x, y) pairs available. The sigmoidal form of the relationship suggests that the following model may be appropriate:

$$f(x, \beta) = \frac{\beta_1}{1 + \exp\{e^{\beta_3}(\log x - \beta_2)\}}, \quad x = D - d, \ 0 \le d \le D. \tag{2.44}$$

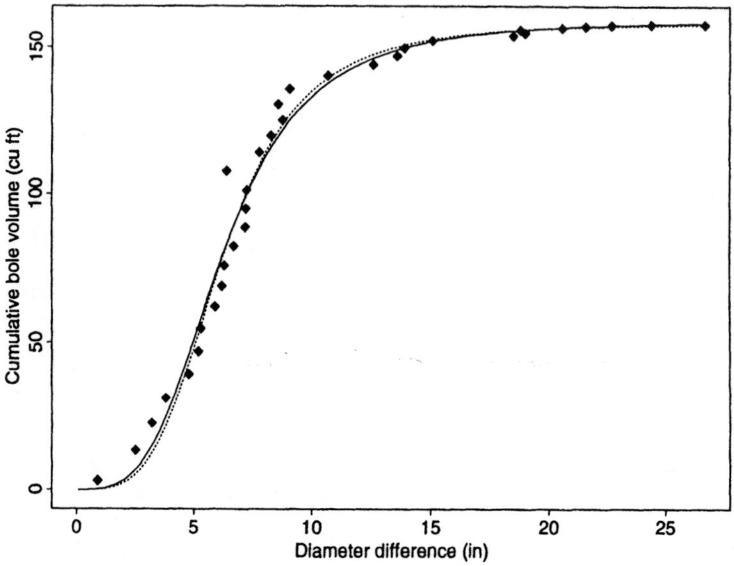

Figure 2.4. *Data for sweetgum tree 10. Fits of model (2.44) are superimposed: The dashed line is the OLS fit ($\theta = 0$), the solid line is the GLS fit with α estimated by PL ($\alpha = 0.37$).*

In (2.44), we have parameterized the exponential multiplier as e^{β_3} to ensure positivity.

Figure 2.4 shows no evidence of within-tree heterogeneity of variance; however, it is reasonable to expect possible spatial correlation of measurements along the bole. The results of an initial OLS fit of (2.44) are summarized in Table 2.3; predicted values for this fit are given by the dashed line in Figure 2.4. This assumes observations within the tree are uncorrelated; thus, a sensible diagnostic for assessing spatial correlation may be based on examination of associations among residuals from the fit. For example, a plot of residuals against suitably lagged versions allows a visual assessment; a general discussion of graphical techniques for exploring correlation structure is given by Diggle, Liang and Zeger (1994, section 3.4). It is reasonable to suppose that adjacent measurements may be similarly correlated along the length of the tree, while observations far apart would be unassociated.

Table 2.3. *Parameter estimates for fits of (2.44) for sweetgum tree 10. The three entries for $\hat{\beta}$ are the estimates of the components of β.*

	$\hat{\beta}$ (SE)	$\hat{\sigma}, \hat{\alpha}$
OLS	158.60,1.83,1.31	7.14,–
	(2.62,0.02,0.07)	
GLS-PL	159.47,1.83,1.23	6.59,0.37
	(3.46,0.02,0.09)	

Accordingly, in Figure 2.5, we plot the OLS residuals against themselves, lagged by successively greater distances; for instance, the upper left hand panel of the figure shows r_j plotted against r_{j-1}, $1 < j \leq n$, where r_j is the jth OLS residual. The plot indicates a possible association at lag 1 (correlation coefficient for the plot = 0.47), although this may be in part driven by one outlying residual, with little suggestion of a relationship at larger lags (|correlation| < 0.14 in all cases).

Based on this evidence, we chose the following model for within-tree correlation:

$$\boldsymbol{\Gamma}_i(\alpha) = \begin{bmatrix} 1 & \alpha & 0 & \cdots & 0 \\ & 1 & \alpha & \cdots & 0 \\ & & \ddots & \ddots & \vdots \\ & & & \ddots & \alpha \\ & & & & 1 \end{bmatrix}.$$

This correlation structure is that of a moving average process of order one and requires $|\alpha| < 0.5$ (see Seber and Wild, 1989, chapter 6). This choice implies, for the jth and kth measurements, $\mathrm{Var}(y_j) = \sigma^2$, $\mathrm{Corr}(y_j, y_k) = \alpha$, $k = j + 1$, and $\mathrm{Corr}(y_j, y_k) = 0$ otherwise.

To carry out the fit of this model, we used the GLS algorithm of section 2.3.4, with the PL objective function for estimation of σ and α. The algorithm converged after 10 iterations; the results are displayed in Table 2.3, and predicted values from this fit are represented by the solid line in Figure 2.4. The estimated value for the correlation parameter, $\alpha = 0.37$, is in moderate agreement with the crude estimate obtained from the plots; the weighting imposed

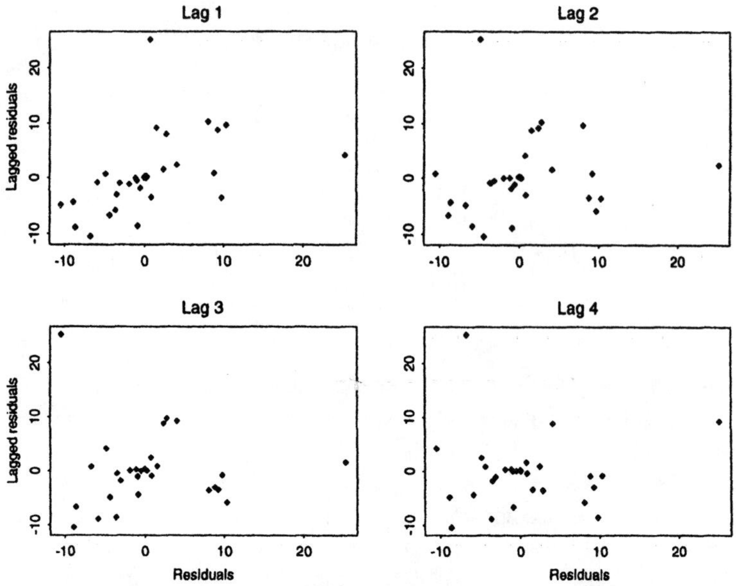

Figure 2.5. *Lagged residuals from the OLS fit of model (2.44). Each panel displays a plot of the residuals, lagged by the number of observations indicated, against the residuals themselves.*

by the PL procedure likely downplayed the influence of the outlying residual. The sign and magnitude of the estimate support the contention that there may be positive association among adjacent measurements.

The OLS and GLS fits are quite similar, with slightly larger standard errors for the latter. Those for the OLS fit were calculated based on the assumption of uncorrelated errors; thus, if within-tree correlation is nonnegligible, these standard errors provide an inappropriate assessment of estimation error. This suggests that, although fits may differ little, adoption of a more complicated covariance structure may have implications for inference on model components; however, the tradeoff between a gain in this direction and additional complexity of implementation is unclear, and may vary for different problems. The inherent difficulty in estimating second moment parameters must be borne in mind in interpretation of results.

2.6 Related approaches

2.6.1 Generalized linear models

The class of generalized linear models (McCullagh and Nelder, 1989) offers a flexible framework for regression analysis where the assumptions of normality and constant variance do not hold. In this approach, the classical linear regression assumptions are relaxed in several ways. The data are assumed to arise from a distribution in a scaled exponential family, so that a natural and specific dependence of variance on the mean response is imposed. For example, in the case of the gamma distribution, the mean and variance are related according to the constant coefficient of variation model (2.4); other distributions in this class are the normal with constant variance, and the Poisson, negative binomial, and inverse Gaussian distributions, all of which imply known mean-variance relationships. In addition, the mean response is assumed to depend on the covariate through a function of a linear combination of the covariate vector, which may be written in our notation as

$$\mathrm{E}(y_j) = f(\boldsymbol{x}_j'\boldsymbol{\beta}), \tag{2.45}$$

where $\boldsymbol{x}_j$ is $(p \times 1)$. There is no penalty for relaxing the requirement that the dependence on $\boldsymbol{x}_j$ and $\boldsymbol{\beta}$ be through a linear combination. Thus, the class of generalized linear models is a special case of the nonlinear regression framework with variance a known function of the mean response ($\boldsymbol{\theta}$ known).

For the exponential family distributions noted above, the maximum likelihood estimator of $\boldsymbol{\beta}$ is identical to the estimator that would be obtained by iterating the GLS algorithm in section 2.3.2 to convergence. That is, for the particular mean-variance specifications imposed by the exponential family, maximum likelihood and GLS coincide; this is shown explicitly in section 2.4 of Carroll and Ruppert (1988). Green (1984) also discusses the correspondence between ML and iteratively-reweighted least squares. This is in contrast to the case of the normal distribution with variance a function of the mean response, which is *not* an exponential family, as discussed above. Thus, the GLS procedure arises naturally for the class of generalized (non)linear models. GLS-type estimation has been proposed in the literature as an omnibus method under the name of *quasilikelihood* (Wedderburn, 1974; McCullagh, 1983) in cases where the exponential family does not hold exactly The class of generalized (non)linear models can be restrictive in that the exponential family requires variance to depend on mean response in a very specific way. Thus, it may not be possible to capture the exact pattern of variance heterogeneity with such a model. For example, in the case of assay data, there may be two components of variation, for which the variance function (2.8) may be an appropriate characterization. This variance function is not among the possibilities encompassed by the exponential family. Furthermore, the classical generalized (non)linear model approach does not allow for the situation where one or more of the parameters in this function are unknown ($\boldsymbol{\theta}$ unknown). Recently, Nelder and Pregibon (1987) have proposed extending the class to allow for more general variance functions and unknown variance parameters; see also Efron (1986) and Davidian and Carroll (1988).

2.6.2 Generalized estimating equations

Liang and Zeger (1986) propose an extension of the generalized linear model to multivariate response; see also Diggle *et al.* (1994). As above, dependence on $\boldsymbol{x}_j$ and $\boldsymbol{\beta}$ through a linear combination, as in (2.45), may be relaxed, so the approach is relevant to more general nonlinear models. Recently, the term *generalized estimating equations* (GEEs) has come to refer broadly in the statistical literature to the class of inferential techniques associated with this model framework and its extensions. The common theme underlying these methods is that no assumptions on the distribution of the data need be made beyond those on moments. The GEE approach is essentially similar to the ideas of M-estimation (e.g. Huber, 1981). These methods have generated considerable interest; here, we note the connection between generalized least squares estimation and techniques of variance function estimation based on residuals for univariate response, discussed in this chapter, and the techniques referred to as GEE methods.

Specifically, these approaches are essentially the same, as they are moment methods (e.g. Judge *et al.*, 1985). In the setting of univariate, rather than multivariate, response, the mean-variance specification given in (2.5),

$$\mathrm{E}(y_j) = f(\boldsymbol{x}_j, \boldsymbol{\beta}), \quad \mathrm{Var}(y_j) = \sigma^2 g^2(\mu_j, \boldsymbol{z}_j, \boldsymbol{\theta}), \quad \mu_j = f(\boldsymbol{x}_j, \boldsymbol{\beta}), \tag{2.46}$$

is of the type considered by Liang and Zeger (1986): the second moment is a function of the mean and additional parameters to be estimated. Under this model, the GEE for the mean parameter $\boldsymbol{\beta}$ is the weighted least squares estimating equation (2.18) with known weights replaced by the variance function, as in the GLS algorithm in sections 2.3.2 and 2.3.3. For estimation of parameters in the second moment, σ and $\boldsymbol{\theta}$ in this context, Prentice (1988), Zhao and Prentice (1990), Prentice and Zhao (1991), and Liang, Zeger and Qaqish (1992) discuss quadratic estimating equations. For (2.46), the form of these equations is essentially that of the PL estimating equations (2.22); this may be seen by comparing, for example, the form of equation (2) of Prentice and Zhao (1991) in the case of univariate response to that of (2.22). Thus, iteration to convergence of the GLS algorithm with PL estimation of variance parameters for (2.46) is equivalent to the approach discussed by Prentice and Zhao (1991, section 2) and referred to as 'GEE1' by Liang *et al.* (1992).

We revisit the connection between methods in this book and generalized estimating equations in Chapter 6.

2.7 Discussion

In this section we comment on a number of issues: model flexibility, other inferential procedures for nonlinear regression models with heterogeneous variance, and other modeling strategies that may be employed to address the issues of nonconstant variance and nonnormality.

Model flexibility

Mean-variance models of the form (2.5) may result from different considerations. When the data are skewed, for example, a distribution such as the gamma or the lognormal, which naturally implies a mean-variance relationship, may be a more realistic model than the normal. For example, the gamma distribution with parameters a and b has expectation ab and variance ab^2; if $a = 1/\sigma^2$ and $b = \sigma^2 f(\boldsymbol{x}_j, \boldsymbol{\beta})$, then y_j has a gamma distribution with mean and variance given in (2.4); we discuss the class of generalized linear models below. Alternatively, the assumption of a lognormal distribution also leads to the mean-variance specification (2.4). In this case, $e_j = y_j - f(\boldsymbol{x}_j, \boldsymbol{\beta})$ will have mean zero, variance $\text{Var}(e_j) = \sigma^2\{f(\boldsymbol{x}_j, \boldsymbol{\beta})\}^2$, and distribution corresponding to that of a gamma random variable minus its mean.

Alternatively, data may follow a normal distribution for given $\boldsymbol{x}_j$ with heterogeneous variance. For example, y_j may be normally distributed, but with constant coefficient of variation rather than constant variance (2.4), which is sometimes represented in the pharmacokinetics literature as $y_j = f(\boldsymbol{x}_j, \boldsymbol{\beta})(1 + u_j)$, where $u_j \sim \mathcal{N}(0, \sigma^2)$. Here, $e_j = f(\boldsymbol{x}_j, \boldsymbol{\beta})u_j$. Thus, (2.1) is a convenient way to represent the response for general mean-variance specifications, regardless of the distribution of the data.

Normal theory maximum likelihood

We have remarked on the use of maximum likelihood assuming the normal distribution as an alternative to the generalized least squares methods we advocate. Joint normal theory maximum likelihood estimation (ML) of the parameters in the mean-variance model (2.5) has been proposed by Peck *et al.* (1984) and Beal and Sheiner (1985, 1988) in the context of pharmacokinetic analysis and has been called 'extended least squares' (ELS) by these authors.

Differentiation of the normal likelihood with respect to $\boldsymbol{\beta}$, σ, and $\boldsymbol{\theta}$ and comparison of the resulting estimating (score) equations to those for GLS and PL yields insight into how the ML and GLS methods may differ. We focus on uncorrelated responses with variance modeled as $\text{Var}(y_j) = \sigma^2 g^2\{f(\boldsymbol{x}_j, \boldsymbol{\beta}), \boldsymbol{z}_j, \boldsymbol{\theta}\}$, but the same issues arise for more complicated covariance structures. The equations corresponding to σ and $\boldsymbol{\theta}$ have the same form as those obtained by differentiating the PL objective function. Differentiation of the likelihood with respect to $\boldsymbol{\beta}$ yields the estimating equations

$$\sigma^2 \sum_{j=1}^{n} \left(\frac{\{y_j - f(\boldsymbol{x}_j, \boldsymbol{\beta})\}^2}{\sigma^2 g^2\{f(\boldsymbol{x}_j, \boldsymbol{\beta}), \boldsymbol{z}_j, \boldsymbol{\theta}\}} - 1 \right) \boldsymbol{\nu}_\beta\{f(\boldsymbol{x}_j, \boldsymbol{\beta}), \boldsymbol{z}_j, \boldsymbol{\theta}\}$$

$$+ \sum_{j=1}^{n} \frac{y_j - f(\boldsymbol{x}_j, \boldsymbol{\beta})}{g^2\{f(\boldsymbol{x}_j, \boldsymbol{\beta}), \boldsymbol{z}_j, \boldsymbol{\theta}\}} \boldsymbol{f}_\beta(\boldsymbol{x}_j, \boldsymbol{\beta}) = \mathbf{0}, \qquad (2.47)$$

where $\boldsymbol{\nu}_\beta\{f(\boldsymbol{x}_j, \boldsymbol{\beta}), \boldsymbol{z}_j, \boldsymbol{\theta}\}$ is the $(p \times 1)$ vector of derivatives of $\log g\{f(\boldsymbol{x}_j, \boldsymbol{\beta}), \boldsymbol{z}_j, \boldsymbol{\theta}\}$ with respect to $\boldsymbol{\beta}$. In contrast, the corresponding GLS estimating equation for $\boldsymbol{\beta}$ is, from (2.18),

$$\sum_{j=1}^{n} \frac{y_j - f(\boldsymbol{x}_j, \boldsymbol{\beta})}{g^2\{f(\boldsymbol{x}_j, \boldsymbol{\beta}), \boldsymbol{z}_j, \boldsymbol{\theta}\}} \boldsymbol{f}_\beta(\boldsymbol{x}_j, \boldsymbol{\beta}) = \mathbf{0}. \qquad (2.48)$$

From (2.47) and (2.48), when the variance function g does not depend on the regression parameter $\boldsymbol{\beta}$, $\boldsymbol{\nu}_\beta\{f(\boldsymbol{x}_j, \boldsymbol{\beta}), \boldsymbol{z}_j, \boldsymbol{\theta}\} = \mathbf{0}$, so that the ML and GLS approaches are identical in that the form of the estimating equations is the same. In this case, full iteration of the GLS algorithm with PL estimation will yield the ML estimates of $\boldsymbol{\beta}$, σ, and $\boldsymbol{\theta}$. When g does depend on $\boldsymbol{\beta}$, $\boldsymbol{\nu}_{\beta}\{f(\boldsymbol{x}_j, \boldsymbol{\beta}), \boldsymbol{z}_j, \boldsymbol{\theta}\} \neq \mathbf{0}$, and the equations differ because of the first term in (2.47). It follows that the methods do not lead to the same estimation scheme. This has led to some debate about the relative performance of the estimators in this situation.

In many instances, even if g depends on the regression parameters, maximum likelihood and generalized least squares will give virtually identical results. This is especially true in the 'small-σ' case, where the magnitude of response variance is small relative to the range of the response (Davidian and Carroll, 1987; Carroll and Ruppert, 1988, Chapter 3); under this condition, the first term in (2.47) is negligible. If this is not the case, the methods may diverge. If the data are normal, ML estimation is superior. This advantage is lost very quickly in the case of nonnormal data, however; because the first term in (2.47) is a quadratic function of e_j, the ML estimator is highly sensitive to outlying observations, while the GLS estimator, which is based on an estimating equation linear in e_j, is robust (Carroll and Ruppert, section 2.4).

Another issue is robustness to misspecification of the variance function g. The ML estimator is nonrobust in the sense that severe bias may result if g is improperly characterized. GLS does not suffer from this deficiency (Carroll and Ruppert, 1982b; Carroll and Ruppert, 1988, section 2.4; van Houwelingen, 1988).

For readers familiar with the literature on generalized estimating equations, in the case of univariate response, joint normal theory maximum likelihood is essentially equivalent to the 'GEE2' approach outlined in Liang *et al.* (1992) and Prentice and Zhao (1991, section 3), with the Gaussian 'working matrix' assumption.

Based on robustness to nonnormality and model misspecification, we prefer GLS methods to ML in general.

Variance function estimation

The methods for variance function estimation we discuss in section 2.3.3 are based on transformations of absolute residuals from a preliminary fit. The methods are appropriate when a functional form for variance can be specified. When variance is heterogeneous, but a smooth relationship between variance and the mean or other factors is not evident, methods for estimation of weights based on replication at each covariate value $\boldsymbol{x}_j$ can be used; see Fuller and Rao (1978) and Carroll and Cline (1988). Even when it is possible to specify a model for variance, other methods for estimation of variance parameters $\boldsymbol{\theta}$, which again require replication at each $\boldsymbol{x}_j$, have been proposed, mainly in the context of the analysis of assay data, see Rodbard and Frazier (1975), Raab (1981), and Sadler and Smith (1985). These methods have been found to be inefficient relative to those we have discussed (Davidian and Carroll, 1987; Carroll and Ruppert, 1988, Chapter 3). For the other applications considered in this book, replication at each covariate value is unlikely or impossible. Thus, we restrict our attention to methods based on residuals.

Robust methods

In order to reduce the sensitivity of the GLS algorithm, including the variance function estimation step, to outlying observations or the 'high leverage' of responses taken at extreme values of the covariate vector $\boldsymbol{x}_j$, various robust modifications of the algorithm have been proposed, see Carroll and Ruppert (1982a), Giltinan, Carroll and Ruppert (1986), and Chapter 6 of Carroll and Ruppert (1988). A discussion of these ideas is beyond the scope of this book; however, we note that many of the procedures we discuss for estimation in the hierarchical nonlinear model introduced in Chapter 4 can be made robust by applying the same principles used for data from a single individual.

Transformations

A popular approach to accommodating heterogeneous variance and departures from normality is by transformation of the response. Originally, transformation was proposed both as a means of achieving homoscedasticity and approximate normality and for inducing a simpler linear model for the transformed response (Box and Cox, 1964). When the model for the mean response is nonlinear, however, it is often because of a meaningful empirical or theoretical relationship between the response and covariate, and it is desirable that this relationship be preserved in an analysis. Consequently, a recent approach to using transformations in the context of nonlinear regression has been the so-called *transform-both-sides* (TBS) model advocated by Carroll and Ruppert (1984). In our notation, it is assumed that there exists a transformation function t depending on a transformation parameter $\boldsymbol{\lambda}$, such that

$$t(y_j, \boldsymbol{\lambda}) = t\{f(\boldsymbol{x}_j, \boldsymbol{\beta}), \boldsymbol{\lambda}\} + e_j, \tag{2.49}$$

where the $e_j \sim \mathcal{N}(0, \sigma^2)$ and are independent. A common choice for t is the Box–Cox transformation family $t(y, \lambda) = (y^\lambda - 1)/\lambda$; note that $\lambda = 0$ gives the logarithmic transformation. A complete discussion of the TBS approach is given in Chapter 4 of Carroll and Ruppert (1988).

In some situations, there is a direct correspondence between mean-variance specifications on the original scale and the approximate mean-variance relationship implied by a TBS model (Carroll and Ruppert, 1988, section 4.2). For example, if t is the Box–Cox transformation, it is straightforward to show that if σ is 'small,' (2.49) is approximately equivalent to the nonlinear regression model on the original scale with

$$\mathrm{E}(y_j) = f(\boldsymbol{x}_j, \boldsymbol{\beta}), \quad \mathrm{Var}(y_j) = \sigma^2 f^{2(1-\lambda)}(\boldsymbol{x}_j, \boldsymbol{\beta}). \tag{2.50}$$

Thus, a common alternative to modeling variance as proportional to a power θ of mean response is the TBS model with Box–Cox transformation function, with approximate correspondence $\theta = 1 - \lambda$.

A limitation of the TBS approach is that the implied mean-variance relationship may not be adequate to model the features of the data, as its form is fairly restrictive. This feature has been relaxed somewhat by models incorporating both transformation and weighting (Carroll and Ruppert, 1988, Chapter 5; Ruppert, Cressie and Carroll, 1989).

For individual data, the explicit modeling framework we have discussed and the TBS model often lead to qualitatively equivalent inferential approaches, and complexity of implementation is similar. In our view, it is more straightforward to extend the mean and variance model on the original scale for data from a single individual to a hierarchical model for data from several individuals.

2.8 Bibliographic notes

Early studies of the nonlinear regression model (Jennrich, 1969; Gallant, 1975) are based on the classical assumptions given in section 2.2.2. The inappropriateness of one or more of these assumptions in a number of important fields of application prompted development of methods to handle departures from them for linear regression models; a small sampling of relevant work in addition to that already referenced in this chapter includes Jacquez *et al.* (1968), Glejser (1969), Amemiya (1973), Wedderburn (1974), Harvey (1976), Jobson and Fuller (1980), and Box and Meyer

(1986a,b). All of these methods are easily generalized to the nonlinear case. Because situations where nonlinear models are appropriate are often those where such departures arise, a number of techniques for handling features such as heterogeneity of variance were proposed directly in the context of nonlinear regression. Much of this work appears in the subject-matter literature. Notable examples include Box and Hill (1974), Gallant and Goebel (1976), Pritchard, Downie and Bacon (1977), work in assay analysis (Rodbard and Frazier, 1975; Finney, 1976; Raab, 1981) and pharmacokinetic modeling of individual data (Beal and Sheiner, 1982; Peck *et al.* 1984).

Although there is an extensive array of reference texts for linear modeling and inference, relatively few comprehensive treatments of nonlinear regression are available. Recent texts include Ratkowsky (1983), Gallant (1987), Bates and Watts (1988), Carroll and Ruppert (1988), Seber and Wild (1989), and Ross (1990). The books of Gallant, Carroll and Ruppert, and Seber and Wild pay particular attention to relaxation of the standard assumptions; the first reference pursues a rigorous theoretical treatment, while the latter two offer practical discussion.

In this chapter, we have focused on classical least squares approaches to estimation and inference for nonlinear models. Bayesian techniques have also been proposed; because there is no accepted strategy for choosing priors for nonlinear model parameters, relatively few references are available. A short list includes Katz, Azen, and Schumitzky (1981), Eaves (1983), and Ye and Berger (1991).

CHAPTER 3

Hierarchical linear models

3.1 Introduction

The main focus of this book is modeling and inference for repeated measurement data where the relationship between response and covariates may be nonlinear in the parameters. A natural framework for inference is the hierarchical nonlinear model and its extensions, discussed in Chapter 4.

In this chapter, we review the hierarchical *linear* model, with particular emphasis on the case of two levels of variability, such as that of repeated measurements on each of several experimental units. Section 3.2 describes the modeling framework, both from a classical mixed effects viewpoint and from a Bayesian perspective. Inference for hierarchical linear models is discussed in section 3.3. Particular attention is paid to the simplest case of a one-way classification with random classes. A thorough understanding of the issues in this simple case is helpful in more complicated examples. Bayesian inferential methods are also discussed in section 3.3. In the setting of hierarchical linear models, there is a strong similarity between sensible inferential procedures arrived at from both a frequentist or a Bayesian point of view. Readers unfamiliar with Bayesian methods may omit sections 3.2.3 and 3.3.3 on a first reading without loss of continuity.

In all but the simplest cases, computational feasibility becomes a practical concern in carrying out inference for hierarchical models. Section 3.4 discusses computational aspects in more detail, including fitting hierarchical models using existing standard software packages. In section 3.5 we provide a brief discussion of limitations and extensions of normal-theory-based hierarchical linear models.

This chapter is intended only to provide enough background to understand subsequent material. Coverage of these topics is in no way exhaustive. There is an extensive literature on hierarchical modeling. Section 3.6 contains bibliographic notes.

3.2 Model specification

In this section we review the (normal theory) hierarchical linear model. In biostatistical applications, this model arises frequently in situations where several measurements are made on a number of individuals. We use the term 'individual' quite broadly; for instance, it might refer to human or animal subjects, experimental runs, litters, laboratories, or devices. Repeated measurements on an individual may be taken over time, at different analyte concentrations, or over some other set of experimental conditions. The existence of repeated measurements requires particular care in characterizing the random variation in the data. In particular, it is important to recognize explicitly two levels of variability: random variation among measurements within a given individual (*intra*-individual variation) and random variation among individuals (*inter*-individual variation). The existence of these different variance components complicates inference on fixed effects of interest.

3.2.1 Examples

Example: Rat pup weights in a reproductive toxicology study

Dempster *et al.* (1984) report data on pup birth weights in grams in a reproductive toxicology study on rats. In this experiment, ten dams were randomly allocated to each of three treatment groups; control, low, and high dose of the test substance. Data were available from only seven litters in the high dose group. Table 3.1 contains data for the first three litters in each group.

For each pup, there are two sources of random variation: the pup-to-pup variability (within-litter) and the litter-to-litter variation. A simple model for pup weights for the kth treatment group is

$$y_{ij} = \mu + \tau_k + b_i + e_{ij}, \tag{3.1}$$

where y_{ij} is the birth weight of the jth pup from the ith litter, μ represents the overall mean, and τ_k is a (fixed) treatment effect for the kth treatment. In (3.1), b_i represent the random litter effects, with mean zero and variance σ_b^2 (inter-litter variability), and e_{ij} correspond to the random pup effects with mean zero and variance σ_e^2 (intra-litter variation).

Other factors may influence birth weights. For instance, pup weights tend to vary inversely with litter size, and males generally weigh more than females. The model (3.1) may be extended

Table 3.1. *Partial data listing: rat pup weights (g) in a reproductive study.*

Litter	*Males*	*Females*
	Controls	
1	6.60,7.40,7.15,7.24,7 10,6.04, 6.98,7.05	6.95,6.29,6.77, 6.57
2	6.37,6.37,6.90,6.34,6.50,6.10, 6.44,6.94,6.41	5.92,6.04,5.82,6.04,5 96
3	7.50,7.08	7.57,7.27
	Low dose	
11	5.65,5.78,6.23,5.70,5.73,6.10, 5.55,5.71,5.81,6.10	5.54,5.72,5.50,5.64,5.42,5,42
12		6.89,7.73
13	5.83,5.97,6.39,5.69,5.69,5.97, 6.04,5.46	6.09,5.39,5.89,5.14
	High dose	
21	5.09,5.57,5.69,5.50,5.45,5.24, 5.36,5.26,5.36,5.01,5.03	5.23,5.13,4.48
22	5.30,5.40,5.55,6.02,5.27	5.19,5.42,5.40,5.12,5.40
23	7.70	7.68,6.33

to incorporate litter size and sex as additional covariates:

$$y_{ij} = \mu + \tau_k + l_i\lambda + x_{ij}\delta + b_i + e_{ij}. \tag{3.2}$$

Here, the interpretation of μ, τ_k, b_i, and e_{ij} is as before; x_{ij} is an indicator variable with value 1 for female pups and 0 for males, and l_i is the size of the ith litter. Thus, in this model, δ represents a term for the (fixed) effect of sex and λ is a regression coefficient for the covariate litter size. Incorporating the effect of litter size on birth weight by means of the linear term $l_i\lambda$ implies a straight line relationship, which is assumed to be parallel across treatment groups and sex. Referring to litter size as a covariate is a slight abuse of terminology, as it may itself be affected by treatment.

Example: Tracer age data

Figure 3.1 shows control data from several runs of a radioimmunoassay for the protein insulin-like growth factor (IGF-I) for

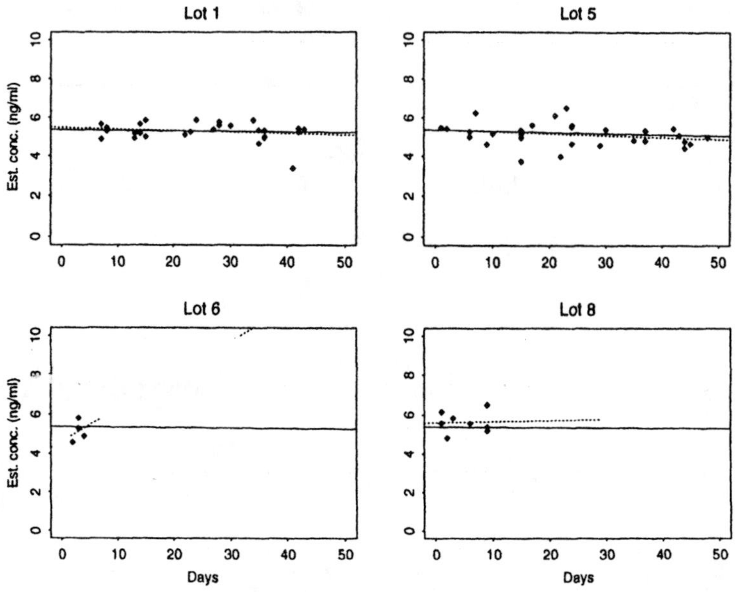

Figure 3.1. *Control data for several runs of a radioimmunoassay for IGF-I for four lots of radioactive tracer; superimposed are OLS fits (dashed lines) and empirical Bayes fits of (3.3) (solid lines).*

each of four lots of radioactive tracer. Each day the assay is run, control samples at three known nominal concentrations are included for the purposes of assay quality control. These samples are aliquots taken from a large stock of IGF-I of known concentration, stored at $-70°$ C. For an assay in good control, calibrated concentrations for control samples should remain stable from day to day. Unusual behavior in control values may be used to diagnose problems in assay performance.

The radioactive tracer used in the assay generates a signal that forms the basis for calibration (a more detailed description of radioimmunoassay procedure is given in section 10.2.1). Approximately every 50 days, the tracer lot is changed, because using a lot for too long could compromise assay performance due to tracer decay. The data in Figure 3.1 represent concentrations of the medium control plotted against tracer age for four lots of tracer. Similar data were available for 10 lots in all. Based on these data, it was of interest to determine if use of tracer lots for up to 50 days is justified.

In statistical terms, this corresponds to evaluating possible trend in control values with tracer age. Formal assessment of trend, accounting correctly for within-lot and lot-to-lot variability, may be achieved within the framework of a random coefficient regression model. Letting y_{ij} denote the jth response (control value) at time t_{ij} for the ith lot, we may postulate a regression relationship specific to the ith lot:

$$\mathrm{E}(y_{ij}|\beta_{0i},\beta_{1i};t_{ij}) = \beta_{0i} + \beta_{1i}t_{ij}, \qquad \mathrm{Var}(y_{ij}|\beta_{0i},\beta_{1i};t_{ij}) = \sigma_e^2. \tag{3.3}$$

Here, β_{0i} and β_{1i} represent the lot-specific regression intercepts and slopes, respectively. The term σ_e^2 corresponds to within-lot variability of the measurements for a subject about the postulated regression line for that lot. Assumption of homogeneous error variance seems reasonable, as the range of response values is relatively narrow.

It proves useful to incorporate a second stage into the model, assuming that the regression coefficients β_{0i} and β_{1i} come from a distribution such that

$$\mathrm{E}\begin{bmatrix}\beta_{0i}\\ \beta_{1i}\end{bmatrix} = \begin{bmatrix}\beta_{0}\\ \beta_{1}\end{bmatrix}, \qquad \mathrm{Cov}\begin{bmatrix}\beta_{0i}\\ \beta_{1i}\end{bmatrix} = \boldsymbol{D} \tag{3.4}$$

The covariance matrix $\boldsymbol{D}$ quantifies the inter-lot variation in the regression coefficients. Within this model formulation, assessment of overall evidence for a change in control values with tracer age may be accomplished by inference on the slope parameter β_1. In addition, it may sometimes be of interest to characterize the slope for a specific tracer lot, β_{1i}.

In the next two sections, we describe a general framework for hierarchical linear models and show how the preceding examples may be viewed as special cases. Inference for these two examples is developed further in section 3.4.4.

3.2.2 General linear mixed effects model

In this section, we give the general parametric form of the linear mixed effects model. Our exposition is based closely on that in Laird and Ware (1982). We consider the following two-stage model:

Stage 1 (intra-individual variation)

Suppose that, for the ith of m individuals, n_i responses have been observed, so that a total of $N = \sum_{i=1}^{m} n_i$ data values are available. Let $\boldsymbol{y}_i$, the $(n_i \times 1)$ data vector for the ith individual, $i = 1, \ldots, m$, satisfy

$$\boldsymbol{y}_i = \boldsymbol{X}_i\boldsymbol{\beta} + \boldsymbol{Z}_i\boldsymbol{b}_i + \boldsymbol{e}_i. \tag{3.5}$$

Here, $\boldsymbol{\beta}$ represents a $(p \times 1)$ vector of fixed effects, $\boldsymbol{X}_i$ is an $(n_i \times p)$ design matrix specific to the ith individual, $\boldsymbol{b}_i$ is a $(k \times 1)$ vector of random effects, $\boldsymbol{Z}_i$ is an $(n_i \times k)$ design matrix linking $\boldsymbol{y}_i$ to the random effects, and $\boldsymbol{e}_i$ is a vector of 'within-individual' errors. Assume that $\boldsymbol{e}_i \sim \mathcal{N}(\boldsymbol{0}, \boldsymbol{R}_i)$, where $\boldsymbol{R}_i$, the within-individual covariance matrix, may depend on i through its dimension, but not otherwise. Typically, $\boldsymbol{R}_i$ is parameterized in terms of a relatively small number of intra-individual variance parameters. The simplest case is that where $\boldsymbol{R}_i = \sigma^2 \boldsymbol{I}_{n_i}$, where $\boldsymbol{I}_{n_i}$ is the $(n_i \times n_i)$ identity matrix. Other possibilities for $\boldsymbol{R}_i$ are discussed later.

Conditional on $\boldsymbol{b}_i$, (3.5) implies

$$\begin{aligned} \mathrm{E}(\boldsymbol{y}_i|\boldsymbol{b}_i) &= \boldsymbol{X}_i\boldsymbol{\beta} + \boldsymbol{Z}_i\boldsymbol{b}_i, \\ \mathrm{Cov}(\boldsymbol{y}_i|\boldsymbol{b}_i) &= \boldsymbol{R}_i. \end{aligned}$$

Stage 2 (inter-individual variation)

Suppose that the random effects, $\boldsymbol{b}_i$, come from a normal distribution with zero mean and $(k \times k)$ dispersion matrix $\boldsymbol{D}$, independently of each other and of the $\boldsymbol{e}_i$. Under these assumptions, it is straightforward to calculate the marginal mean and covariance for $\boldsymbol{y}_i$:

$$\begin{aligned} \mathrm{E}(\boldsymbol{y}_i) &= \mathrm{E}\{\mathrm{E}(\boldsymbol{y}_i|\boldsymbol{b}_i)\} = \boldsymbol{X}_i\boldsymbol{\beta}; \\ \mathrm{Cov}(\boldsymbol{y}_i) &= \mathrm{E}\{\mathrm{Cov}(\boldsymbol{y}_i|\boldsymbol{b}_i)\} + \mathrm{Cov}\{\mathrm{E}(\boldsymbol{y}_i|\boldsymbol{b}_i)\} \\ &= \boldsymbol{R}_i + \boldsymbol{Z}_i\boldsymbol{D}\boldsymbol{Z}_i' \\ &= \boldsymbol{V}_i. \end{aligned} \tag{3.6}$$

Thus, under the assumptions of normality and independence for $\boldsymbol{b}_i$ and $\boldsymbol{e}_i$, unconditionally, $\boldsymbol{y}_i$ are independently and normally distributed with mean $\boldsymbol{X}_i\boldsymbol{\beta}$ and covariance matrix $\boldsymbol{V}_i = \boldsymbol{R}_i + \boldsymbol{Z}_i\boldsymbol{D}\boldsymbol{Z}_i'$. This ability to specify the marginal distribution in closed form plays a key role in guiding sensible inference and depends critically on the assumed linear, additive dependence of the response on the fixed effects (through the design matrices $\boldsymbol{X}_i$) and the random effects (through the design matrices $\boldsymbol{Z}_i$). If instead, $\mathrm{E}(\boldsymbol{y}_i)$ depends on the $\boldsymbol{b}_i$ in a nonlinear fashion, it is not, in general, possible to specify the marginal distribution in closed forms. As discussed in subsequent chapters, this complicates inference considerably.

It is typically the case that $\boldsymbol{R}_i + \boldsymbol{Z}_i\boldsymbol{D}\boldsymbol{Z}_i'$ depends on a vector of distinct variance parameters $\boldsymbol{\omega}$ of low dimension. We shall use w to denote the dimension of $\boldsymbol{\omega}$.

Rat pup weight data

For the rat pup birth weight data, the model in equation (3.2) may be seen to be a particular case of that in equation (3.5). Set

$$\boldsymbol{X}_i = \begin{bmatrix} 1 & t_{i11} & t_{i12} & t_{i13} & l_i & x_{i1} \\ \vdots & \vdots & \vdots & \vdots & \vdots & \vdots \\ 1 & t_{in_i1} & t_{in_i2} & t_{in_i3} & l_i & x_{in_i} \end{bmatrix};$$

$$\begin{aligned} t_{ijk} &= 1,\ \text{if pup } (i,j) \text{ is in group } k, \\ &= 0,\ \text{otherwise;} \\ l_i &= \text{size of the } i\text{th litter;} \\ x_{ij} &= 1,\ \text{if pup } (i,j) \text{ is female,} \\ &= 0,\ \text{otherwise;} \\ \boldsymbol{b}_i &= b_i \text{ (scalar litter effect);} \end{aligned}$$

$$\boldsymbol{\beta} = \begin{bmatrix} \mu \\ \tau_1 \\ \tau_2 \\ \tau_3 \\ \lambda \\ \delta \end{bmatrix}; \quad \boldsymbol{Z}_i = \begin{bmatrix} 1 \\ \vdots \\ 1 \end{bmatrix}; \quad \boldsymbol{e}_i = \begin{bmatrix} e_{i1} \\ \vdots \\ e_{in_i} \end{bmatrix},$$

where $\boldsymbol{Z}_i$ is $(n_i \times 1)$. For this example, then, $p = 6$.

Assume that $\boldsymbol{e}_i \sim \mathcal{N}(\boldsymbol{0}, \sigma_e^2 \boldsymbol{I}_{n_i})$, so that $\boldsymbol{R}_i = \sigma_e^2 \boldsymbol{I}_{n_i}$. Assume that $b_i \sim \mathcal{N}(0, \sigma_b^2)$. Then, unconditionally,

$$\text{Cov}(\boldsymbol{y}_i) = \boldsymbol{R}_i + \boldsymbol{Z}_i\boldsymbol{D}\boldsymbol{Z}_i' = \sigma_e^2 \boldsymbol{I}_{n_i} + \sigma_b^2 \boldsymbol{J}_{n_i},$$

where $\boldsymbol{J}_{n_i}$ is an $(n_i \times n_i)$ matrix of ones, so that

$$\begin{aligned} \text{Var}(y_{ij}) &= \sigma_e^2 + \sigma_b^2, \\ \text{Cov}(y_{ij_1}, y_{ij_2}) &= \sigma_b^2. \end{aligned}$$

Note that in this example, the marginal covariance matrix, $\boldsymbol{R}_i + \boldsymbol{Z}_i\boldsymbol{D}\boldsymbol{Z}_i'$, is a function of a very small number (two) of variance parameters. Thus, $\omega = [\sigma_e^2, \sigma_b^2]'$, and $w = 2$.

Tracer lot data

To see that the random coefficient regression model (3.3) described in section 3.2.1 is a special case of the general model (3.5), write

$$y_{ij} = \beta_{0i} + \beta_{1i}t_{ij} + e_{ij}, \tag{3.7}$$

where $\mathrm{E}(e_{ij}) = 0$ and $\mathrm{Var}(e_{ij}) = \sigma_e^2$, which implies equation (3.3) when the e_{ij} are independent of β_{0i} and β_{1i}. Write

$$\boldsymbol{b}_i = \begin{bmatrix} b_{0i} \\ b_{1i} \end{bmatrix} = \begin{bmatrix} \beta_{0i} - \beta_0 \\ \beta_{1i} - \beta_1 \end{bmatrix}.$$

Define $\beta = [\beta_0, \beta_1]'$, so that $p = 2$, and set

$$\boldsymbol{X}_i = \begin{bmatrix} 1 & t_{i1} \\ \vdots & \vdots \\ 1 & t_{in_i} \end{bmatrix}.$$

With these definitions, (3.7) may be put into the form of equation (3.5), where $\boldsymbol{Z}_i = \boldsymbol{X}_i$ in this case.

The simplest specification for $\boldsymbol{R}_i$ is $\boldsymbol{R}_i = \sigma_e^2 \boldsymbol{I}_{n_i}$. For a random intercepts-and-slopes model, $\boldsymbol{D}$, the lot-to-lot covariance matrix

$$\boldsymbol{D} = \mathrm{Cov}(\boldsymbol{b}_i),$$

has form

$$\begin{bmatrix} \sigma_0^2 & \sigma_{01} \\ \sigma_{01} & \sigma_1^2 \end{bmatrix}$$

and involves three variance parameters. Thus, in the simplest instance, the marginal covariance matrix $\boldsymbol{V}_i = \boldsymbol{R}_i + \boldsymbol{Z}_i \boldsymbol{D} \boldsymbol{Z}_i'$ depends on a vector of variance parameters $\omega = [\sigma_0^2, \sigma_{01}, \sigma_1^2, \sigma_e^2]'$ of dimension $w = 4$. It is possible, and may be appropriate, to allow more complicated within-individual covariance structures. We defer discussion of this issue to the next chapter, but remark that, even if more complicated forms for $\boldsymbol{R}_i$ are allowed, it is still generally desirable to keep the dimension of the variance parameter vector ω relatively small.

3.2.3 Bayesian model specification

In this section, we describe a Bayesian formulation of the general hierarchical linear model. Further details of this approach to linear modeling may be found in Lindley and Smith (1972), Louis (1991), and Searle *et al.* (1992, Chapter 9).

We again consider the case where response vectors $\boldsymbol{y}_1, \ldots, \boldsymbol{y}_m$ for each of m individuals are assumed to depend linearly upon individual-specific regression parameters, which in turn represent samples from a prior distribution. This may be accommodated by the following hierarchical Bayesian model.

Write

$$\boldsymbol{y}_i = \boldsymbol{X}_i\boldsymbol{\beta} + \boldsymbol{Z}_i\boldsymbol{b}_i + \boldsymbol{e}_i \tag{3.8}$$

as before. As we have seen in section 3.2.2, the classical mixed model interpretation of (3.8) regards $\boldsymbol{\beta}$ as an unknown, fixed parameter and $\boldsymbol{b}_i$ as an unknown random vector. $\boldsymbol{X}_i$ and $\boldsymbol{Z}_i$ are known, fixed design matrices, and $\boldsymbol{e}_i$ is an unknown random vector. In the Bayesian interpretation of (3.8), the distribution of $\boldsymbol{e}_i$, which characterizes intra-individual variation, is sometimes referred to as the sampling distribution. Variation in $\boldsymbol{\beta}$ and $\boldsymbol{b}_i$ is accommodated at higher levels of the modeling hierarchy, in the following scheme:

Stage 1 (intra-individual variation)

$$\boldsymbol{y}_i = \boldsymbol{X}_i\boldsymbol{\beta} + \boldsymbol{Z}_i\boldsymbol{b}_i + \boldsymbol{e}_i, \quad \boldsymbol{e}_i \sim \mathcal{N}(\boldsymbol{0}, \boldsymbol{R}_i).$$

Stage 2 (inter-individual variation)

$$\boldsymbol{b}_i \sim \mathcal{N}(\boldsymbol{0}, \boldsymbol{D}).$$

Stage 3 (hyperprior distribution)

$$\boldsymbol{\beta} \sim \mathcal{N}(\boldsymbol{\beta}^*, \boldsymbol{H}),$$

$$\boldsymbol{D}^{-1} \sim \text{Wishart}, \quad \boldsymbol{R}_i^{-1} \sim \text{Wishart}.$$

The parameters $\boldsymbol{\beta}^*$, $\boldsymbol{H}$, and those characterizing the independent Wishart distributions for $\boldsymbol{D}^{-1}$ and $\boldsymbol{R}_i^{-1}$ are assumed to be known. Generally, choices are made to reflect weak knowledge at this stage of the hierarchy; a convenient assumption is that $\boldsymbol{H}^{-1} = \boldsymbol{0}$. Choice of normal and Wishart priors is an example of the common Bayesian strategy of using *conjugate* priors; a conjugate prior is one for which the resulting posterior distributions of interest come from the same distributional family. Inference within this Bayesian model framework is discussed in section 3.3.3.

3.3 Inference

This section provides a brief review of inferential techniques for the linear hierarchical model. A key element is the correct incorporation of both fixed and random effects in estimation and inference. In the case of repeated measurements on an individual, both inter- and intra-individual covariance structures must be correctly taken into account. To understand the interplay between estimation of fixed and random effects, we begin in section 3.3.1 by considering the simplest case of a single fixed and a single random effect. These ideas are generalized in section 3.3.2. In the Bayesian setting, the important distinction is that between *observable* and *unobservable* quantities. The Bayesian approach is summarized in section 3.3.3. Despite broad differences in philosophy between the Bayesian and classical approaches, we shall see that both lead to very similar inferential procedures in this setting.

3.3.1 One-way classification with random effects

For the pup weight data, consider the problem of estimating average body weight for the low dose males based on data for that group only. For simplicity of exposition, we omit the covariate litter size in this section. The model for body weight thus becomes particularly simple:

$$y_{ij} = \mu + b_i + e_{ij}, \tag{3.9}$$

where μ represents mean body weight, b_i is the random effect for the ith litter, and e_{ij} is a within-litter random error term. Let $\boldsymbol{b} = [b_1, \ldots, b_m]'$ and $\boldsymbol{e}_i = [e_{i1}, \ldots, e_{in_i}]'$. As before, assume that

$$\boldsymbol{b} \sim \mathcal{N}(\boldsymbol{0}, \sigma_b^2 \boldsymbol{I}_m), \qquad \boldsymbol{e}_i \sim \mathcal{N}(\boldsymbol{0}, \sigma_e^2 \boldsymbol{I}_{n_i}), \qquad \text{Cov}(\boldsymbol{b}, \boldsymbol{e}_i) = \boldsymbol{0}.$$

Three types of inference may be of interest:

(i) Inference regarding fixed effects – in this simple case, inference on the overall group mean μ.

(ii) Inference regarding variance components – estimation of and inference on the between-litter variance σ_b^2 and the within-litter variance σ_e^2.

(iii) Inference regarding the litter means – prediction of the quantities $\mu_i = \mu + b_i$.

These might appear to be separate questions. In fact, they are interrelated, as the following development shows.

Estimation of fixed effect assuming variance components are known

Suppose for the moment that the variance components σ_b^2 and σ_e^2 are *known*. Consider the problem of estimating μ. Three possibilities are immediate:

(i) Take the simple average of all pup weights.

(ii) Take the simple average of the litter means.

(iii) Do something intermediate between (i) and (ii).

Which of the three options is 'best?' One answer is the following. Denote the mean for the ith litter (of size n_i) by $\bar{y}_{i\cdot}$. Then it is straightforward to show that

$$\text{Var}(\bar{y}_i) = \sigma_b^2 + \sigma_e^2/n_i.$$

This suggests an estimate of μ based on the following weighted combination of litter means:

$$\hat{\mu}_w = \frac{\sum_{i=1}^m w_i \bar{y}_i}{\sum_{i=1}^m w_i},$$

where $w_i = (\sigma_b^2 + \sigma_e^2/n_i)^{-1}$, the inverse of $\text{Var}(\bar{y}_i)$. This is the *generalized least squares* (GLS) estimator of μ.

Other choices of w_i result in different estimates of μ. However, it may be shown (Searle *et al.* 1992) that

$$\text{Var}\{GLS(\mu)\} = \left(\sum_{i=1}^m \frac{n_i}{n_i\sigma_b^2 + \sigma_e^2} \right)^{-1}$$

achieves minimum variance among the class of estimators consisting of weighted averages of the litter means. Note that if $\sigma_b^2 = 0$ (no litter-to-litter variation), the GLS estimator reduces to the simple average of pup weights. If $\sigma_e^2 = 0$ (no pup-to-pup variation), the GLS estimator is a simple average of the litter means. In the general case, the GLS estimator represents an intermediate strategy, depending on the relative magnitudes of the variance components σ_b^2 and σ_e^2 and the values of n_i. For the balanced case, $n_i \equiv n$, options (i) and (ii) coincide.

In practice, of course, σ_b^2 and σ_e^2 will not be known. If estimates of the variance components are available, then GLS estimation of μ may be carried out using weights based on these estimates.

This simple example illustrates a key feature of inference in mixed effects models. Estimation of fixed effects and variance components are inextricably linked. In many applications, inference regarding variance components is of interest *per se*, but even in cases where variance component estimation is not of primary interest, it is a necessary prerequisite for sensible inference on fixed effects.

Estimation of variance components

There is an enormous literature devoted to variance component estimation. A comprehensive review is given by Searle *et al.* (1992) In this book, we shall restrict attention to two methods, maximum likelihood (ML) and restricted maximum likelihood (REML).

Normal theory maximum likelihood. Given the specification of a normal distribution for both random effects and the consequent ability to write down the marginal likelihood, maximum likelihood (ML) represents an appealing approach to variance component estimation. One advantage is that the estimates are constrained to be non-negative; a second is that ML techniques extend readily to more complex mixed models. Iterative computation is required, but as we discuss in section 3.4, several algorithms are available for the general mixed effects model. One disadvantage of maximum likelihood estimation is that it makes no allowance for the loss of degrees of freedom associated with estimation of fixed effects when estimating variance components. As a result, variance component estimates are generally biased downward.

Restricted maximum likelihood. Restricted maximum likelihood estimation (REML) of variance components corrects for the loss of degrees of freedom due to estimating fixed effects. In the simple case considered here, this amounts to estimating σ_e^2 by dividing the within-litter sum of squares by $N - m$ rather than N as in the case of maximum likelihood. A more precise definition of restricted maximum likelihood for the general setting is given in the next section. Iterative computation is typically needed; details are discussed in sections 3.3.2 and 3.4.

It is reasonable to ask which of these two methods is preferable for estimating variance components. REML estimation techniques are generally favored, as it is considered desirable to correct for loss of degrees of freedom for fixed effects where appropriate.

Both methods are quite nonrobust to extreme data points, being based on sums of squared deviations. In our discussion of the hierarchical nonlinear model, we shall also consider estimation of variance parameters based on absolute rather than squared deviations, as in section 2.3.3. Such methods can help alleviate the susceptibility of variance parameter estimation to a small fraction of outlying observations.

Sensible estimation of μ, the fixed effect in this example, is predicated upon knowledge of the variance components σ_e^2 and σ_b^2. Estimation of these, in turn, is typically based on deviations involving the current estimate of μ. Thus, estimation of $(\mu, \sigma_e^2, \sigma_b^2)$ can proceed according to an iterative strategy, wherein the estimate of μ is updated to reflect current estimates of (σ_e^2, σ_b^2) and conversely. Details are presented for the general case in section 3.3.2; a similar generalized least squares strategy for estimation of regression parameters in nonlinear regression models for data from a single individual was considered in section 2.3.

Best linear unbiased prediction

Before proceeding to the general model, we consider the problem of estimating $\mu_i = \mu + b_i$ in the simplest case (3.9). An example where this might be of interest is the case where several test measurements are made on each of a number of subjects. In this situation, we may be interested in both inference on μ, the average score for the population, and on $\mu + b_i$, the average score for the ith subject.

Information on b_i, the deviation of the ith subject's average from the overall mean, is provided by the data through the quantity $\bar{y}_{i\cdot}$, the average of the n_i scores for that subject. Intuitively, if $\bar{y}_{i\cdot}$ is 'high' (e.g. higher than the overall average $\bar{y}_{\cdot\cdot}$), it seems likely that b_i is positive. A reasonable predictor of b_i is given by

$$\tilde{b}_i = \mathrm{E}(b_i|\bar{y}_{i\cdot}).$$

(Note that $\mathrm{E}(b_i)$ is not a sensible predictor; it is zero by assumption and makes no use of the data.)

Under the assumptions of (3.9), it is straightforward to derive an explicit expression for $\mathrm{E}(b_i|\bar{y}_{i\cdot})$. It is straightforward to show that

$$\begin{bmatrix} b_i \\ \bar{y}_{i\cdot} \end{bmatrix} \sim \mathcal{N}\left(\begin{bmatrix} 0 \\ \mu \end{bmatrix}, \begin{bmatrix} \sigma_b^2 & \sigma_b^2 \\ \sigma_b^2 & \sigma_b^2 + \sigma_e^2/n_i \end{bmatrix} \right),$$

from which it follows that

$$\mathrm{E}(b_i|\bar{y}_{i\cdot}) = \frac{n_i\sigma_b^2}{\sigma_e^2 + \sigma_b^2 n_i}(\bar{y}_{i\cdot} - \mu).$$

If μ is replaced by its GLS estimate in this regression, the resulting predictor of b_i is known as the *best linear unbiased predictor* (BLUP) of b_i:

$$BLUP(b_i) = \frac{n_i\sigma_b^2}{\sigma_e^2 + n_i\sigma_b^2}\{\bar{y}_{i\cdot} - GLS(\mu)\}.$$

As the name implies, this expression is unbiased (i.e. it has mean zero) and has minimum variance among all linear functions of the observations that are unbiased. The natural estimate of $\mu_i = \mu + b_i$, the mean for the ith subject, is then

$$\begin{aligned} BLUP(\mu + b_i) &= GLS(\mu) + BLUP(b_i) \\ &= GLS(\mu) + \frac{n_i\sigma_b^2}{\sigma_e^2 + n_i\sigma_e^2}\{\bar{y}_{i\cdot} - GLS(\mu)\}. \end{aligned}$$

A little algebra shows that this may be rewritten as

$$BLUP(\mu_i) = \bar{y}_{i\cdot} - \frac{\sigma_e^2}{n_i\sigma_b^2 + \sigma_e^2}\{\bar{y}_{i\cdot} - GLS(\mu)\}.$$

That is, the estimate shrinks $\bar{y}_{i\cdot}$ closer to the overall mean by an amount depending on n_i and the relative magnitudes of the variance components σ_b^2 and σ_e^2. This type of estimate, wherein an individual average is 'shrunk' closer to an overall population value, is often referred to as a James–Stein (1961) or *empirical Bayes* estimate. The connection with Bayesian inference is elucidated further in section 3.3.3. In practice, σ_b^2 and σ_e^2 will not be known, so implementation of BLUP requires point estimation of the variance components.

3.3.2 General linear mixed effects model

We now generalize the preceding arguments. The notation becomes more complicated, but the basic ideas are the same. Inference on three questions is of primary interest: the fixed effect parameters $\boldsymbol{\beta}$, the random effects $\boldsymbol{b}_i$, and the covariance components, $\boldsymbol{R}_i$ and $\boldsymbol{D}$. As before, it proves conceptually useful to separate estimation of $\boldsymbol{\beta}$ and $\boldsymbol{b}_i$ from that of $\boldsymbol{R}_i$ and $\boldsymbol{D}$.

Estimation of effects if covariance structure is known

The marginal distribution of y_i is $\mathcal{N}(\boldsymbol{X}_i\boldsymbol{\beta}, \boldsymbol{V}_i)$, where $\boldsymbol{V}_i = \boldsymbol{R}_i + \boldsymbol{Z}_i\boldsymbol{D}\boldsymbol{Z}_i'$. In the case where the covariance matrices $\boldsymbol{R}_i$ and $\boldsymbol{D}$ are known, one can base inference regarding $\boldsymbol{\beta}$ and $\boldsymbol{b}_i$ on the marginal likelihood.

To streamline the presentation, it is customary to define the *combined* model for the data from all m individuals as follows. Let

$$\boldsymbol{y} = \begin{bmatrix} \boldsymbol{y}_1 \\ \vdots \\ \boldsymbol{y}_m \end{bmatrix}, \quad \boldsymbol{b} = \begin{bmatrix} \boldsymbol{b}_1 \\ \vdots \\ \boldsymbol{b}_m \end{bmatrix}, \quad \boldsymbol{X} = \begin{bmatrix} \boldsymbol{X}_1 \\ \vdots \\ \boldsymbol{X}_m \end{bmatrix},$$

so that $\boldsymbol{y}$ is $(N \times 1)$, $\boldsymbol{b}$ is $(km \times 1)$, and $\boldsymbol{X}$ is an $(N \times p)$ matrix. Define the block diagonal matrices $\tilde{\boldsymbol{D}} = \text{diag}(\boldsymbol{D}, \ldots, \boldsymbol{D})$ $(km \times km)$, $\boldsymbol{Z} = \text{diag}(\boldsymbol{Z}_1, \ldots, \boldsymbol{Z}_m)$ $(N \times km)$, and $\boldsymbol{R} = \text{diag}(\boldsymbol{R}_1, \ldots, \boldsymbol{R}_m)$ $(N \times N)$. Then $\boldsymbol{V} = \text{diag}(\boldsymbol{V}_1, \ldots, \boldsymbol{V}_m) = \boldsymbol{R} + \boldsymbol{Z}\tilde{\boldsymbol{D}}\boldsymbol{Z}'$ $(N \times N)$. The combined model is written as

$$\boldsymbol{y} = \boldsymbol{X}\boldsymbol{\beta} + \boldsymbol{Z}\boldsymbol{b} + \boldsymbol{e}, \tag{3.10}$$

and the marginal distribution for the combined data vector $\boldsymbol{y}$ is $\mathcal{N}(\boldsymbol{X}\boldsymbol{\beta}, \boldsymbol{V})$.

For *known* $\boldsymbol{D}$ and $\boldsymbol{R}_i$, inference for $\boldsymbol{\beta}$ and $\boldsymbol{b}_i$ may be based on the marginal likelihood. Estimates may be obtained by jointly minimizing in $\boldsymbol{\beta}$ and $\boldsymbol{b}_i$ the objective function

$$\begin{aligned} &\log|\tilde{\boldsymbol{D}}| + \boldsymbol{b}'\tilde{\boldsymbol{D}}^{-1}\boldsymbol{b} + \log|\boldsymbol{R}| \\ &+(\boldsymbol{y} - \boldsymbol{X}\boldsymbol{\beta} - \boldsymbol{Z}\boldsymbol{b})'\boldsymbol{R}^{-1}(\boldsymbol{y} - \boldsymbol{X}\boldsymbol{\beta} - \boldsymbol{Z}\boldsymbol{b})'. \end{aligned} \tag{3.11}$$

(3.11) is twice the negative log of the posterior density of $\boldsymbol{b}_i$ for fixed $\boldsymbol{\beta}$ and twice the negative loglikelihood for $\boldsymbol{\beta}$ for fixed $\boldsymbol{b}_i$. Minimization may be accomplished by implementing a so-called 'pseudo-data' regression approach. Consider the linear least squares problem with $(N + km \times 1)$ augmented data vector

$$\begin{bmatrix} \boldsymbol{R}^{-1/2}\boldsymbol{y} \\ \boldsymbol{0}_{km\times 1} \end{bmatrix}$$

and linear regression function

$$\begin{bmatrix} \boldsymbol{R}^{-1/2}\boldsymbol{X} & \boldsymbol{R}^{-1/2}\boldsymbol{Z} \\ \boldsymbol{0}_{km\times p} & \tilde{\boldsymbol{D}}^{-1/2} \end{bmatrix} \begin{bmatrix} \boldsymbol{\beta} \\ \boldsymbol{b} \end{bmatrix} = \begin{bmatrix} \boldsymbol{R}^{-1/2}(\boldsymbol{X}\boldsymbol{\beta} + \boldsymbol{Z}\boldsymbol{b}) \\ \tilde{\boldsymbol{D}}^{-1/2}\boldsymbol{b} \end{bmatrix}.$$

Solution of this pseudo-data regression problem minimizes the required objective function, (3.11). It is straightforward to show that the resulting estimators for $\boldsymbol{\beta}$ and $\boldsymbol{b}$ solve the so-called *mixed model equations*

$$\begin{bmatrix} \boldsymbol{X}'\boldsymbol{R}^{-1}\boldsymbol{X} & \boldsymbol{X}'\boldsymbol{R}^{-1}\boldsymbol{Z} \\ \boldsymbol{Z}'\boldsymbol{R}^{-1}\boldsymbol{X} & \boldsymbol{Z}'\boldsymbol{R}^{-1}\boldsymbol{Z} + \tilde{\boldsymbol{D}}^{-1} \end{bmatrix} \begin{bmatrix} \hat{\boldsymbol{\beta}} \\ \hat{\boldsymbol{b}} \end{bmatrix} = \begin{bmatrix} \boldsymbol{X}'\boldsymbol{R}^{-1}\boldsymbol{y} \\ \boldsymbol{Z}'\boldsymbol{R}^{-1}\boldsymbol{y} \end{bmatrix}. \tag{3.12}$$

Without the $\tilde{\boldsymbol{D}}^{-1}$ in the matrix on the left hand side of (3.12), these would supply the maximum likelihood estimates for the model treating $\boldsymbol{b}$ as fixed effects.

Using the matrix identity

$$\boldsymbol{R}^{-1} - \boldsymbol{R}^{-1}\boldsymbol{Z}(\boldsymbol{Z}'\boldsymbol{R}^{-1}\boldsymbol{Z} + \tilde{\boldsymbol{D}}^{-1})^{-1}\boldsymbol{Z}'\boldsymbol{R}^{-1} = (\boldsymbol{R} + \boldsymbol{Z}\tilde{\boldsymbol{D}}\boldsymbol{Z}')^{-1} ,$$

it may be shown that the estimator for $\boldsymbol{\beta}$ that solves (3.12) is the generalized least squares estimator

$$\begin{aligned} \hat{\boldsymbol{\beta}} &= (\boldsymbol{X}'\boldsymbol{V}^{-1}\boldsymbol{X})^{-1}\boldsymbol{X}'\boldsymbol{V}^{-1}\boldsymbol{y} \\ &= \Big(\sum_{i=1}^{m} \boldsymbol{X}_i'\boldsymbol{V}_i^{-1}\boldsymbol{X}_i\Big)^{-1} \sum_{i=1}^{m} \boldsymbol{X}_i'\boldsymbol{V}_i^{-1}\boldsymbol{y}_i; \end{aligned} \tag{3.13}$$

(3.13) is the maximum likelihood estimator for known $\tilde{\boldsymbol{D}}$ and $\boldsymbol{R}$. The estimator $\hat{\boldsymbol{b}}$ for $\boldsymbol{b}$ which solves (3.12) satisfies

$$\hat{\boldsymbol{b}} = (\boldsymbol{Z}'\boldsymbol{R}^{-1}\boldsymbol{Z} + \tilde{\boldsymbol{D}}^{-1})^{-1}\boldsymbol{Z}'\boldsymbol{R}^{-1}(\boldsymbol{y} - \boldsymbol{X}\hat{\boldsymbol{\beta}}).$$

Applying the matrix identity

$$(\tilde{\boldsymbol{D}}^{-1} + \boldsymbol{Z}'\boldsymbol{R}^{-1}\boldsymbol{Z})^{-1}\boldsymbol{Z}'\boldsymbol{R}^{-1} = \tilde{\boldsymbol{D}}\boldsymbol{Z}'\boldsymbol{V}^{-1},$$

it may be shown that this estimator may be written more simply as

$$\hat{\boldsymbol{b}} = \tilde{\boldsymbol{D}}\boldsymbol{Z}'\boldsymbol{V}^{-1}(\boldsymbol{y} - \boldsymbol{X}\hat{\boldsymbol{\beta}}), \tag{3.14}$$

so that the individual estimates $\hat{\boldsymbol{b}}_i$ are given by

$$\hat{\boldsymbol{b}}_i = \boldsymbol{D}\boldsymbol{Z}_i'\boldsymbol{V}_i^{-1}(\boldsymbol{y}_i - \boldsymbol{X}_i\hat{\boldsymbol{\beta}}).$$

This estimate is BLUP for $\boldsymbol{b}$ (Searle *et al.*, 1992, section 7.6). $\hat{\boldsymbol{b}}$ may also be derived as the posterior mean $\mathrm{E}(\boldsymbol{b}|\boldsymbol{y}, \hat{\boldsymbol{\beta}}; \boldsymbol{R}, \tilde{\boldsymbol{D}})$, and thus has an empirical Bayes interpretation, as a weighted average of $\boldsymbol{0}$ $(= \mathrm{E}(\boldsymbol{b}))$ and the weighted least squares estimate obtained by regarding $\boldsymbol{b}$ as a fixed effect.

Because both $\boldsymbol{\beta}$ and $\boldsymbol{b}_i$ are linear functions of $\boldsymbol{y}$, it is straightforward to derive standard errors. Specifically,

$$\mathrm{Cov}(\hat{\boldsymbol{\beta}}) = \boldsymbol{X}'\boldsymbol{V}^{-1}\boldsymbol{X} = \Big(\sum_{i=1}^{m} \boldsymbol{X}_i'\boldsymbol{V}_i^{-1}\boldsymbol{X}_i\Big)^{-1}, \tag{3.15}$$

and

$$\begin{aligned} \mathrm{Cov}(\hat{\boldsymbol{b}}_i - \boldsymbol{b}_i) = \boldsymbol{D} &- \boldsymbol{D}(\boldsymbol{Z}_i'\boldsymbol{V}_i^{-1}\boldsymbol{Z}_i)\boldsymbol{D} + \boldsymbol{D}(\boldsymbol{Z}'\boldsymbol{V}_i^{-1}\boldsymbol{X}_i) \\ &\times \Big(\sum_{i=1}^{m} \boldsymbol{X}_i'\boldsymbol{V}_i^{-1}\boldsymbol{X}_i\Big)^{-1}(\boldsymbol{X}_i'\boldsymbol{V}_i^{-1}\boldsymbol{Z}_i)\boldsymbol{D}. \end{aligned} \tag{3.16}$$

As pointed out by Laird and Ware (1982), (3.16) is used to assess the error of estimation of $\boldsymbol{b}_i$ rather than $\text{Cov}(\hat{\boldsymbol{b}}_i)$ because the latter expression ignores the variation of $\boldsymbol{b}_i$.

Estimation of effects if covariance structure is unknown

If the covariance matrices are unknown, but a point estimate of $\boldsymbol{\omega}$, and hence of $\boldsymbol{R}_i$ and $\boldsymbol{D}$, is available, a natural extension of the estimator given above is to replace $\boldsymbol{V}_i$ by $\hat{\boldsymbol{V}}_i = \hat{\boldsymbol{R}}_i + \boldsymbol{Z}_i\hat{\boldsymbol{D}}\boldsymbol{Z}_i$, where, in obvious notation, $\hat{\boldsymbol{R}}_i$ and $\hat{\boldsymbol{D}}$ are these matrices evaluated at the estimate of $\boldsymbol{\omega}$. Expressions for standard errors of the resulting estimators may be obtained by substitution of $\hat{\boldsymbol{V}}_i$ and $\hat{\boldsymbol{D}}$ in (3.15) and (3.16). These standard errors are optimistic, as they fail to reflect the uncertainty due to estimation of $\boldsymbol{\omega}$.

Estimation of variance components

Here we focus on methods based on maximum likelihood (ML) and restricted (or residual) maximum likelihood (REML) techniques. An underlying assumption of normality is made; this leads to mathematically tractable solutions, but may restrict the suitability of these methods for some data sets.

Maximum likelihood estimation of the covariance parameter vector $\boldsymbol{\omega}$ involves maximization of the marginal loglikelihood

$$\log L = -\frac{1}{2}N\log 2\pi - \frac{1}{2}\log|\boldsymbol{V}| - \frac{1}{2}(\boldsymbol{y} - \boldsymbol{X\beta})'\boldsymbol{V}^{-1}(\boldsymbol{y} - \boldsymbol{X\beta}) \tag{3.17}$$

with respect to the components of $\boldsymbol{\omega}$. Joint maximization with respect to $\boldsymbol{\beta}$ and $\boldsymbol{\omega}$ results in the GLS estimates of $\boldsymbol{\beta}$, described above.

Restricted maximum likelihood estimation adjusts for loss of degrees of freedom due to estimating fixed effects (e.g. $\boldsymbol{\beta}$). One way to think of REML estimation is that of estimating variance components based on residuals calculated after fitting fixed effects only. Algebraically, this corresponds to replacing $\log L$ in (3.17) by

$$\log L_R = \log L + \frac{1}{2}p\log 2\pi - \frac{1}{2}\log|\boldsymbol{X}'\boldsymbol{V}^{-1}\boldsymbol{X}|, \tag{3.18}$$

where $\log L$ is evaluated at $\hat{\boldsymbol{\beta}}$. Another view of REML estimation, in terms of marginal likelihood, is presented in section 3.3.3.

Discussion of computational aspects is deferred until section 3.4.

3.3.3 Bayesian inference

We recall the fundamental approach of Bayesian inference. Given a random variable $\boldsymbol{Y}$ with density $p(\boldsymbol{y}|\boldsymbol{\tau})$ depending on a parameter $\boldsymbol{\tau}$ and a prior distribution $\pi(\boldsymbol{\tau})$ for $\boldsymbol{\tau}$, inference regarding $\boldsymbol{\tau}$ is based on the posterior distribution of $\boldsymbol{\tau}$ given the data:

$$\pi(\boldsymbol{\tau}|\boldsymbol{y}) = \frac{p(\boldsymbol{y}|\boldsymbol{\tau})\pi(\boldsymbol{\tau})}{\int p(\boldsymbol{y}|\boldsymbol{\tau})\pi(\boldsymbol{\tau})\,d\boldsymbol{\tau}}.$$

For instance, an estimate of $\boldsymbol{\tau}$ might be computed as the mean or the mode of the posterior distribution. This approach is conceptually straightforward, though in practice calculation of the relevant posterior distributions can be difficult. The normal two-stage hierarchy described in section 3.2.3 does allow relatively tractable methods, however. As in previous sections, we begin by considering inference on model effects $\boldsymbol{\beta}$ and $\boldsymbol{b}$ assuming variance components are known, and then turn to the issue of covariance estimation.

Our presentation here is based on that in Searle *et al.* (1992). Further details may be found in Chapter 9 of that reference.

Estimation of effects if covariance structure is known

(i) *Estimating $\boldsymbol{\beta}$.* The posterior distribution of interest is given by

$$p_{\beta}(\boldsymbol{\beta}|\boldsymbol{y}) = \frac{\int p(\boldsymbol{y}|\boldsymbol{\beta},\boldsymbol{b})p_{\beta}(\boldsymbol{\beta})p_{b}(\boldsymbol{b})\,d\boldsymbol{b}}{\int\int p(\boldsymbol{y}|\boldsymbol{\beta},\boldsymbol{b})p_{\beta}(\boldsymbol{\beta})p_{b}(\boldsymbol{b})\,d\boldsymbol{b}\,d\boldsymbol{\beta}}, \tag{3.19}$$

where

$$\boldsymbol{y}|\boldsymbol{\beta},\boldsymbol{b} \sim \mathcal{N}(\boldsymbol{X\beta}+\boldsymbol{Zb},\boldsymbol{R}),$$

$$\boldsymbol{\beta} \sim \mathcal{N}(\boldsymbol{\beta}^*,\boldsymbol{H}), \qquad \boldsymbol{b} \sim \mathcal{N}(\boldsymbol{0},\tilde{\boldsymbol{D}}),$$

so that $p(\boldsymbol{y}|\boldsymbol{\beta},\boldsymbol{b})$, $p_{\beta}(\boldsymbol{\beta})$, and $p_b(\boldsymbol{b})$ are normal densities. For simplicity, we have suppressed the dependence on dispersion matrices in (3.19).

Under the assumptions of the normal theory hierarchy, the joint distribution of

$$\begin{bmatrix} \boldsymbol{\beta} \\ \boldsymbol{b} \\ \boldsymbol{y} \end{bmatrix}$$

is given by

$$\mathcal{N}\left(\begin{bmatrix} \boldsymbol{\beta}^* \\ \mathbf{0}_{km} \\ \boldsymbol{X\beta}^* \end{bmatrix}, \begin{bmatrix} \boldsymbol{H} & \mathbf{0}_{p\times km} & \boldsymbol{HX}' \\ \mathbf{0}_{km\times p} & \tilde{\boldsymbol{D}} & \tilde{\boldsymbol{D}}\boldsymbol{Z}' \\ \boldsymbol{XH} & \boldsymbol{Z}\tilde{\boldsymbol{D}} & \boldsymbol{XHX}' + \boldsymbol{Z}\tilde{\boldsymbol{D}}\boldsymbol{Z}' + \boldsymbol{R} \end{bmatrix}\right),$$

where $\mathbf{0}_\ell$ and $\mathbf{0}_{\ell\times u}$ are $(\ell\times 1)$ and $(\ell\times u)$ matrices of zeroes, respectively. Using standard results for conditional multivariate normal distributions and exploiting certain matrix identities, it can be shown that

$$\boldsymbol{\beta}|\boldsymbol{y} \sim \mathcal{N}\{\boldsymbol{C}^{-1}(\boldsymbol{X}'\boldsymbol{V}^{-1}\boldsymbol{y} + \boldsymbol{H}^{-1}\boldsymbol{\beta}^*), \boldsymbol{C}^{-1}\}, \qquad (3.20)$$

where $\boldsymbol{C} = \boldsymbol{X}'\boldsymbol{V}^{-1}\boldsymbol{X} + \boldsymbol{H}^{-1}$.

This result may be used as a basis for inference on $\boldsymbol{\beta}$. For instance, a possible point estimate for $\boldsymbol{\beta}$ is

$$\mathrm{E}(\boldsymbol{\beta}|\boldsymbol{y}) = \boldsymbol{C}^{-1}(\boldsymbol{X}'\boldsymbol{V}^{-1}\boldsymbol{y} + \boldsymbol{H}^{-1}\boldsymbol{\beta}^*),$$

with variance given by

$$\mathrm{Cov}\{E(\boldsymbol{\beta}|\boldsymbol{y})\} = \boldsymbol{C}^{-1}(\boldsymbol{X}'\boldsymbol{V}^{-1}\boldsymbol{X})\boldsymbol{C}^{-1}.$$

In the event that a noninformative prior is chosen for $\boldsymbol{\beta}$, $\boldsymbol{H}^{-1} = \mathbf{0}$,

$$\mathrm{E}(\boldsymbol{\beta}|\boldsymbol{y}) = (\boldsymbol{X}'\boldsymbol{V}^{-1}\boldsymbol{X})^{-1}\boldsymbol{X}'\boldsymbol{V}^{-1}\boldsymbol{y},$$

$$\mathrm{Cov}\{E(\boldsymbol{\beta}|\boldsymbol{y})\} = (\boldsymbol{X}'\boldsymbol{V}^{-1}\boldsymbol{X})^{-1},$$

the GLS estimate of $\boldsymbol{\beta}$ and its variance. Thus, the generalized least squares estimator for $\boldsymbol{\beta}$ may be viewed as a posterior estimate obtained from a prior distribution with infinite variance.

(ii) ***Estimating b.*** Just as we did for $\boldsymbol{\beta}$, we can derive the posterior distribution of $\boldsymbol{b}$ given $\boldsymbol{y}$ from the joint distribution of $\boldsymbol{\beta}$, $\boldsymbol{b}$, and $\boldsymbol{y}$. It is multivariate normal with mean

$$\mathrm{E}(\boldsymbol{b}|\boldsymbol{y}) = (\boldsymbol{Z}'\boldsymbol{L}\boldsymbol{Z}+\tilde{\boldsymbol{D}})^{-1}\boldsymbol{Z}'\boldsymbol{L}(\boldsymbol{y} - \boldsymbol{X\beta}^*),$$

where $\boldsymbol{L} = (\boldsymbol{R} + \boldsymbol{XHX}')^{-1}$, and dispersion matrix

$$\mathrm{Cov}(\boldsymbol{b}|\boldsymbol{y}) = (\boldsymbol{Z}'\boldsymbol{L}\boldsymbol{Z}+\tilde{\boldsymbol{D}})^{-1}.$$

Assuming vague prior information on $\boldsymbol{\beta}$ by setting $\boldsymbol{H}^{-1} = \mathbf{0}$, recalling that $\boldsymbol{V} = \boldsymbol{Z}\tilde{\boldsymbol{D}}\boldsymbol{Z}' + \boldsymbol{R}$ and $\hat{\boldsymbol{\beta}} = (\boldsymbol{X}'\boldsymbol{V}^{-1}\boldsymbol{X})^{-1}\boldsymbol{X}'\boldsymbol{V}^{-1}\boldsymbol{y}$, and invoking standard results from matrix theory, it may be shown that

$$\mathrm{E}(\boldsymbol{b}|\boldsymbol{y}) = \tilde{\boldsymbol{D}}\boldsymbol{Z}'\boldsymbol{V}^{-1}(\boldsymbol{y} - \boldsymbol{X}\hat{\boldsymbol{\beta}}),$$

the best linear unbiased predictor (BLUP) of $\boldsymbol{b}$.

Empirical Bayes estimation

The expressions for the posterior quantities $\mathrm{E}(\boldsymbol{\beta}|\boldsymbol{y})$, $\mathrm{Cov}(\boldsymbol{\beta}|\boldsymbol{y})$, and $\mathrm{E}(\boldsymbol{b}|\boldsymbol{y})$ involve the unobservable covariance parameters $\tilde{\boldsymbol{D}}$ and $\boldsymbol{R}$. A fully Bayesian approach would require specifying prior distributions for $\boldsymbol{D}$ and $\boldsymbol{R}$ and then integrating out these parameters. However, as this can be quite cumbersome, a simpler strategy is to substitute point estimates for $\tilde{\boldsymbol{D}}$ and $\boldsymbol{R}$ when evaluating the expression for $\mathrm{E}(\boldsymbol{\beta}|\boldsymbol{y})$ and $\mathrm{E}(\boldsymbol{b}|\boldsymbol{y})$. This general approach of substituting point estimates for certain parameters in the posterior distribution of interest is usually referred to as *empirical Bayes estimation*. In the context of hierarchical models in particular, empirical Bayes estimation is most often applied to the 'random effects' $\boldsymbol{b}_i$.

Estimation of covariance components

Conceptually, within the Bayesian framework, inference regarding the dispersion matrices $\boldsymbol{D}$ and $\boldsymbol{R}$ could be carried out by computation of the relevant posterior distributions. In practice, however, the high-dimensional integrations that such derivations entail can be prohibitively complicated. Recent advances in computational techniques, such as Gibbs sampling methods, may prove useful in this context. These methods are considered further in Chapter 8. For the moment, we confine our attention to simpler strategies for estimating variances. In particular, we show the connection between the ML and REML estimation techniques considered previously and Bayesian estimation.

The integrations over the next few pages may seem forbidding at first sight, but are actually quite straightforward, involving nothing more than standard techniques of multivariate integration and matrix algebra. If desired, the reader may omit the details of this section without significant loss of continuity.

Consider the hierarchical specification that uses a point mass prior for $\boldsymbol{\beta}$:

$$\boldsymbol{y}|\boldsymbol{\beta}, \boldsymbol{b} \sim \mathcal{N}(\boldsymbol{X}\boldsymbol{\beta} + \boldsymbol{Z}\boldsymbol{b}, \boldsymbol{R}),$$

$$\boldsymbol{b} \sim \mathcal{N}(\boldsymbol{0}, \tilde{\boldsymbol{D}}). \tag{3.21}$$

We may obtain the usual likelihood function for $\boldsymbol{\beta}, \boldsymbol{R}$, and $\tilde{\boldsymbol{D}}$ by integrating over $\boldsymbol{b}$:

$$L(\boldsymbol{\beta}, \tilde{\boldsymbol{D}}, \boldsymbol{R}|\boldsymbol{y}) = \int L(\boldsymbol{b}, \boldsymbol{\beta}, \tilde{\boldsymbol{D}}, \boldsymbol{R}|\boldsymbol{y})\, d\boldsymbol{b},$$

where

$$L(\boldsymbol{b}, \boldsymbol{\beta}, \tilde{\boldsymbol{D}}, \boldsymbol{R}|\boldsymbol{y}) = \frac{1}{(2\pi)^{N/2}|\boldsymbol{R}|^{1/2}}$$

$$\times \exp\Big[-\frac{1}{2}\{\boldsymbol{y} - (\boldsymbol{X\beta} + \boldsymbol{Zb})\}'\boldsymbol{R}^{-1}\{\boldsymbol{y} - (\boldsymbol{X\beta} + \boldsymbol{Zb})\}\Big]$$

$$\times \frac{1}{(2\pi)^{km/2}|\tilde{\boldsymbol{D}}|^{1/2}} \exp(-\frac{1}{2}\boldsymbol{b}'\tilde{\boldsymbol{D}}^{-1}\boldsymbol{b})$$

$$= \frac{1}{(2\pi)^{N/2}|\boldsymbol{R}|^{1/2}} \times \frac{1}{(2\pi)^{km/2}|\tilde{\boldsymbol{D}}|^{1/2}}$$

$$\times \exp\Big[-\frac{1}{2}(\boldsymbol{y} - \boldsymbol{X\beta})'\boldsymbol{R}^{-1}(\boldsymbol{y} - \boldsymbol{X\beta})$$

$$-\frac{1}{2}\boldsymbol{b}'(\tilde{\boldsymbol{D}}^{-1} + \boldsymbol{Z}'\boldsymbol{R}^{-1}\boldsymbol{Z})\boldsymbol{b} + \boldsymbol{b}'\boldsymbol{Z}'\boldsymbol{R}^{-1}(\boldsymbol{y} - \boldsymbol{X\beta})\Big].$$

Writing $\boldsymbol{A} = \tilde{\boldsymbol{D}}^{-1} + \boldsymbol{Z}'\boldsymbol{R}^{-1}\boldsymbol{Z}$, and completing the square in the exponent, we obtain

$$L(\boldsymbol{b}, \boldsymbol{\beta}, \tilde{\boldsymbol{D}}, \boldsymbol{R}|\boldsymbol{y}) = \frac{1}{(2\pi)^{N/2}|\boldsymbol{R}|^{1/2}} \times \frac{1}{(2\pi)^{km/2}|\tilde{\boldsymbol{D}}|^{1/2}}$$

$$\times \exp\Big[-\frac{1}{2}(\boldsymbol{y} - \boldsymbol{X\beta})'\boldsymbol{R}^{-1}(\boldsymbol{y} - \boldsymbol{X\beta}) - \frac{1}{2}\boldsymbol{d}'\boldsymbol{A}\boldsymbol{d}$$

$$-\frac{1}{2}(\boldsymbol{y} - \boldsymbol{X\beta})'\boldsymbol{R}^{-1}\boldsymbol{Z}\boldsymbol{A}^{-1}\boldsymbol{Z}'\boldsymbol{R}^{-1}(\boldsymbol{y} - \boldsymbol{X\beta})\Big]$$

$$= \frac{1}{(2\pi)^{N/2}|\boldsymbol{R}|^{1/2}} \times \frac{1}{(2\pi)^{km/2}|\tilde{\boldsymbol{D}}|^{1/2}}$$

$$\times \exp\Big[-\frac{1}{2}(\boldsymbol{y} - \boldsymbol{X\beta})'\boldsymbol{V}^{-1}(\boldsymbol{y} - \boldsymbol{X\beta}) - \frac{1}{2}(\boldsymbol{d}'\boldsymbol{A}\boldsymbol{d})\Big],$$

where $\boldsymbol{d} = \boldsymbol{b} - \boldsymbol{A}^{-1}\boldsymbol{Z}'\boldsymbol{R}^{-1}(\boldsymbol{y} - \boldsymbol{X\beta})$ and using the identity

$$\boldsymbol{R}^{-1} - \boldsymbol{R}^{-1}\boldsymbol{Z}\boldsymbol{A}^{-1}\boldsymbol{Z}'\boldsymbol{R}^{-1} = (\boldsymbol{Z}\tilde{\boldsymbol{D}}\boldsymbol{Z}' + \boldsymbol{R})^{-1} = \boldsymbol{V}^{-1}.$$

Noting that

$$\int \exp\Big[-\frac{1}{2}\boldsymbol{d}'\boldsymbol{A}\boldsymbol{d}\Big]\,d\boldsymbol{b} = (2\pi)^{km/2}|\boldsymbol{A}|^{-1},$$

we have

$$\int L(\boldsymbol{b}, \boldsymbol{\beta}, \tilde{\boldsymbol{D}}, \boldsymbol{R}|\boldsymbol{y})\,d\boldsymbol{b} = \frac{1}{(2\pi)^{N/2}|\boldsymbol{R}|^{1/2}|\tilde{\boldsymbol{D}}|^{1/2}|\boldsymbol{A}|^{1/2}}$$

$$\times \exp\Big[-\frac{1}{2}(\boldsymbol{y} - \boldsymbol{X\beta})'\boldsymbol{V}^{-1}(\boldsymbol{y} - \boldsymbol{X\beta})\Big]$$

$$= \frac{1}{(2\pi)^{N/2}|\boldsymbol{V}|^{1/2}} \exp\Big[-\frac{1}{2}(\boldsymbol{y} - \boldsymbol{X\beta})'\boldsymbol{V}^{-1}(\boldsymbol{y} - \boldsymbol{X\beta})\Big],$$

using the identity $|\boldsymbol{R}||\tilde{\boldsymbol{D}}||\boldsymbol{A}| = |\boldsymbol{V}|$.

Thus, maximum likelihood estimation of $\boldsymbol{R}$ and $\tilde{\boldsymbol{D}}$ may be viewed as estimation where $\boldsymbol{\beta}$ is taken to be a fixed unknown constant and $\boldsymbol{b}$ is integrated out according to the hierarchy in (3.21).

Suppose, instead, we begin with a hierarchy in which $\boldsymbol{\beta}$ has a noninformative prior $\boldsymbol{\beta} \sim \text{uniform}(-\infty, \infty)$ or $\boldsymbol{\beta} \sim \mathcal{N}(\boldsymbol{\beta}^*, \boldsymbol{H})$, with $\boldsymbol{H}^{-1} = \boldsymbol{0}$:

$$\boldsymbol{y}|\boldsymbol{\beta}, \boldsymbol{b} \sim \mathcal{N}(\boldsymbol{X\beta} + \boldsymbol{Zb}, \boldsymbol{R}),$$

$$\boldsymbol{\beta} \sim \text{uniform}(-\infty, \infty),$$

$$\boldsymbol{b} \sim \mathcal{N}(\boldsymbol{0}, \tilde{\boldsymbol{D}}).$$

Then consider the marginal likelihood obtained by integrating out *both* $\boldsymbol{\beta}$ and $\boldsymbol{b}$:

$$L(\tilde{\boldsymbol{D}}, \boldsymbol{R}|\boldsymbol{y}) = \int\int L(\boldsymbol{b}, \boldsymbol{\beta}, \tilde{\boldsymbol{D}}, \boldsymbol{R}|\boldsymbol{y})\, d\boldsymbol{b}\, d\boldsymbol{\beta}.$$

By the derivation above, we may substitute for the first of these integrals to obtain

$$L(\tilde{\boldsymbol{D}}, \boldsymbol{R}|\boldsymbol{y}) = \int \frac{1}{(2\pi)^{N/2}|\boldsymbol{V}|^{1/2}} \times \exp\left[-\frac{1}{2}(\boldsymbol{y} - \boldsymbol{X\beta})'\boldsymbol{V}^{-1}(\boldsymbol{y} - \boldsymbol{X\beta})\right] d\boldsymbol{\beta}.$$

The exponent in this expression may be written as

$$\begin{aligned}(\boldsymbol{y} - \boldsymbol{X\beta})'\boldsymbol{V}^{-1}(\boldsymbol{y} - \boldsymbol{X\beta}) &= \boldsymbol{y}'\{\boldsymbol{V}^{-1} - \boldsymbol{V}^{-1}\boldsymbol{X}(\boldsymbol{X}'\boldsymbol{V}^{-1}\boldsymbol{X})^{-1}\boldsymbol{X}'\boldsymbol{V}^{-1}\}\boldsymbol{y} \\ &\quad + (\boldsymbol{\beta} - \hat{\boldsymbol{\beta}})'(\boldsymbol{X}'\boldsymbol{V}^{-1}\boldsymbol{X})(\boldsymbol{\beta} - \hat{\boldsymbol{\beta}}).\end{aligned}$$

$\hat{\boldsymbol{\beta}} = (\boldsymbol{X}'\boldsymbol{V}^{-1}\boldsymbol{X})^{-1}\boldsymbol{X}'\boldsymbol{V}^{-1}\boldsymbol{y}$, so that, integrating over $\boldsymbol{\beta}$, we obtain

$$\begin{aligned}L(\tilde{\boldsymbol{D}}, \boldsymbol{R}|\boldsymbol{y}) &= \frac{(2\pi)^{p/2}|\boldsymbol{X}'\boldsymbol{V}^{-1}\boldsymbol{X}|^{-1/2}}{(2\pi)^{N/2}|\boldsymbol{V}|^{1/2}} \\ &\quad \times \exp\left[-\frac{1}{2}\boldsymbol{y}'\{\boldsymbol{V}^{-1} - \boldsymbol{V}^{-1}\boldsymbol{X}(\boldsymbol{X}'\boldsymbol{V}^{-1}\boldsymbol{X})^{-1}\boldsymbol{X}'\boldsymbol{V}^{-1}\}\boldsymbol{y}\right] \\ &= \frac{(2\pi)^{p/2}|\boldsymbol{X}'\boldsymbol{V}^{-1}\boldsymbol{X}|^{-1/2}}{(2\pi)^{N/2}|\boldsymbol{V}|^{1/2}} \exp\left[-\frac{1}{2}\boldsymbol{y}'\boldsymbol{P}_V\boldsymbol{y}\right],\end{aligned}$$

where $\boldsymbol{P}_V$ is the projection matrix

$$\boldsymbol{P}_V = \boldsymbol{V}^{-1} - \boldsymbol{V}^{-1}\boldsymbol{X}(\boldsymbol{X}'\boldsymbol{V}^{-1}\boldsymbol{X})^{-1}\boldsymbol{X}'\boldsymbol{V}^{-1}.$$

Thus,

$$L(\tilde{\boldsymbol{D}}, \boldsymbol{R}|\boldsymbol{y}) = (2\pi)^{p/2}|\boldsymbol{X}'\boldsymbol{V}^{-1}\boldsymbol{X}|^{-1/2}L(\hat{\boldsymbol{\beta}}, \tilde{\boldsymbol{D}}, \boldsymbol{R}|\boldsymbol{y}).$$

Comparison with (3.18) shows that maximization of the marginal likelihood $L(\tilde{\boldsymbol{D}}, \boldsymbol{R}|\boldsymbol{y})$ is equivalent to restricted maximum likelihood.

Thus, both ML and REML estimation of variance components arise naturally within a hierarchical Bayes framework. The former maximizes the marginal normal likelihood, integrating over random effects, conditional on the value of $\boldsymbol{\beta}$. In restricted maximum likelihood, both $\boldsymbol{b}$ and $\boldsymbol{\beta}$ are integrated out, the latter integration using a flat prior.

3.4 Computational aspects

3.4.1 EM algorithm

We have seen in section 3.3.2 that fixed effects may be estimated by solving the generalized least squares estimating equations (3.12) if the vector of variance parameters $\boldsymbol{\omega}$ specifying the covariance matrices $\boldsymbol{R}$ and $\boldsymbol{D}$ is known. Use of the EM algorithm (Dempster, Laird and Rubin, 1977) to estimate $\boldsymbol{\omega}$ by maximum likelihood or restricted maximum likelihood is discussed by several authors (Laird and Ware, 1982; Dempster, Rubin and Tsutakawa, 1981; Laird, Lange and Stram, 1987; Searle *et al.*, 1992, section 8.3). Here, we provide a brief overview.

The EM algorithm is generally motivated as being applicable to incomplete data. In the context of unbalanced longitudinal data, one possible view is to regard the case of a fully balanced design (equal number of observations at the same covariate values available for each individual) as constituting 'complete data.' This is not the paradigm which proves useful in implementing the EM algorithm, however. Instead, one regards the situation where one could observe the random effects, or 'unobservables,' $\boldsymbol{b}_i$ and $\boldsymbol{e}_i$, in addition to the responses $\boldsymbol{y}_i$, as corresponding to the complete data case.

ML estimation

If $\boldsymbol{b}_i$ and $\boldsymbol{e}_i$ were observed, we could easily write down closed-form normal theory maximum likelihood estimates of σ^2 and the remaining components of $\boldsymbol{\omega}$. For $\boldsymbol{R}_i$ of the form $\sigma^2\boldsymbol{I}_{n_i}$, $\boldsymbol{D}$ arbitrary, these would be

$$\hat{\sigma}^2 = \frac{\sum_{i=1}^{m} e_i' e_i}{\sum_{i=1}^{m} n_i} = t_1/N, \tag{3.22}$$

$$\hat{\boldsymbol{D}} = \sum_{i=1}^{m} \boldsymbol{b}_i \boldsymbol{b}_i'/m = \boldsymbol{T}_2/m. \tag{3.23}$$

Here, t_1 and the $k(k+1)/2$ nonredundant components of $\boldsymbol{T}_2$ are the sufficient statistics for $\boldsymbol{\omega}$, under the assumption of normality.

In practice, $\boldsymbol{b}_i$ and $\boldsymbol{e}_i$ will not be known. The EM algorithm provides a means for substituting alternative values. Given current values for parameter estimates, replace the sufficient statistics t_1 and $\boldsymbol{T}_2$ in (3.22) and (3.23) by their expectations, conditional on the observed data vector $\boldsymbol{y}$. This defines the 'E-step' of the algorithm:

$$\begin{aligned} \mathrm{E}\{t_1|\boldsymbol{y}, \hat{\boldsymbol{\beta}}(\hat{\boldsymbol{\omega}}), \hat{\boldsymbol{\omega}}\} &= \mathrm{E}\Big(\sum_{i=1}^{m} \boldsymbol{e}_i'\boldsymbol{e}_i|\boldsymbol{y}_i, \hat{\boldsymbol{\beta}}(\hat{\boldsymbol{\omega}}), \hat{\boldsymbol{\omega}}\Big) \\ &= \sum_{i=1}^{m}\Big(\boldsymbol{e}_i(\hat{\boldsymbol{\omega}})'\boldsymbol{e}_i(\hat{\boldsymbol{\omega}}) + \mathrm{tr}\ \mathrm{Cov}\{\boldsymbol{e}_i|\boldsymbol{y}_i, \hat{\boldsymbol{\beta}}(\hat{\boldsymbol{\omega}}), \hat{\boldsymbol{\omega}}\}\Big) \end{aligned} \tag{3.24}$$

$$\begin{aligned} \mathrm{E}\{\boldsymbol{T}_2|\boldsymbol{y}, \hat{\boldsymbol{\beta}}(\hat{\boldsymbol{\omega}}), \hat{\boldsymbol{\omega}}\} &= \mathrm{E}\Big(\sum_{i=1}^{m} \boldsymbol{b}_i\boldsymbol{b}_i'|\boldsymbol{y}_i, \hat{\boldsymbol{\beta}}(\hat{\boldsymbol{\omega}}), \hat{\boldsymbol{\omega}}\Big) \\ &= \sum_{i=1}^{m}\Big(\boldsymbol{b}_i(\hat{\boldsymbol{\omega}})\boldsymbol{b}_i(\hat{\boldsymbol{\omega}})' + \mathrm{Cov}\{\boldsymbol{b}_i|\boldsymbol{y}_i, \hat{\boldsymbol{\beta}}(\hat{\boldsymbol{\omega}}), \hat{\boldsymbol{\omega}}\}\Big). \end{aligned} \tag{3.25}$$

In (3.24) and (3.25), $\boldsymbol{e}_i(\hat{\boldsymbol{\omega}})$ and $\boldsymbol{b}_i(\hat{\boldsymbol{\omega}})$ represent the conditional expectations $\mathrm{E}\{\boldsymbol{e}_i|\boldsymbol{y}_i, \hat{\boldsymbol{\beta}}(\hat{\boldsymbol{\omega}}), \hat{\boldsymbol{\omega}}\}$ and $\mathrm{E}\{\boldsymbol{b}_i|\boldsymbol{y}_i, \hat{\boldsymbol{\beta}}(\hat{\boldsymbol{\omega}}), \hat{\boldsymbol{\omega}}\}$, respectively. The second equality in each of these equations follows from standard results on the expectations of quadratic forms.

The relevant conditional distributions for (3.24) and (3.25) are

$$\boldsymbol{e}_i|\boldsymbol{y}, \hat{\boldsymbol{\beta}}(\hat{\boldsymbol{\omega}}) \sim \mathcal{N}\{\boldsymbol{y}_i - \boldsymbol{X}_i\hat{\boldsymbol{\beta}}(\hat{\boldsymbol{\omega}}) - \boldsymbol{Z}_i\boldsymbol{b}_i(\hat{\boldsymbol{\omega}}), \sigma^2(\boldsymbol{I}_{n_i} - \boldsymbol{V}_i^{-1})\}$$

$$\boldsymbol{b}_i|\boldsymbol{y}, \hat{\boldsymbol{\beta}}(\hat{\boldsymbol{\omega}}) \sim \mathcal{N}\Big(\boldsymbol{D}\boldsymbol{Z}_i\boldsymbol{V}^{-1}\{\boldsymbol{y} - \boldsymbol{X}\hat{\boldsymbol{\beta}}(\hat{\boldsymbol{\omega}})\}, \boldsymbol{D} - \boldsymbol{D}\boldsymbol{Z}_i'\boldsymbol{V}_i^{-1}\boldsymbol{Z}_i\boldsymbol{D}\Big)$$

where $\boldsymbol{V}_i = \sigma^2\boldsymbol{I}_{n_i} + \boldsymbol{Z}_i\boldsymbol{D}\boldsymbol{Z}_i'$, as before.

The 'M-step' of the algorithm consists of solving (3.22) and (3.23) with t_1 and $\boldsymbol{T}_2$ being replaced by their current expectations given in (3.24) and (3.25).

REML estimation

Restricted maximum likelihood estimation may also be implemented by the EM algorithm. Because the 'complete data' still correspond to $\boldsymbol{y}$, $\boldsymbol{b}$, and $\boldsymbol{e}$, the equations for the 'M-step' (3.22) and (3.23) remain the same. In the 'E-step,' however, instead of taking expectations conditional on $\boldsymbol{y}$ and $\boldsymbol{\beta}$, they are taken conditional on $\boldsymbol{y}$ only, as one is dealing with the marginal likelihood, having integrated out $\boldsymbol{\beta}$. Thus, t_1 and $\boldsymbol{T}_2$ in (3.22) and (3.23) are replaced by their conditional expectations

$$\begin{aligned} \mathrm{E}\{t_1|\boldsymbol{y},\hat{\omega}\} &= \mathrm{E}\Big(\sum_{i=1}^{m} \boldsymbol{e}_i{}'\boldsymbol{e}_i|\boldsymbol{y}_i,\hat{\omega}\Big) \\ &= \sum_{i=1}^{m}\Big(\boldsymbol{e}_i(\hat{\omega})'\boldsymbol{e}_i(\hat{\omega}) + \mathrm{tr}\ \mathrm{Cov}\{\boldsymbol{e}_i|\boldsymbol{y}_i,\hat{\omega}\}\Big) \end{aligned} \tag{3.26}$$

$$\begin{aligned} \mathrm{E}\{\boldsymbol{T}_2|\boldsymbol{y},\hat{\omega}\} &= \mathrm{E}\Big(\sum_{i=1}^{m} \boldsymbol{b}_i\boldsymbol{b}_i'|\boldsymbol{y}_i,\hat{\omega}\Big) \\ &= \sum_{i=1}^{m}\Big(\boldsymbol{b}_i(\hat{\omega})\boldsymbol{b}_i(\hat{\omega})' + \mathrm{Cov}\{\boldsymbol{b}_i|\boldsymbol{y}_i,\hat{\omega}\}\Big). \end{aligned} \tag{3.27}$$

In (3.26) and (3.27), $\boldsymbol{e}_i(\hat{\omega})$ and $\boldsymbol{b}_i(\hat{\omega})$ are conditional expectations, as before. The relevant conditional distributions are

$$\boldsymbol{e}_i|\boldsymbol{y} \sim \mathcal{N}\{\boldsymbol{y}_i - \boldsymbol{X}_i\hat{\boldsymbol{\beta}}(\hat{\omega}) - \boldsymbol{Z}_i\boldsymbol{b}_i(\hat{\omega}), \sigma^2(\boldsymbol{I}_{n_i} - \boldsymbol{P}_{V_i})\}$$

$$\boldsymbol{b}_i|\boldsymbol{y} \sim \mathcal{N}(\boldsymbol{D}\boldsymbol{Z}_i\boldsymbol{P}_V\boldsymbol{y}, \boldsymbol{D} - \boldsymbol{D}\boldsymbol{Z}_i'\boldsymbol{P}_{V_i}\boldsymbol{Z}_i\boldsymbol{D}),$$

where $\boldsymbol{P}_{V_i}$ is the projection matrix

$$\boldsymbol{P}_{V_i} = \boldsymbol{V}_i^{-1} - \boldsymbol{V}_i^{-1}\boldsymbol{X}_i(\boldsymbol{X}_i'\boldsymbol{V}_i^{-1}\boldsymbol{X}_i)^{-1}\boldsymbol{X}_i'\boldsymbol{V}_i^{-1}$$

corresponding to the ith individual. Laird *et al.* (1987) point out that inversion of the $(n_i \times n_i)$ matrices $\boldsymbol{V}_i$ can be carried out by using the identity

$$\boldsymbol{V}_i^{-1} = \boldsymbol{I}_{n_i} - \boldsymbol{Z}_i(\boldsymbol{Z}_i'\boldsymbol{Z}_i + \boldsymbol{D}^{-1})\boldsymbol{Z}_i'.$$

Lindstrom and Bates (1988) suggest further simplification by the use of Cholesky decompositions.

Existence of closed-form solutions for balanced data

Laird *et al.* (1987) also discuss the special case of complete, balanced data (the same number of observations, taken at the same design points for each individual). In this case, closed-form expressions may be obtained for ML and REML estimates for all parameters. In particular, the generalized least squares estimate of β reduces to ordinary least squares, regardless of the forms of Z and D:

$$\hat{\beta} = (X'X)^{-1}X'y = \left(\sum_{i=1}^{m} X_i'X_i\right)^{-1} \sum_{i=1}^{m} X_i'y_i. \tag{3.28}$$

This is as one might expect: the balance in the design obviates the need for any kind of differential weighting.

This observation provides insight into the validity of a commonly used simple approach to linear repeated measures regression, that of obtaining individual-specific estimates of regression parameters and using these as summary statistics for a subsequent analysis. This approach may be expected to give good results when the degree of imbalance in the data is not too severe. For highly unbalanced data, some form of differential weighting is needed. In Chater 5, we discuss inference based on individual-specific estimates for nonlinear repeated measurement data.

Starting values and convergence issues

Laird *et al.* (1987) suggest starting values for σ^2 and D based on OLS estimates of β and b_i:

$$\hat{\sigma}_0^2 = \frac{1}{\{N - (m-1)k - p\}} \Bigg(\sum_{i=1}^{m} y_i'y_i - \\ \hat{\beta}_{OLS} \sum_{i=1}^{m} X_i'y_i - \sum_{i=1}^{m} \hat{b}_i Z_i'(y_i - X_i\hat{\beta}_{OLS})^2 \Bigg) \tag{3.29}$$

and

$$\hat{D}_0 = \frac{1}{m}\sum_{i=1}^{m} \hat{b}_i\hat{b}_i' - \frac{1}{m}\hat{\sigma}_0^2 \sum_{i=1}^{m} (Z_i'Z_i)^{-1}, \tag{3.30}$$

where $\hat{\beta}_{OLS}$ is the ordinary least squares estimator given by (3.28) and

$$\hat{b}_i = (Z_i'Z_i)^{-1}Z_i'(y_i - X_i\hat{\beta}_{OLS}).$$

If $n_i < k$ for some i, so that $Z_i'Z_i$ is singular, one may exclude these individuals from these calculations.

A general property of the EM algorithm is that the loglikelihood may be shown to increase at each iteration (Dempster *et al.*,

1977). In addition, nonnegativity constraints for variance components and dispersion matrices are automatically satisfied at each iteration, provided that the initial values satisfy these constraints. Slowness of convergence is an issue in implementing the EM algorithm, however. Methods of accelerating convergence in practice, using the techniques of Aitken acceleration, are discussed in Laird *et al.* (1987, section 5.2).

3.4.2 Newton–Raphson and the method of scoring

A general discussion of the Newton–Raphson technique is given in section 2.4.1. Implementation of the Newton–Raphson algorithm in the context of ML and REML estimation in linear mixed effects models is discussed by Jennrich and Schluchter (1986) and Lindstrom and Bates (1988). Details of the necessary matrix derivatives to implement the scheme may be found in these papers and in Searle *et al.* (1992, Chapter 6), and are not presented here.

Lindstrom and Bates (1988) discuss several ways to improve the performance of the Newton–Raphson algorithm in this context. Specifically, they recommend:

(i) Reparameterization in terms of a matrix $\boldsymbol{D}^*$ such that $\boldsymbol{D} = \sigma^2\boldsymbol{D}^*$, so that $\boldsymbol{V}_i = \sigma^2(\boldsymbol{R}_i + \boldsymbol{Z}_i\boldsymbol{D}^*\boldsymbol{Z}_i')$, and then 'profiling out' the intra-individual scale parameter σ^2. That is, substitute the maximizing value $\hat{\sigma}^2$ (which is easily calculated at any given iteration) into the loglikelihood and maximize the resulting *profile* loglikelihood in $\boldsymbol{\beta}$ and the remaining components of $\boldsymbol{\omega}$, where $\boldsymbol{\omega}$ now contains the distinct elements of $\boldsymbol{D}^*$.

(ii) Reduction of the matrices $\boldsymbol{X}_i$ and $\boldsymbol{Z}_i$ to orthogonal-triangular (QR) form (Kennedy and Gentle, 1980).

(iii) Optimization of the loglikelihood as a function of the nonzero entries of the Cholesky factor of $\boldsymbol{D}$. This transforms the constrained optimization problem into an unconstrained problem and ensures positive definite estimates for $\boldsymbol{D}$.

Global comparison of different estimation algorithms is somewhat tricky, as performance may often be dictated by specific model and data configurations. Lindstrom and Bates (1988) compare performance characteristics of the improved Newton–Raphson technique and the EM algorithm incorporating Aitken acceleration for a number of data sets and demonstrate superiority of the Newton–Raphson approach. In general, this appears to be true:

Newton–Raphson iterations are more expensive, but convergence occurs in fewer iterations than with the EM algorithm.

3.4.3 Software implementation

Until recently, widely available commercial software packages such as BMDP and SAS had limited capabilities for inference in general mixed effects models. Within the last several years, this has changed; both packages now offer the capacity to fit mixed effects models using ML or REML techniques. In SAS, this may be accomplished using PROC MIXED. Module 5V in BMDP implements the methods in this chapter. Other packages for mixed model analysis are ML3 (Prosser, Rasbash and Goldstein, 1991) and EPILOG PROC LRE (Epicenter Software, 1994). In addition, NLME (Pinheiro, Bates and Lindstrom, 1993), a collection of Splus functions for fitting nonlinear mixed effects models, may also be used for inference in linear mixed effects models. Each of the packages mentioned above uses some version of the Newton–Raphson algorithm in fitting. The reader is referred to the relevant documentation for details. A comparative review of these and other software packages is given by Kreft, de Leeuw, and Van Der Leeden (1995). In the examples below, we have used SAS PROC MIXED in analyzing the pup weight data, and the linear option of NLME for fitting the tracer data.

3.4.4 Examples revisited

Pup weight data

Table 3.2 summarizes the results of restricted maximum likelihood fitting of the model specified in (3.2) to the pup weight data in Table 3.1. Results presented here were obtained using the MIXED procedure in SAS and are identical to those reported by Dempster *et al.* (1984) in their discussion of this data set. An illustration of the use of BMDP-5V to fit these data may be found in the BMDP manual (1990, Volume 2, p. 1335–8).

For the fit reported in Table 3.2, the model was parameterized in such a way that the intercept corresponded to the mean response for the control group, and the two degrees of freedom for treatment group differences were the low dose versus control and high dose versus control contrasts, respectively.

Table 3.2. *Estimates of fixed and random effects for the pup weight data.*

Parameter	*Estimate*	*Standard Error*
Litter size coefficient (λ)	−0.129	0.019
Sex effect (M-F) (δ)	0.359	0.047
Control (= Intercept)	8.310	0.274
Low dose − control	−0.429	0.150
High dose − control	−0.859	0.182
Litter-to-litter variance (σ_b^2)	0.097	
Pup-to-pup variance (σ^{2e})	0.163	

Each of the fixed effects in the model were statistically significant at the conventional 5% level. Sex and litter effects were in the expected directions, and the effect of the test agent was dose-related, with quite a substantial effect observed at the high dose. As pointed out by Dempster *et al.* (1984), this effect could not be observed by simple inspection of the treatment means because dams in the high dose group generally have fewer pups than dams in the other two groups. Incorporating litter size as a covariate increases sensitivity to detect potential treatment effects.

Tracer lot data

The random coefficients model specified by (3.3) and (3.4) was fitted to the radioimmunoassay tracer data by restricted maximum likelihood. The simple intra-individual correlation structure $\boldsymbol{R}_i = \sigma^2 \boldsymbol{I}_{n_i}$ was assumed for all lots. Table 3.3 summarizes results of the fit using the Splus functions NLME of Pinheiro *et al.* (1993).

The estimated average rate of decline in calibrated control concentration with tracer age was 0.0019 ng/ml/day, with a standard error of 0.0046 ng/ml/day. This would correspond to a loss of less than 2% over a 50 day period, and is neither statistically nor practically significant. The data support continuing to use tracer lots for up to 50 days.

The solid line superimposed on the raw data plots in Figure 3.1 represents predicted values based on empirical Bayes estimation of lot-specific regression parameters. For lots with many observations, empirical Bayes and OLS fits are very close. In contrast, where few points are available for a lot (e.g. lot 6), the fits are quite

Table 3.3. *REML estimation for the random coefficients model, tracer data.*

Parameter	*Estimate*	*Standard Error*
Intercept (β_0)	5.369	0.104
Slope (β_1)	−0.0019	0.0046
D_{11}	$< 10^{-12}$	
D_{12}	$< 10^{-12}$	
D_{22}	0.000029	
σ_e^2	0.68	

different. The empirical Bayes fit 'borrows' information from other lots to shrink the parameter estimates for lot 6 closer to the overall average. Lot-to-lot variation in intercept was effectively zero. One could refit a model assuming a constant intercept; this does not affect the conclusions.

3.5 Limitations of the normal hierarchical linear model

The normal theory hierarchical framework described in this chapter offers considerable flexibility in modeling repeated measurement data. Within this modeling framework, however, there are certain restrictions. We review these briefly here. Extensions of the modeling framework to overcome some of these restrictions are discussed in subsequent chapters.

Normality of the conditional (intra-individual) distribution

The assumption that the within-individual random terms $\boldsymbol{e}_i$ follow a normal distribution may not always be appropriate. For example, suppose that, in a reproductive toxicology study, y_{ij} represents the measured response for the jth pup in the ith litter. In the case where y_{ij} represents birth weight, it may be reasonable to model within the normal theory framework

$$\boldsymbol{y}_i = \boldsymbol{X}_i\boldsymbol{\beta} + \boldsymbol{Z}_i\boldsymbol{b}_i + \boldsymbol{e}_i,$$

with $\boldsymbol{e}_i \sim \mathcal{N}(\boldsymbol{0}, \boldsymbol{R}_i)$, as we did in section 3.2.2. However, if y_{ij} represents a discrete or binary response, for instance, an indicator for presence or absence of a particular birth defect or malformation, the assumption that the e_{ij} are normally distributed is clearly inappropriate. Even for continuous y_{ij}, a distribution with heavier tails, such as the double exponential or Student's t, may be a more realistic model than the normal. It is thus of interest to consider a hierarchical framework that accommodates discrete, binary, or otherwise nonnormal responses.

For the particular case of binary response data, early proposals for hierarchical modeling focused on the beta-binomial model (Williams, 1975; Crowder, 1978):

$$y_{ij}|p_i \sim \text{ independent Bernoulli } (p_i),$$
$$p_i \sim \text{ independent beta } (\alpha_1, \alpha_2).$$

Extensions to this method have been proposed, which adopt the approach of generalized linear models. For example, Stiratelli, Laird, and Ware (1984) proposed a hierarchy in which the individual-specific means p_i are related to fixed and random effects by a logistic link function:

$$y_{ij}|\boldsymbol{b}_i \sim \text{ independent Bernoulli } (p_i),$$
$$\text{logit}(p_i) = \boldsymbol{X}_i\boldsymbol{\beta} + \boldsymbol{Z}_i\boldsymbol{b}_i, \quad \boldsymbol{b}_i \sim \mathcal{N}(\boldsymbol{0}, \boldsymbol{D}).$$

Inference for this type of hierarchy is complicated; the simplifications that allow evaluation of the likelihood in the normal-normal hierarchy no longer apply. Stiratelli, Laird and Ware (1984) propose certain simplifying assumptions. A further complication lies in the interpretation of the regression parameters $\boldsymbol{\beta}$. For the normal-normal hierarchy with scalar β, β represents the change in the mean of y_{ij} associated with a unit change in x_{ij}, both for the marginal *and* conditional distribution of y_{ij} given $\boldsymbol{b}_i$. This follows from linearity, because

$$\text{E}(y_{ij}) = \text{E}\{\text{E}(y_{ij}|\boldsymbol{b}_i)\} = \text{E}(\beta x_{ij} + b_i) = \beta x_{ij}.$$

In the hierarchy considered above, this equality does not hold true, because of the nonlinearity of the logistic link function. β represents the change on the logit scale in the conditional mean of y_{ij}, for a change of one unit in x_{ij}, but this is not true for the marginal mean.

This type of hierarchy is considered further, and extended to allow more general link functions, by Liang and Zeger (1986) and Zeger and Liang (1986). The paper by Zeger, Liang and Albert (1988) discusses the important distinction between what they refer to as *subject-specific* (SS) parameters, which pertain to inference in the conditional distribution for a given subject, and *population-averaged* (PA) models, where attention focuses on inference in the marginal distribution. We defer further discussion of this issue to section 4.4.2; as we shall be concerned with *nonlinear* regression models, the distinction between SS and PA parameters will be important in our development.

Normality of random effects distribution

The ability to carry out likelihood based inference, both in the classical and Bayesian hierarchies, depends strongly on the assumption of a Gaussian distribution for the random effects (or unobservables $\boldsymbol{b}_i$). In practice, this may not be true. Approaches to inference that require less restrictive parametric assumptions are of interest. This issue is discussed further in Chapters 4, 7 and 8.

Linearity

In many applications, the relationship between the conditional mean and covariates may depend on individual-specific parameters $\boldsymbol{\beta}_i$ in a nonlinear manner. Apart from the intrinsic complexity of nonlinear modeling, this nonlinearity complicates inference in two ways. The ability, even under the assumption of normality, to write down the marginal distribution of a response vector, $\boldsymbol{y}_i$, in closed form is lost. Furthermore, difficulties arise, since the individual-specific parameters and their marginal (population) means no longer have the same interpretation. A further possibility is that the dependence of individual-specific parameters on population covariates is also nonlinear in the relevant population parameters. All of these issues are discussed in gory detail in subsequent chapters.

Specification of the intra-individual covariance structure

The methods of section 3.3 are presented in terms of a general intra-individual covariance matrix $\boldsymbol{R}_i$. The simplest possible specification is that $\boldsymbol{R}_i = \sigma^2 \boldsymbol{I}_{n_i}$ for all individuals. In many applications, more complex intra-individual covariance structures are needed, as discussed in Chapter 2. However, if $\boldsymbol{R}_i$ depends on $\boldsymbol{\beta}_i$, the ability to separate estimation of fixed effects and covariance components as cleanly as in section 3.3 is lost. We incorporate the possibility of accommodating heterogeneous variance, serial correlation, or both, in the within-individual measurements in the hierarchical nonlinear model in Chapter 4.

Patterns of missingness

Caution is in order when analyzing unbalanced repeated measurement data. The lack of balance may arise for a number of reasons. Quite frequently, the proposed study design may have called for data that are fully balanced, but imbalance arises due to missing data. One then needs to be concerned with the 'pattern of missingness.' The methods of section 3.4 yield unbiased inference on fixed effects provided that the data are 'missing at random' (MAR) or 'missing completely at random' (MCAR) in the sense of Little and Rubin (1987). In some cases, however, modifications are needed; for instance, if the missing data arise as the result of informative censoring. An example would be the tracking of CD4 counts or other markers in different treatment groups, where treatment has a differential effect on the probability that a subject's count is recorded at a scheduled visit. Methods for handling these issues have generated much recent interest; some relevant references include Wu and Carroll (1988), Wu and Bailey (1989), Little (1993), and Diggle and Kenward (1994).

3.6 Bibliographic notes

There is a considerable literature on variance components and linear mixed effects models. The book by Searle, Casella and McCulloch (1992) provides a comprehensive review. In several instances in this chapter, we have adapted or shortened derivations of key results from Searle *et al.*; the reader is referred to that book for full details. The book by Crowder and Hand (1990) provides a basic introduction to the analysis of repeated measurement data and includes an extensive bibliography. Four other recent texts deal extensively with the analysis of repeated measurement data. Jones (1993) adopts a state-space approach to the analysis of hierarchical linear models. Longford (1993) covers random coefficient models and extensions in detail. The books by Lindsey (1993) and Diggle *et al.* (1994) offer a broad treatment of methods for repeated measurement data.

Early papers on variance component estimation include those by Henderson (1953), Hartley and Rao (1967), and Patterson and Thompson (1971). A series of papers in the 1970s discusses computational aspects; these include Hemmerle and Hartley (1973), Corbeil and Searle (1976), and Jennrich and Sampson (1976). Harville (1976) provides a review of methods to that date. The connection between REML estimation and Bayesian inference was elucidated by Harville (1974). Other approaches to variance component estimation, based on quadratic estimation, are discussed by Lamotte (1973) and Rao (1973, Chapter 4). Best linear unbiased prediction is discussed by Henderson (1984). Applications to animal breeding are discussed by Harville (1990) and Henderson (1990); Robinson (1991) provides a very readable review.

Lindley and Smith (1972) is a seminal reference for the Bayesian approach to hierarchical linear modeling. Implementation of the EM algorithm (Dempster *et al.*, 1977) in this context is discussed by Rubin (1980), Dempster, Rubin and Tsutakawa (1981), Hui and Berger (1983), and Dempster *et al.* (1984). An integrated review of the application of these methods to longitudinal data is provided by Laird and Ware (1982) and by Ware (1985). Kass and Steffey (1989) offer a general treatment of Bayesian inference in hierarchical models. Louis (1991) considers applications of empirical Bayes methods to problems in biopharmaceutical research. Escobar and West (1992) discuss a nonparametric Bayesian approach.

Computational aspects are discussed in the papers by Laird and Ware (1982), Laird, Lange and Stram (1987), Jennrich and Schluchter (1986), and Lindstrom and Bates (1988). Treatment of general covariance structures may be found in the papers by Reinsel (1984, 1985), Jennrich and Schluchter (1986), and Chi and Reinsel (1989).

The case of nonnormal intra-individual error distributions is considered by Stiratelli, Laird and Ware (1984) and in papers by Liang and Zeger (1986), Zeger and Liang (1986), and Zeger, Liang and Albert (1988).

CHAPTER 4

Hierarchical nonlinear models

4.1 Introduction

The objective of this book is to provide a description of the broad range of techniques available for the analysis of nonlinear repeated measurement data. Central to this objective is a unified presentation of the various modeling and inferential strategies within a single general model framework. In this chapter, we introduce the *hierarchical nonlinear model* that forms the basis for the inferential methods discussed in the rest of the book.

The hierarchical nonlinear model may be regarded as both an extension of the nonlinear regression models discussed in Chapter 2 to handle data from several individuals and of the linear model reviewed in Chapter 3 to the case of a *nonlinear* response function. This extension is straightforward in principle. As in the linear case, intra- and inter-individual variation are accommodated within the framework of a two-stage model. At the first stage, *intra-individual* variation is characterized by a nonlinear regression model with a model for the individual covariance structure, as in Chapter 2. *Inter-individual* variability is represented in the second stage through individual-specific regression parameters, which may incorporate both systematic and random effects.

The basic model framework is described in section 4.2. Different assumptions about the inter-individual variation lead to different model specifications. These are described in section 4.3. The final section of this chapter discusses specific difficulties arising for the hierarchical nonlinear case beyond those encountered in the linear setting.

There is a wide but scattered literature on nonlinear models for repeated measurement data. Because most of the references are connected with specific inferential strategies, we defer compilation of an extensive bibliography in this chapter. Rather, we point out a few relevant references in connection with each of the model specifications discussed in section 4.3, and give a more detailed listing of related work when we discuss the associated inferential techniques in Chapters 5–8.

4.2 General hierarchical nonlinear model

In this section we describe the basic hierarchical nonlinear model. As in the linear case, the model involves two stages. After introduction of the basic framework in section 4.2.1, each stage is considered in detail in sections 4.2.2 and 4.2.3. A formal summary of the model is given in section 4.2.4.

The use of this modeling framework was pioneered by Sheiner, Rosenberg and Melmon (1972), who recognized the need to accommodate both variation within and among subjects in pharmacokinetic analysis. In contrast to the hierarchical linear model, the nonlinear case has only recently received widespread attention in the statistical literature. References include Racine-Poon (1985), Mallet (1986), Lindstrom and Bates (1990), Vonesh and Carter (1992), Davidian and Giltinan (1993a,b), Davidian and Gallant (1993), and Wakefield (1995); this list is in no way exhaustive.

4.2.1 Basic model

Let y_{ij} denote the jth response, $j = 1, \ldots, n_i$, for the ith individual, $i = 1, \ldots, m$, taken at a set of conditions summarized by the vector of covariates $\boldsymbol{x}_{ij}$, so that a total of $N = \sum_{i=1}^{m} n_i$ responses have been observed. The vector $\boldsymbol{x}_{ij}$ incorporates variables such as time, dose, analyte concentration, etc. Suppose that a (nonlinear) function $f(\boldsymbol{x}, \boldsymbol{\beta})$ may be specified to model the relationship between y and $\boldsymbol{x}$, where $\boldsymbol{\beta}$ is a $(p \times 1)$ vector of parameters.

Although the form of f is common to all individuals, the parameter $\boldsymbol{\beta}$ may vary across individuals. This possibility is taken into account by specification of a separate $(p \times 1)$ vector of parameters $\boldsymbol{\beta}_i$ for the ith individual. For example, for the cefamandole data considered in section 1.1.1, $\boldsymbol{\beta}_i$ would be the (4×1) vector whose components are the parameters of the biexponential function (1.1) for the ith individual.

The mean response for individual i depends on the regression parameter $\boldsymbol{\beta}_i$ specific to that individual. This may be written as

$$\mathrm{E}(y_{ij}|\boldsymbol{\beta}_i) = f(\boldsymbol{x}_{ij}, \boldsymbol{\beta}_i).$$

We may define the following two-stage model:

Stage 1 (intra-individual variation)

Assume that, for individual i, the jth response follows the model

$$y_{ij} = f(\boldsymbol{x}_{ij}, \boldsymbol{\beta}_i) + e_{ij}, \tag{4.1}$$

where e_{ij} is a random error term reflecting uncertainty in the response, given the ith individual, with $\mathrm{E}(e_{ij}|\boldsymbol{\beta}_i) = 0$. Collecting the responses and errors for the ith individual into the $(n_i \times 1)$ vectors $\boldsymbol{y}_i = [y_{i1}, \ldots, y_{in_i}]'$ and $\boldsymbol{e}_i = [e_{i1}, \ldots, e_{in_i}]'$, respectively, and defining the $(n_i \times 1)$ vector

$$\boldsymbol{f}_i(\boldsymbol{\beta}_i) = \begin{bmatrix} f(\boldsymbol{x}_{i1}, \boldsymbol{\beta}_i) \\ \vdots \\ f(\boldsymbol{x}_{in_i}, \boldsymbol{\beta}_i) \end{bmatrix},$$

we may summarize the data for the ith individual as

$$\boldsymbol{y}_i = \boldsymbol{f}_i(\boldsymbol{\beta}_i) + \boldsymbol{e}_i, \tag{4.2}$$

where $\mathrm{E}(\boldsymbol{e}_i|\boldsymbol{\beta}_i) = \boldsymbol{0}$.

The model given in (4.1) and (4.2) describes the systematic and random variation associated with measurements on the ith individual. Systematic variation is characterized through the regression function f, while random variation is represented by an assumption on the random errors $\boldsymbol{e}_i$. Specification of a model for the distribution of the $\boldsymbol{e}_i$ completes the description of intra-individual variation for the ith individual. We consider this issue in section 4.2.2.

Stage 2 (inter-individual variation)

Variation among different individuals is accounted for through the individual-specific regression parameters $\boldsymbol{\beta}_i$. Part of the inter-individual variation in parameter values may be due to systematic dependence on individual characteristics, such as treatment group, weight, or age. Parameters may also differ due to unexplained variation in the population of individuals, for example, due to natural biological or physical variability among individuals or run-to-run variation in assay procedure. To account for these possibilities, this stage consists of a *model* for the dependence of $\boldsymbol{\beta}_i$ on systematic and random components. A formal development of modeling the regression parameters $\boldsymbol{\beta}_i$ is given in section 4.2.3.

4.2.2 Intra-individual variation

In Chapter 2, we emphasized that, for a given individual, variability in the y_{ij} may be a systematic function of the mean response for that individual, other known constants, and additional, possibly unknown parameters. Correlation among measurements on a given individual may also arise. Thus, a useful description of the random intra-individual variation represented by the errors $\boldsymbol{e}_i$ should allow for the possibility of heterogeneous within-individual variance as well as within-individual correlation.

In many contexts, it is reasonable to expect a comparable pattern of intra-individual variation across individuals. For example, in the development of a new immunoassay or bioassay, once assay procedure has stabilized, one would expect similar intra-assay variation from one run to the next. In the pharmacokinetic context, if plasma assay procedures remain consistent, then it may be acceptable to assume that the mean-variance relationship is the same for all subjects. The pattern of correlation among measurements taken at the same time points for all individuals would also be likely to remain constant across individuals. In the following model for intra-individual variation, we adopt this perspective; we comment on exceptions to this idea in section 4.4.

Paralleling the development in section 2.2.3, we may write a general specification for the common intra-individual variance structure as

$$\text{Cov}(\boldsymbol{e}_i|\boldsymbol{\beta}_i) = \boldsymbol{R}_i(\boldsymbol{\beta}_i, \boldsymbol{\xi}), \quad \boldsymbol{\xi} = [\sigma, \boldsymbol{\theta}', \boldsymbol{\alpha}']' \tag{4.3}$$

allowing for variance heterogeneity and correlation within individuals. In accordance with the assumption of a common within-individual pattern of variation, the functional form of $\boldsymbol{R}_i(\boldsymbol{\beta}_i, \boldsymbol{\xi})$ and the covariance parameter $\boldsymbol{\xi}$ are the same across individuals. As in section 2.3.4, we will often find it convenient to write

$$\boldsymbol{R}_i(\boldsymbol{\beta}_i, \boldsymbol{\xi}) = \sigma^2 \boldsymbol{S}_i(\boldsymbol{\beta}_i, \boldsymbol{\gamma}), \quad \boldsymbol{\gamma} = [\boldsymbol{\theta}', \boldsymbol{\alpha}']',$$

where $\boldsymbol{\gamma}$ is the $(v \times 1)$, $v = q + s$, vector of parameters in the functional part of the covariance model. As in the case of data from a single individual, $\boldsymbol{R}_i(\boldsymbol{\beta}_i, \boldsymbol{\xi})$ may be chosen to reflect heterogeneity of variance, within-individual correlation, or both.

The matrix $\boldsymbol{R}_i(\boldsymbol{\beta}_i, \boldsymbol{\xi})$ is analogous to the matrix $\boldsymbol{R}_i$ defined in section 3.2.2 to represent intra-individual covariance in the hierarchical linear model. In the linear case, this matrix was assumed to depend on i through its dimension, but not otherwise. In contrast, we consider here a more flexible model, also allowing dependence on

i to be through the individual-specific information and individual mean response, given $\boldsymbol{\beta}_i$. As we have seen, this greater generality may be necessary in order to characterize the complicated patterns of intra-individual variation likely to arise with nonlinear repeated measurement data.

To complete the description of intra-individual variation, an assumption is made about the conditional distribution of $\boldsymbol{e}_i$ given $\boldsymbol{\beta}_i$. The most common distributional assumption is that of intra-individual normality of the response, which follows from the error specification

$$\boldsymbol{e}_i|\boldsymbol{\beta}_i \sim \mathcal{N}\{\mathbf{0}, \boldsymbol{R}_i(\boldsymbol{\beta}_i, \boldsymbol{\xi})\}.$$

Other distributional assumptions are also possible; for example, in pharmacokinetic analysis, the response is often assumed to follow a lognormal distribution.

Often, the assumption about the conditional distribution may be made with a good deal of confidence. In some situations, rich information may be available on each individual, allowing assessment of the appropriateness of an assumption at the individual level. In other settings, despite the availability of only sparse individual data, an assumption may be justified based on more complete information gathered from previous studies. For a number of the inferential techniques we consider in Chapters 5 and 6, which incorporate the idea of generalized least squares estimation described in sections 2.3.2–2.3.4, an explicit assumption about the conditional distribution is not required beyond specification of the first two conditional moments $\mathrm{E}(\boldsymbol{e}_i|\boldsymbol{\beta}_i)$ and $\mathrm{Cov}(\boldsymbol{e}_i|\boldsymbol{\beta}_i)$. Other procedures, which are predicated on the likelihood principle or Bayesian ideas, require an explicit assumption about the intra-individual response distribution, see Chapters 6–8.

In fact, in nonlinear repeated measurement studies, far more emphasis typically is placed on understanding the nature of inter-individual systematic and random variation. This issue is addressed in detail in sections 4.2.3 and 4.3.

4.2.3 Inter-individual variation

To account for inter-individual variation among the $\boldsymbol{\beta}_i$, the standard approach is to specify a model for the $\boldsymbol{\beta}_i$. The degree of complexity of this model will depend on the nature of the data. To illustrate the range of possibilities, we refer to the four examples introduced in section 1.1.

Consider the relaxin assay data described in section 1.1.4. For repeated runs of the assay, the values of the parameters of the four-parameter logistic model (1.6) vary across runs. Assuming that assay procedures are consistent from run to run, it is unlikely that there is a systematic, identifiable basis for this variation. Rather, differences in parameter values are for the most part due to random variation in the assay procedure. Similarly, the subjects in the study of the pharmacokinetics of cefamandole (section 1.1.1) were chosen from a relatively homogeneous population of healthy volunteers. Thus, variation among the parameter values of the bi-exponential function (1.1) across subjects is probably attributable mainly to random variation among subjects rather than to differences in individual demographic and physiological characteristics that would be more pronounced in a heterogeneous patient population.

In both of these situations, a model for the $\boldsymbol{\beta}_i$ that assumes inter-individual variation is due entirely to unexplained phenomena would be appropriate:

$$\boldsymbol{\beta}_i = \boldsymbol{\beta} + \boldsymbol{b}_i. \tag{4.4}$$

In (4.4), $\boldsymbol{\beta}$ is a $(p \times 1)$ vector of fixed parameters and $\boldsymbol{b}_i$ is a $(p \times 1)$ random effect assumed to arise from a population with mean $\boldsymbol{0}$ and covariance matrix $\boldsymbol{D}$. (4.4) implies that the individual-specific regression parameters have mean $\boldsymbol{\beta}$. The covariance matrix $\boldsymbol{D}$ quantifies the random inter-individual variation.

A more complicated model allows for dependence on both random and systematic effects. To fix ideas, consider the soybean growth study described in section 1.1.3, for which

$$f(\boldsymbol{x}_{ij}, \boldsymbol{\beta}_i) = \frac{\beta_{1i}}{1 + \beta_{2i}\exp(\beta_{3i}x_{ij})}, \tag{4.5}$$

$\boldsymbol{\beta}_i = [\beta_{1i}, \beta_{2i}, \beta_{3i}]'$. Growth patterns may be thought to vary among plots in part because of genotype (F or P) and weather condition (in the years 1988, 1989, or 1990), and in part because of natural variation expected to occur among plots. Both possibilities are taken into account by assuming that the regression parameter for plot i arises from a distribution with mean depending on the particular genotype-weather condition combination under which that plot was observed. If the six possible combination groups are indexed by g, where $g = 1$ corresponds to genotype F in 1988, $g = 2$ corresponds to genotype P in 1988, and so on, then the regression parameter for individual i appearing in the kth combination group may be represented according to the model

$$\boldsymbol{\beta}_i = \boldsymbol{\beta}^{(g)} + \boldsymbol{b}_i. \tag{4.6}$$

Here, $\boldsymbol{b}_i$ is a (3×1) random effect assumed to arise from a population with mean $\mathbf{0}$ and covariance matrix $\boldsymbol{D}$, and $\boldsymbol{\beta}^{(g)}$ is a (3×1) vector of fixed parameters representing the mean value for the gth group. The specification for $\boldsymbol{b}_i$ implies that $\boldsymbol{D}$, and hence the pattern of random variation in the $\boldsymbol{\beta}_i$, is the same across all groups. This assumption is common, but not necessary.

The model given in (4.6) may be written more compactly in the form

$$\boldsymbol{\beta}_i = \boldsymbol{A}_i\boldsymbol{\beta} + \boldsymbol{b}_i, \tag{4.7}$$

where $\boldsymbol{\beta} = [\boldsymbol{\beta}^{(1)\prime}, \ldots, \boldsymbol{\beta}^{(6)\prime}]'$, that is, the (18×1) vector with the six group mean vectors $\boldsymbol{\beta}^{(g)}$ stacked vertically; and $\boldsymbol{A}_i$ is a (3×18) 'design' matrix such that

$$\boldsymbol{A}_i = [\boldsymbol{I}_3 | \mathbf{0}_{3\times3} | \cdots | \mathbf{0}_{3\times3}]$$

if $g = 1$,

$$\boldsymbol{A}_i = [\mathbf{0}_{3\times3} | \boldsymbol{I}_3 | \cdots | \mathbf{0}_{3\times3}]$$

if $g = 2$, and so on, where $\mathbf{0}_{3\times3}$ is a (3×3) matrix of zeroes. Note that under this model, comparison of growth patterns can be made by investigating differences among the $\boldsymbol{\beta}^{(g)}$. For example, the comparison of the overall growth pattern between genotypes F and P within the year 1988 could be made by contrasting the values of $\boldsymbol{\beta}^{(1)}$ and $\boldsymbol{\beta}^{(2)}$. More complicated issues, such as evaluation of the genotype-weather condition interaction effect and main effects, could be investigated by considering appropriate (linear) functions of the combined mean parameter vector $\boldsymbol{\beta}$.

In this example, the $\boldsymbol{\beta}_i$ may vary systematically according to treatment group, so that the design matrix $\boldsymbol{A}_i$ is an indicator matrix of zeroes and ones. Systematic dependence on continuous variables is also possible. In the population study of the pharmacokinetics of quinidine, described in section 1.1.2, several continuous individual-specific variables are available, such as height, weight, and age. If the pharmacokinetic parameters in the model given in (1.2)–(1.4) for the ith individual are collected in the vector $\boldsymbol{\beta}_i = [Cl_i, V_i, k_{ai}]'$, one could model linear dependence of each of Cl_i, V_i, and k_{ai} on height h_i using (4.7), with (3×6) design matrix

$$\boldsymbol{A}_i = \begin{bmatrix} 1 & h_i & 0 & 0 & 0 & 0 \\ 0 & 0 & 1 & h_i & 0 & 0 \\ 0 & 0 & 0 & 0 & 1 & h_i \end{bmatrix}$$

and (6×1) parameter vector

$$\boldsymbol{\beta} = \begin{bmatrix} \beta_{0Cl} \\ \beta_{1Cl} \\ \beta_{0V} \\ \beta_{1V} \\ \beta_{0k_a} \\ \beta_{1k_a} \end{bmatrix},$$

where $(\beta_{0Cl}, \beta_{1Cl})$, (β_{0V}, β_{1V}), and $(\beta_{0k_a}, \beta_{1k_a})$ are the intercepts and slopes, of the assumed straight-line relationships of each of Cl_i, V_i, and k_{ai}, respectively, with height. We consider the more complex models typically used to model pharmacokinetic parameters in the context of this example shortly.

In some situations, it is appropriate to assume that certain components of the $\boldsymbol{\beta}_i$ do not vary across individuals. Such an assumption might be made based on theoretical considerations. For example, if the data arise from a study of chemical kinetics, a physically meaningful unknown parameter in a theoretical model of the process might be a function of a fixed universal constant. With this knowledge, it may be sensible to regard this parameter as fixed across individual runs of the process. An initial analysis of data may also lead to the assumption that one or more components of the $\boldsymbol{\beta}_i$ do not vary across individuals. If n_i are relatively large for each individual, parameter estimates can be obtained for each individual using generalized least squares methods. An examination of these individual fits may reveal that certain parameter values are very consistent across individuals. In general, unless overwhelming theoretical or empirical evidence exists, we favor adopting the assumption that all components are random across individuals in an initial analysis; we discuss this issue in more detail in section 5.6.

To illustrate how this possibility might be incorporated into a model for $\boldsymbol{\beta}_i$, consider again the soybean growth study discussed in section 1.1.3. Relatively complete profiles are available from each plot to fit the logistic model (4.5) to the data from each plot separately. Suppose that, from a preliminary analysis, although individual estimates of β_{1i} and β_{2i} vary substantially across plots, estimates of the relative rate of growth β_{3i} are quite consistent from plot to plot, regardless of genotype or weather pattern; this scenario is used for illustrative purposes only and is not consistent with the actual data. This observation may be accommodated in the model as follows. For an individual from the gth group, let the elements of $\boldsymbol{\beta}_i$ be given by

$$\beta_{1i} = \beta_1^{(g)} + b_{1i}, \quad \beta_{2i} = \beta_2^{(g)} + b_{2i}, \quad \beta_{3i} = \beta_3^{(g)}, \tag{4.8}$$

where $\boldsymbol{\beta}^{(g)} = [\beta_1^{(g)}, \beta_2^{(g)}, \beta_3^{(g)}]'$, and $\boldsymbol{b}_i = [b_{1i}, b_{2i}]'$ is a random vector with mean zero and covariance matrix $\boldsymbol{D}$. In (4.8), β_{3i} is the same for all plots and equal to the fixed value $\beta_3^{(g)}$, with no associated random effect.

The situation in (4.8) may be expressed compactly by a generalization of the expression in (4.7). Let

$$\boldsymbol{\beta}_i = \boldsymbol{A}_i \boldsymbol{\beta} + \boldsymbol{B}_i \boldsymbol{b}_i, \tag{4.9}$$

where the (3×18) design matrix $\boldsymbol{A}_i$ corresponding to the systematic part of the model is as before, and the (3×2) design matrix $\boldsymbol{B}_i$ associated with the random effect $\boldsymbol{b}_i$ is given by

$$\boldsymbol{B}_i = \begin{bmatrix} 1 & 0 \\ 0 & 1 \\ 0 & 0 \end{bmatrix}.$$

A linear specification of the form in (4.9) offers a considerable range of possibilities for modeling inter-individual variation in the $\boldsymbol{\beta}_i$. In some situations, however, one may wish to model the inter-individual variation as a *nonlinear* function of systematic and random components. Such an approach to modeling of inter-individual variation is common in the area of population pharmacokinetic modeling.

Consider again the study of the population pharmacokinetics of quinidine, described in section 1.1.2. Typically, pharmacokinetic parameters such as clearance Cl and apparent volume of distribution V exhibit skewed distributions with constant coefficient of variation in patient populations, so a standard approach is to allow log-linear dependence of the parameters on the random effects (e.g. Verme *et al.*, 1992).

For the quinidine study, $\boldsymbol{\beta}_i = [Cl_i, V_i, k_{ai}]' = [\beta_{1i}, \beta_{2i}, \beta_{3i}]'$. Consider first a simple log-linear model for each of the elements of $\boldsymbol{\beta}_i$, assuming no systematic dependence on individual attributes:

$$\beta_{1i} = \beta_1 \exp(b_{1i}), \quad \beta_{2i} = \beta_2 \exp(b_{2i}), \quad \beta_{3i} = \beta_3 \exp(b_{3i}).$$

Here, $\boldsymbol{b}_i = [b_{1i}, b_{2i}, b_{3i}]'$ has mean zero and covariance matrix $\boldsymbol{D}$.

Note that if the random vector were normally distributed, the distributions of the elements of β_i would be lognormal, which is skewed with constant CV.

Incorporation of systematic dependence of individual pharmacokinetic parameters on subject-specific variables in a log-linear model is straightforward. For example, suppose that Cl_i is thought to depend on a subject's weight w_i, α_1-acid glycoprotein concentration g_i, and creatinine clearance (categorized as $<$ or $\geq$ 50 ml/min), and V_i is thought to depend on weight. Then, letting c_i be an indicator variable equal to 1 if creatinine clearance is $\geq$ 50, zero otherwise, consider the following model:

$$\beta_{1i} = (\beta_1 + \beta_4 w_i + \beta_5 g_i + \beta_6 c_i) \exp(b_{1i}),$$

$$\beta_{2i} = (\beta_2 + \beta_7 w_i) \exp(b_{2i}), \quad \beta_{3i} = \beta_3 \exp(b_{3i}). \qquad (4.10)$$

Other model forms are common; a model similar to (4.10) is the log-linear form

$$\beta_{1i} = \exp(\beta_1 + \beta_4 w_i + \beta_5 g_i + \beta_6 c_i + b_{1i}),$$

$$\beta_{2i} = \exp(\beta_2 + \beta_7 w_i + b_{2i}), \quad \beta_{3i} = \exp(\beta_3 + b_{3i}). \qquad (4.11)$$

As an alternative to nonlinear models for $\boldsymbol{\beta}_i$ such as (4.10) and (4.11), one might consider a reparameterization of the individual model f so that appropriate modeling of the individual parameters $\boldsymbol{\beta}_i$ may be accomplished by a simpler, linear form. For example, if one parameterizes the model for quinidine pharmacokinetics (1.2)–(1.4) in terms of $(\log Cl_i, \log V_i, \log k_{ai})$, so that $\boldsymbol{\beta}_i = [\log Cl_i, \log V_i, \log k_{ai}]'$, (4.11) may be represented by the linear specification

$$\beta_{1i} = \beta_1 + \beta_4 w_i + \beta_5 g_i + \beta_6 c_i + b_{1i},$$

$$\beta_{2i} = \beta_2 + \beta_7 w_i + b_{2i}, \quad \beta_{3i} = \beta_3 + b_{3i}.$$

Depending on the context and application, the relative simplicity of a linear model for $\boldsymbol{\beta}_i$ as well as the flexibility of available software may make a linear specification preferable.

Our discussion suggests that it makes sense to consider a quite general model for variation among individuals. Thus, we specify the following general form for a model for inter-individual variation as a function of fixed parameters, individual-specific characteristics, and random effects.

Let $\boldsymbol{\beta}_i$ be a $(p \times 1)$ vector of regression parameters specific to the ith individual. Let $\boldsymbol{a}_i$ be an $(a \times 1)$ covariate vector corresponding to individual attributes for individual i, let $\boldsymbol{b}_i$ be a $(k \times 1)$ vector of random effects associated with the ith individual, and let $\boldsymbol{\beta}$ be a $(r \times 1)$ vector of fixed parameters, or *fixed effects*. Then a general model for $\boldsymbol{\beta}_i$ is given by

$$\boldsymbol{\beta}_i = \boldsymbol{d}(\boldsymbol{a}_i, \boldsymbol{\beta}, \boldsymbol{b}_i), \tag{4.12}$$

where $\boldsymbol{d}$ is a p-dimensional vector-valued function. Each element of $\boldsymbol{d}$ is associated with the corresponding element of $\boldsymbol{\beta}_i$, so that the functional relationship may be of a different form for each element. For example, (4.10) may be put in this form by defining the elements of $\boldsymbol{d}$ to be

$$d_1(\boldsymbol{a}_i, \boldsymbol{\beta}, \boldsymbol{b}_i) = (\beta_1 + \beta_4 w_i + \beta_5 g_i + \beta_6 c_i) \exp(b_{1i}),$$

$$d_2(\boldsymbol{a}_i, \boldsymbol{\beta}, \boldsymbol{b}_i) = (\beta_2 + \beta_7 w_i) \exp(b_{2i}),$$

$$d_3(\boldsymbol{a}_i, \boldsymbol{\beta}, \boldsymbol{b}_i) = \beta_3 \exp(b_{3i}),$$

where $\boldsymbol{a}_i = [w_i, g_i, c_i]'$.

In a model of the form (4.12), a complete characterization of inter-individual variation requires an assumption about the distribution of the random effects $\boldsymbol{b}_i$. In our discussion, we have assumed that $\boldsymbol{b}_i$ has mean $\boldsymbol{0}$ and covariance matrix $\boldsymbol{D}$. As in the hierarchical linear model, the assumption of a constant covariance matrix may be relaxed, and we discuss this issue in section 4.4. The mean zero assumption is also not required in some model specifications; see Chapter 7.

By analogy to the linear case, normality is the most common assumption for the distribution of the random effects; that is,

$$\boldsymbol{b}_i \sim \mathcal{N}(\boldsymbol{0}, \boldsymbol{D}).$$

In the case of a general specification for the systematic component of variation as in (4.12), it does not follow that the individual regression parameters $\boldsymbol{\beta}_i$ themselves need be normally distributed; the implied distribution will follow from the form of the function $\boldsymbol{d}$ in (4.12). Distributional assumptions for the random effects are discussed in section 4.3.

In the preceding development, we have seen that the standard characterization of inter-individual variation involves a parametric model $\boldsymbol{d}(\boldsymbol{a}_i, \boldsymbol{\beta}, \boldsymbol{b}_i)$. This represents an assumption about the $\boldsymbol{\beta}_i$ that may not always be desirable. An alternative is to make no parametric assumptions about inter-individual variation. This is discussed further in section 4.3.2.

4.2.4 Summary

For convenience, we summarize the usual form of the general hierarchical nonlinear model.

Let $\boldsymbol{y}_i$ and $\boldsymbol{e}_i$ be the $(n_i \times 1)$ vectors of responses and random intra-individual errors for individual i. Define the $(n_i \times 1)$ mean response vector $\boldsymbol{f}_i(\boldsymbol{\beta}_i) = [f(\boldsymbol{x}_{i1}, \boldsymbol{\beta}_i), \ldots, f(\boldsymbol{x}_{in_i}, \boldsymbol{\beta}_i]'$ for the ith individual, depending on the $(p \times 1)$ individual-specific regression parameter $\boldsymbol{\beta}_i$. Suppose that

$$\mathrm{E}(\boldsymbol{e}_i|\boldsymbol{\beta}_i) = \boldsymbol{0},$$

$$\begin{aligned} \mathrm{Cov}(\boldsymbol{e}_i|\boldsymbol{\beta}_i) &= \sigma^2 \boldsymbol{G}_i^{1/2}(\boldsymbol{\beta}_i, \boldsymbol{\theta})\boldsymbol{\Gamma}_i(\boldsymbol{\alpha})\boldsymbol{G}_i^{1/2}(\boldsymbol{\beta}_i, \boldsymbol{\theta}) \\ &= \boldsymbol{R}_i(\boldsymbol{\beta}_i, \boldsymbol{\xi}), \;\; \boldsymbol{\xi} = [\sigma, \boldsymbol{\theta}', \boldsymbol{\alpha}']', \end{aligned}$$

where the $(n_i \times n_i)$ diagonal matrix $\boldsymbol{G}_i^{1/2}(\boldsymbol{\beta}_i, \boldsymbol{\theta})$ characterizes intra-individual variance and the $(n_i \times n_i)$ matrix $\boldsymbol{\Gamma}_i(\boldsymbol{\alpha})$ describes the correlation pattern within the ith individual. These matrices in turn depend on variance parameters σ and $\boldsymbol{\theta}$ $(q \times 1)$, and correlation parameters $\boldsymbol{\alpha}$ $(s \times 1)$, common to all individuals. Suppose that the $\boldsymbol{b}_i$ is a $(k \times 1)$ vector of random effects, and let $\boldsymbol{\beta}$ be a $(r \times 1)$ vector of fixed effects and $\boldsymbol{a}_i$ an $(a \times 1)$ vector of individual characteristics. Let $\boldsymbol{d}$ be a p-dimensional function of $\boldsymbol{a}_i$, $\boldsymbol{\beta}$, and $\boldsymbol{b}_i$.

With these definitions, the general hierarchical nonlinear model may be written, for $i = 1, \ldots, m$,

Stage 1 (intra-individual variation)

$$\boldsymbol{y}_i = \boldsymbol{f}_i(\boldsymbol{\beta}_i) + \boldsymbol{e}_i, \quad \boldsymbol{e}_i|\boldsymbol{\beta}_i \sim \Big(\boldsymbol{0}, \boldsymbol{R}_i(\boldsymbol{\beta}_i, \boldsymbol{\xi})\Big), \tag{4.13}$$

Stage 2 (inter-individual variation)

$$\boldsymbol{\beta}_i = \boldsymbol{d}(\boldsymbol{a}_i, \boldsymbol{\beta}, \boldsymbol{b}_i), \tag{4.14}$$

$\boldsymbol{b}_i$ are independent and identically distributed; often,

$$\boldsymbol{b}_i \sim (\boldsymbol{0}, \boldsymbol{D}),$$

where $\boldsymbol{D}$ is a $(k \times k)$ covariance matrix. Thus, in this general description of Stage 2, we allow for the possibility that there may be situations in which the assumption of zero mean for the random effects is not made.

Note that $\boldsymbol{\beta}_i$ in (4.14) is specific to individual i through $\boldsymbol{b}_i$ and the known individual characteristics $\boldsymbol{a}_i$. Thus, by substitution of the expression $\boldsymbol{d}$ for $\boldsymbol{\beta}_i$ in f and $\boldsymbol{R}_i$, we may express the conditional moments of $\boldsymbol{e}_i$ as functions of $\boldsymbol{b}_i$; that is, conditional on $\boldsymbol{b}_i$

$$\boldsymbol{e}_i|\boldsymbol{b}_i \sim \Big(\boldsymbol{0}, \boldsymbol{R}_i(\boldsymbol{\beta}_i, \boldsymbol{\xi})\Big).$$

In this notation, the dependence of the conditional moments on $\boldsymbol{a}_i$ is suppressed.

The intra-individual covariance parameter $\boldsymbol{\xi}$ and the distinct elements of the inter-individual covariance matrix $\boldsymbol{D}$ may be collected into a single covariance parameter vector $\boldsymbol{\omega}$, of dimension w. If the covariance matrix $\boldsymbol{D}$ is completely unstructured, $w = v + k(k+1)/2$. If one imposes some structure on $\boldsymbol{D}$, for example, assuming that the elements of $\boldsymbol{\beta}_i$ are uncorrelated, $\boldsymbol{D}$ may be taken to be a diagonal matrix, in which case $w = v + k$.

Within the framework of (4.13) and (4.14), two types of inference may be of interest. Estimation of the population parameters $\boldsymbol{\beta}$ and $\boldsymbol{D}$ allows inference regarding population characteristics. In pharmacokinetic or growth studies, for example, interest may focus on average kinetic behavior or growth characteristics as well as on inter-individual variation about the average. Inference about individual $\boldsymbol{\beta}_i$ and quantities derived from them are the primary focus in other settings, such as assay development and the individualization of dosage regimens in pharmacokinetic analysis. The strategies for addressing these issues given in the remainder of this book are based on different sets of assumptions about the model, described in section 4.3.

4.2.5 Time-dependent covariates

In the development in section 4.2.3, it is assumed that the covariate vector $\boldsymbol{a}_i$ summarizing individual characteristics is constant across the observations on individual i. Consequently, the general model given in section 4.2.4 above specifies implicitly that the value of the regression parameter $\boldsymbol{\beta}_i$ for individual i remains fixed for that individual over the course of observation. In some settings, particularly in the area of population pharmacokinetics, individual-specific information may change during the course of observation, and it may be reasonable to expect pharmacokinetic parameters to exhibit corresponding changes. For example, one might anticipate within-individual changes in clearance of drug to be associated with changing levels of circulating binding proteins. This is the case for the quinidine data, and, accordingly, several measurements on

α_1-acid glycoprotein concentration are available for most patients. Similarly, worsening renal function, tracked by frequent creatinine clearance measurements, may be expected to affect overall clearance for drugs that are primarily renally eliminated.

The standard approach to handling time-varying individual attributes is to permit the individual regression parameters to depend on changing individual-specific information (Beal and Sheiner, 1992; Davidian and Gallant, 1992; Wakefield, 1995). Formally, let $\boldsymbol{a}_{ij}$ represent the vector of covariate values for individual i corresponding to the jth condition of measurement $\boldsymbol{x}_{ij}$, and let $\boldsymbol{\beta}_{ij}$ be the value of the regression parameter for individual i at conditions j. The adaptation of the standard two-stage hierarchical model to this situation is given as follows:

Stage 1 (intra-individual variation)

$$y_{ij} = f(\boldsymbol{x}_{ij}, \boldsymbol{\beta}_{ij}) + e_{ij}, \tag{4.15}$$

$$\mathrm{E}(e_{ij}|\boldsymbol{\beta}_{ij}) = 0, \quad \mathrm{Var}(e_{ij}|\boldsymbol{\beta}_{ij}) = \sigma^2 g^2\{f(\boldsymbol{x}_{ij}, \boldsymbol{\beta}_{ij}), \boldsymbol{z}_{ij}, \boldsymbol{\theta}\},$$

with the possibility of correlation among the e_{ij} in the usual fashion.

Stage 2 (inter-individual variation)

$$\boldsymbol{\beta}_{ij} = \boldsymbol{d}(\boldsymbol{a}_{ij}, \boldsymbol{\beta}, \boldsymbol{b}_i). \tag{4.16}$$

For simplicity of exposition, our description of the various model specifications in section 4.3 and the inferential techniques in Chapters 5–8 is in the context of the general framework given in section 4.2.4 with time-*independent* covariates $\boldsymbol{a}_i$. The model specifications extend easily to accommodate (4.15) and (4.16), as do many of the inferential methods; we note such extensions in our subsequent development and consider examples where time-dependent covariates arise in Chapter 9 and 11.

4.2.6 Accommodation of multiple responses

In some instances,. more than one type of response may be measured repeatedly on each individual. For example, in a pharmacokinetic/pharmacodynamic study, the measurements on each subject over time following administration of drug may consist of both plasma concentrations and measurements of drug effect. Moreover, the number and placement of time points at which each type of measurement is taken may vary among response types. A description of such a study is given in section 9.5. In this situation, a separate mean response model, with associated regression parameters, and covariance model, with response-specific parameters, may be available for each variable. One approach would be to proceed with a separate model and analysis for each variable. An obvious drawback of this procedure is failure to take into account possible correlations among responses of different types. The hierarchical nonlinear model may be modified easily to handle this situation. For simplicity, we focus our discussion on the case of two response variables, but extension to more than two is straightforward.

Denote the two response variables by y_1 and y_2, and write the jth response for the ith individual on each as y_{1ij}, $j = 1, \ldots, n_{1i}$, and y_{2ij}, $j = 1, \ldots, n_{2i}$, respectively, taken at associated covariate values $\boldsymbol{x}_{1ij}$ and $\boldsymbol{x}_{2ij}$. Let $\boldsymbol{y}_{1i}$ and $\boldsymbol{y}_{2i}$ denote the $(n_{1i} \times 1)$ and $(n_{2i} \times 1)$ vectors of observations of each type, respectively.

Suppose that for the kth response type, $k = 1, 2$, we have specified a hierarchical nonlinear model of the form (4.13), (4.14). Subscripting quantities pertaining to the kth response in an obvious fashion, we may write a *combined* hierarchical model. To allow a simpler description, suppose that there are no shared elements among the two individual-specific regression parameters $\boldsymbol{\beta}_{1i}$ and $\boldsymbol{\beta}_{2i}$ or the fixed effects vectors $\boldsymbol{\beta}^{(1)}$ and $\boldsymbol{\beta}^{(2)}$; relaxation of these assumptions is straightforward. Let $n_i = n_{1i} + n_{2i}$, the total number of measurements on individual i, and let the $(n_i \times 1)$ vector $\boldsymbol{y}_i$ denote the vector consisting of $\boldsymbol{y}_{1i}$ and $\boldsymbol{y}_{2i}$ 'stacked,' with jth element y_{ij}, so that j now indexes position in this vector, and let $\boldsymbol{e}_i$ and its elements denote the associated within-individual errors. Index the vectors $\boldsymbol{x}_{ij}$ similarly. Form the combined individual-specific regression parameter $\boldsymbol{\beta}_i = [\boldsymbol{\beta}_{1i}', \boldsymbol{\beta}_{2i}']'$ for individual i, and similarly, let $\boldsymbol{\beta} = [\boldsymbol{\beta}^{(1)\prime}, \boldsymbol{\beta}^{(2)\prime}]'$ be the combined vector of fixed effects. Similarly, let $\boldsymbol{a}_i$ denote the collection of individual-specific covariates appearing in both models. Let $\boldsymbol{\xi}$ be the combined vector of intra-individual covariance parameters. Then the combined hierarchy is as follows:

Stage 1 (intra-individual variation)

$$y_{ij} = f_k(\boldsymbol{x}_{ij}, \boldsymbol{\beta}_i) + e_{ij}$$

if observation j on subject i is of type k, $k = 1, 2$, where the kth regression model, f_k, has been written as a function of the combined parameter vector, $\boldsymbol{\beta}_i$, and

$$\boldsymbol{e}_i|\boldsymbol{\beta}_i \sim \Big(\mathbf{0}, \boldsymbol{R}_i(\boldsymbol{\beta}_i, \boldsymbol{\xi})\Big).$$

Here, $\boldsymbol{R}_i(\boldsymbol{\beta}_i, \boldsymbol{\xi})$ is the $(n_i \times n_i)$ covariance matrix whose upper left $(n_{1i} \times n_{1i})$ submatrix corresponds to the covariance structure for $\boldsymbol{y}_{1i}$, whose lower right $(n_{2i} \times n_{2i})$ submatrix corresponds to the covariance structure for $\boldsymbol{y}_{2i}$, and whose other entries may correspond to a model for within-individual correlations between the elements of $\boldsymbol{y}_{1i}$ and $\boldsymbol{y}_{2i}$.

Stage 2 (inter-individual variation)

$$\boldsymbol{\beta}_i = \boldsymbol{d}(\boldsymbol{a}_i, \boldsymbol{\beta}, \boldsymbol{b}_i), \quad \boldsymbol{b}_i \sim (\mathbf{0}, \boldsymbol{D}),$$

where the components of the inter-individual model $\boldsymbol{d}$ correspond to those in the separate models $\boldsymbol{d}_k$ for each response, and $\boldsymbol{b}_i = [\boldsymbol{b}_{1i}', \boldsymbol{b}_{2i}']'$.

In Stage 1 of the combined model, possible scaling differences between variable types are accommodated by the possibility of different intra-individual covariance parameters for each type, as in the separate models above. In Stage 2, the individual-specific regression parameters are collected into a single vector, as are the associated random effects. The matrix $\boldsymbol{D}$, is the *joint* covariance matrix for all random effects, thus allowing correlations among the individual-specific regression parameters for the two different response types to be taken into account. By combining the two models in this way, associations among factors corresponding to the two responses are accommodated, and elements common to both models may be estimated with greater precision, using information from both sets of measurements. Note that there is no requirement that the number of observations of each type be the same; by 'stacking' all responses into a single vector, one may achieve the objective of accounting for correlations without resorting to the response matrix formulations used in traditional multivariate analysis, which have this requirement.

An analysis based on such a combined hierarchical model is given in section 9.5.

4.3 Model specification

In setting forth the general model framework in section 4.2, we have mainly emphasized accommodation of the systematic elements of intra- and inter-individual variation. As noted in section 4.2.2, it is often the case that an assumption about the conditional distribution of intra-individual errors may be based on theoretical considerations or on previously gathered empirical evidence. A standard assumption is that of normality, which we may summarize in terms of the general hierarchical nonlinear model of section 4.2.4 as

Stage 1 (intra-individual variation)

$$\begin{aligned} \boldsymbol{y}_i &= \boldsymbol{f}_i(\boldsymbol{\beta}_i) + \boldsymbol{e}_i, \\ \boldsymbol{e}_i|\boldsymbol{\beta}_i &\sim \mathcal{N}\{\mathbf{0}, \boldsymbol{R}_i(\boldsymbol{\beta}_i, \boldsymbol{\xi})\}. \end{aligned} \quad (4.17)$$

Other possibilities exist; for example, the conditional distribution of the intra-individual random component may be taken to be lognormal or gamma in situations where the normal is inappropriate. In the model specifications described in the next three sections, Stage 1 of the hierarchical nonlinear model is assumed to be of the above form, where conditional normality of the within-individual errors may be replaced by another distributional assumption if desired.

In contrast, specification of a distribution to characterize inter-individual variation is fundamentally more difficult. In this section, we discuss different approaches to this issue. These approaches lead to the various model specifications described below.

4.3.1 Fully parametric model specification

As in the case of the hierarchical linear model, a standard approach to inference for the nonlinear case is based on full distributional assumptions for both the intra- and inter-individual random components $\boldsymbol{e}_i$ and $\boldsymbol{b}_i$, respectively. By far the most common assumption is that in which both random components are taken to be normally distributed. That is, in the general hierarchical nonlinear model summarized in section 4.2.4, assume that

$$\begin{aligned} \boldsymbol{e}_i|\boldsymbol{\beta}_i \sim \mathcal{N}\{\mathbf{0}, \boldsymbol{R}_i(\boldsymbol{\beta}_i, \boldsymbol{\xi})\}, \ \ \boldsymbol{\beta}_i &= \boldsymbol{d}(\boldsymbol{a}_i, \boldsymbol{\beta}, \boldsymbol{b}_i), \\ \boldsymbol{b}_i &\sim \mathcal{N}(\mathbf{0}, \boldsymbol{D}). \end{aligned} \quad (4.18)$$

This set of distributional assumptions is used by Beal and Sheiner (1982) in their work on the analysis of population pharmacokinetic studies.

Under this assumption, Stage 1 is given by (4.17), and Stage 2 of the fully parametric model specification may be summarized as

Stage 2 (inter-individual variation)

$$\boldsymbol{\beta}_i = \boldsymbol{d}(\boldsymbol{a}_i, \boldsymbol{\beta}, \boldsymbol{b}_i),$$

$$\boldsymbol{b}_i \sim \mathcal{N}(\boldsymbol{0}, \boldsymbol{D}).$$

As an alternative to normality, a distribution such as the multivariate t (Wakefield, 1995) or a mixture of normal distributions (Beal and Sheiner, 1992) may be assumed for the inter-individual random effects. The t distribution, with its heavier tails, may provide a robust alternative to handle outlying individuals. A mixture of normals accommodates the possibility of multimodality of the distribution of the random effects; multiple modes are discussed in detail in the next two sections. However, in general, if a fully parametric distributional assumption is made for the random effects $\boldsymbol{b}_i$, it is almost always taken to be the normal model (4.18), see for example, Beal and Sheiner (1982) and Lindstrom and Bates (1990). Many of the inferential methods discussed in Chapters 5 and 6 rely heavily on this assumption.

4.3.2 Nonparametric model specification

Specification of a parametric model, $\boldsymbol{d}(\boldsymbol{a}_i, \boldsymbol{\beta}, \boldsymbol{b}_i)$, for the $\boldsymbol{\beta}_i$ is a natural approach when potentially relevant covariate information is available. If one wishes, however, to avoid making this kind of assumption, an alternative strategy is to characterize the variation in the $\boldsymbol{\beta}_i$ in less restrictive ways. Mallet (1986) proposes a model formulation that is completely nonparametric in the following sense. No parametric form is assumed for the $\boldsymbol{\beta}_i$ in terms of fixed effects nor is any assumption made about their distribution.

Under the nonparametric model specification, then, the conditional distribution, given $\boldsymbol{\beta}_i$, of the within-individual errors $\boldsymbol{e}_i$ is specified by a parametric form, as in (4.17). The distribution of the $\boldsymbol{\beta}_i$ themselves is left completely unspecified. The nonparametric model specification is as follows.

Stage 2 (inter-individual variation)

$$\beta_i \sim H,$$

where H is a completely unrestricted distribution function. Thus, the inter-individual variation (Stage 2) is accommodated solely through the distribution H. This formulation does not facilitate incorporation of covariate information in a straightforward manner. This issue and inferential techniques for this model are described in Chapter 7.

4.3.3 Semiparametric model specification

The nonparametric model specification allows the random parameters to arise from virtually any distribution H, but does not provide a natural framework for accommodation of covariates. In contrast, the distributional assumptions underlying the fully parametric model, such as normality, may be unduly restrictive. The actual distribution of the random component might be skewed, or have heavier tails than the normal, and misspecification of this distribution may adversely affect inference. This suggests a compromise between these two specifications, borrowing features from both. An intermediate possibility is to assume a parametric model for the β_i but to allow a more flexible distributional form for the random effects. In particular, assume that random effects arise from a *class* of probability densities that includes the normal density, densities with multiple modes, skewed densities, and densities with excess dispersion, but excludes distributions with unusual features unlikely to be sensible models for real biological and physical phenomena. This is the approach taken by Davidian and Gallant (1992, 1993), who suggested that the density could then be estimated by appropriate methods for density estimation.

Another motivation for greater flexibility in the assumed distribution for the random component of β_i is potential misspecification of the systematic component, as the following development shows. Consider the study of the population pharmacokinetics of quinidine, described in section 1.1.2. Suppose that quinidine clearance Cl_i for individual i depends on that individual's smoking status; for example, suppose that clearance is faster among smokers than nonsmokers. Thus, if the true Cl_i values were available, it would be reasonable to expect them to cluster roughly into two groups corresponding to smokers and nonsmokers, once other systematic factors were taken into account. Suppose further that this dependence on smoking status was not expected *a priori*, so that smoking status is not included in the model $\boldsymbol{d}$ for systematic dependence of Cl_i on individual characteristics. For definiteness, suppose that the model for Cl_i is

$$\log Cl_i = \boldsymbol{\beta}'\boldsymbol{a}_i + b_i, \tag{4.19}$$

where $\boldsymbol{a}_i$ is a vector of individual characteristic information not including smoking status, and b_i is a random effect corresponding to clearance. Because smoking status in fact does have a systematic effect on Cl_i, this model will assign to the random component b_i all unexplained variation, including that due to the systematic effect of smoking omitted from the model. The 'random' component of variation in this admittedly misspecified model would be best represented by a bimodal distribution, in order to capture the embedded systematic effect of smoking status. The normal distribution, or indeed any unimodal density, would thus be an inadequate representation of the apparent nature of random variation accounted for in (4.19) by b_i.

The implication is that, in order to allow detection of possible omissions of relevant information from the description of systematic variation in the $\boldsymbol{\beta}_i$, it is desirable that the distributional specification for the random effects $\boldsymbol{b}_i$ be flexible enough to capture multimodality, as well as other departures from the usual normality assumption.

Formally, the semiparametric model specification that embodies this idea is given as follows:

Stage 2 (inter-individual variation)

$$\boldsymbol{\beta}_i = \boldsymbol{d}(\boldsymbol{a}_i, \boldsymbol{\beta}, \boldsymbol{b}_i),$$

$$\boldsymbol{b}_i \sim h, \quad h \in \mathcal{H},$$

where h is a density belonging to a class $\mathcal{H}$ of 'smooth' densities.

As in the nonparametric case, the normal distribution has been used to model the random intra-individual errors (4.17), but other parametric forms may be specified.

A number of possibilities exist for defining the class of densities $\mathcal{H}$. Davidian and Gallant (1992, 1993) describe a particular class of densities $\mathcal{H}$ which has the desirable properties noted above.

Inferential techniques appropriate for this class of densities, as well as alternative approaches, are described in Chapter 7.

4.3.4 Bayesian model specification

In the preceding three sections, the hierarchical nonlinear model framework parallels that for linear mixed models, with the degree of restrictiveness regarding the assumption on the distribution of the random effects being the difference among models. Alternatively, one could impose a complete structure derived from Bayesian ideas. Bayesian models for the hierarchical nonlinear framework have been discussed by Racine *et al.* (1986), Gelfand *et al.* (1990), Zeger and Karim (1991), and Wakefield (1995).

As in the linear case, outlined in section 3.2.3, in the Bayesian formulation, the natural hierarchy involves three stages:

Stage 1 (intra-individual variation)

$$\boldsymbol{y}_i = \boldsymbol{f}_i(\boldsymbol{\beta}_i) + \boldsymbol{e}_i,$$

$$\boldsymbol{e}_i|\boldsymbol{\beta}_i \sim \Big(\boldsymbol{0}, \boldsymbol{R}_i(\boldsymbol{\beta}_i, \boldsymbol{\xi})\Big).$$

Stage 2 (inter-individual variation)

$$\boldsymbol{\beta}_i = \boldsymbol{d}(\boldsymbol{a}_i, \boldsymbol{\beta}, \boldsymbol{b}_i), \quad \boldsymbol{b}_i \sim (\boldsymbol{0}, \boldsymbol{D}).$$

Stage 3 (hyperprior distribution)

The model specification is completed by assumption of a distribution for all parameters in Stages 1 and 2: $\boldsymbol{\beta}$, $\boldsymbol{\xi}$, and $\boldsymbol{D}$.

The first two stages are often referred to as the *sampling distribution* and the *prior distribution*, respectively. The most common distributional choice for both Stages 1 and 2 is that of the multivariate normal distribution. The hyperprior is generally chosen to be noninformative. For $\boldsymbol{\beta}$, this is usually accomplished by the assumption that $\boldsymbol{\beta} \sim \mathcal{N}(\boldsymbol{\beta}^*, \boldsymbol{H})$, where $\boldsymbol{H}^{-1} = \boldsymbol{0}$. The assumption of a Wishart distribution for $\boldsymbol{D}^{-1}$ is convenient. A prior distribution is also assumed for some or all of the components of the within-individual variance model $\boldsymbol{R}_i(\boldsymbol{\beta}_i, \boldsymbol{\xi})$. For example, if individual errors are thought to be independent, with individual variance following the power model (2.6), that is,

$$\text{Var}(e_{ij}|\boldsymbol{\beta}_i) = \sigma^2 f^{2\theta}(\boldsymbol{x}_{ij}, \boldsymbol{\beta}_i),$$

one might assume prior distributions on the scale parameter σ, the power parameter θ, or both. A common specification for the scale parameter is to assume a gamma prior for σ^{-2}. If θ is taken to be fixed but unknown, a prior specification may be employed for θ; Wakefield (1995) uses a uniform prior over an interval of likely values for θ.

Departures from the standard normal paradigm are possible within the Bayesian framework. Wakefield (1995) uses a multivariate t distribution in place of a normal model for the random effects $\boldsymbol{b}_i$ in the second stage and a lognormal distribution for the sampling distribution of the response in the first stage to model the data from the study of quinidine described in section 1.1.2. Wakefield *et al.* (1994) describe further nonnormal modeling strategies in a hierarchical Bayesian nonlinear model.

We discuss inference for the Bayesian model in Chapter 8.

4.4 Discussion

Throughout our discussion of the hierarchical nonlinear model, we have pointed out several instances in which nonlinearity of the mean response function introduces complications not encountered in the linear case. We have also indicated that the model given in section 4.2.4 may be generalized in a number of ways. In this section, we address these and other issues.

Form of the marginal distribution

As we have discussed, a fundamental difference between the linear and nonlinear versions of the hierarchical model is in the ability to derive explicitly the marginal distribution of the response $\boldsymbol{y}_i$. This difficulty arises for all model formulations. To illustrate, consider the situation where both the (conditional) distribution of the intra-individual errors and that of the random effects are assumed to be appropriate normal distributions (the fully parametric case of section 4.3.1). Write the conditional density of $\boldsymbol{y}_i$ given $\boldsymbol{b}_i$ as

$$p_{y|b}(\boldsymbol{y}_i|\boldsymbol{x}_{i1},\ldots,\boldsymbol{x}_{in_i},\boldsymbol{a}_i,\boldsymbol{\beta},\boldsymbol{\xi},\boldsymbol{b}_i),$$

where we have emphasized the dependence on all individual-specific information and the fixed effects. Write the density of $\boldsymbol{b}_i$ as $p_b(\boldsymbol{b}_i|\boldsymbol{D})$, noting the dependence on fixed parameters such as the elements of

$\boldsymbol{D}$. Then the marginal distribution of $\boldsymbol{y}_i$ is given by

$$\int p_{y|b}(\boldsymbol{y}_i|\boldsymbol{x}_{i1},\ldots,\boldsymbol{x}_{in_i},\boldsymbol{a}_i,\boldsymbol{\beta},\boldsymbol{\xi},\boldsymbol{b})\, p_b(\boldsymbol{b}|\boldsymbol{D})\, d\boldsymbol{b}. \tag{4.20}$$

For the hierarchical linear model discussed in Chapter 3, both $p_{y|b}$ and p_b are normal distributions, and the intra-individual covariance matrix $\boldsymbol{R}_i$ does not depend on $\boldsymbol{b}_i$. In this case, it is possible to evaluate the integral in (4.20) explicitly to obtain the form of the (normal) marginal distribution. In contrast, for the hierarchical nonlinear model, under the same conditions except for the nonlinearity of the mean response function f, it is not possible to evaluate this integral in general. To see this explicitly, it is necessary to consider only the simplest case where

$$\boldsymbol{e}_i|\boldsymbol{\beta}_i \sim \mathcal{N}(\boldsymbol{0},\sigma^2\boldsymbol{I}_{n_i}),$$

$$\boldsymbol{\beta}_i = \boldsymbol{\beta}+\boldsymbol{b}_i, \quad \boldsymbol{b}_i \sim \mathcal{N}(\boldsymbol{0},\boldsymbol{D}).$$

In this situation,

$$p_{y|b}(\boldsymbol{y}_i|\boldsymbol{x}_{i1},\ldots,\boldsymbol{x}_{in_i},\boldsymbol{a}_i,\boldsymbol{\beta},\boldsymbol{\xi},\boldsymbol{b}_i)$$
$$= \frac{1}{(2\pi\sigma^2)^{n_i/2}}\exp\left[-\frac{1}{2\sigma^2}\{\boldsymbol{y}_i - \boldsymbol{f}_i(\boldsymbol{\beta}+\boldsymbol{b}_i)\}'\{\boldsymbol{y}_i - \boldsymbol{f}_i(\boldsymbol{\beta}+\boldsymbol{b}_i)\}\right],$$

and

$$p_b(\boldsymbol{b}_i|\boldsymbol{D}) = \frac{1}{(2\pi)^{n_i/2}|\boldsymbol{D}|^{1/2}}\exp\left(-\frac{1}{2}\boldsymbol{b}_i'\boldsymbol{D}^{-1}\boldsymbol{b}_i\right).$$

From these expressions, it is clear that if f is a *linear* function of $\boldsymbol{\beta}+\boldsymbol{b}_i$, as in the case, for example, of a linear random coefficient model with

$$\boldsymbol{f}_i(\boldsymbol{\beta}_i) = \boldsymbol{X}_i\boldsymbol{\beta}_i = \boldsymbol{X}_i\boldsymbol{\beta} + \boldsymbol{X}_i\boldsymbol{b}_i,$$

then, using standard arguments as in section 3.3.3 (completing the square), it is straightforward to calculate the integral (4.20) explicitly. For a general *nonlinear* function f of $\boldsymbol{\beta}+\boldsymbol{b}_i$, however, these arguments do not apply; it is impossible to find a general strategy to 'complete the square' or to identify a general transformation to allow analytic evaluation of the integral. There are, of course, specific exceptions; for example, if f is the monoexponential function

$$f(x,\boldsymbol{\beta}) = \beta_1\exp(-\beta_2 x),$$

along with the assumptions of constant intra-individual variance and independent within-individual responses as above, it is possible to carry out the integration, as the integral has the form of the moment generating function for the normal distribution (Beal, 1984a).

Note that this difficulty arises even in this most simple of cases. When the intra-individual covariance structure depends on β_i, and hence on b_i, the integral is generally intractable even in the case of a linear mean response function. Nonlinearity of the relationship d between β_i and b_i adds an additional level of complication in either the linear or nonlinear case as well.

In general, then, the form of the integral necessary to obtain the marginal distribution can be quite complex, making implementation of classical likelihood techniques difficult. In principle, the integral can be evaluated numerically, as discussed in Chapter 7. Alternatively, other approaches to deriving general inferential strategies for complex versions of the hierarchical nonlinear model are based on approximations to the marginal distribution or appeal to large sample theory results. These methods are described in Chapters 5 and 6.

Interpretation of model parameters

The discussion in the preceding section highlights an additional complication that arises in the hierarchical model framework when the mean response model is nonlinear. Consider again the case of the fully parametric model specification with normal intra- and inter-individual error, as above.

Under the hierarchical linear model specification for an *individual* response given in (3.5), that is,

$$y_i = X_i\beta + Z_i b_i + e_i,$$

the marginal mean of y_i is $X_i\beta$, with marginal covariance structure of the form $V_i = R_i + Z_i D Z_i'$. This model starts with consideration of a specific individual – the mean response for the specific individual is $X_i\beta + Z_i b_i$ – with the marginal model for the population of response vectors y_i arising as a consequence.

Alternatively, one might choose to start with consideration of behavior in the population as a whole, and model the marginal moments directly. If a linear model were deemed adequate for the marginal mean response, one would likely use the mean response function $X_i\beta$, with a suitably chosen marginal covariance model to characterize the variation among the response vectors y_i.

In the case of the linear model, whether one chooses to start at the individual level or to consider the population as a whole, one is led to the same model for the marginal mean. Thus, from either standpoint, the parameter vector of fixed effects, $\boldsymbol{\beta}$, has the same interpretation as the value of $\boldsymbol{\beta}$ producing the 'typical' response vector at the covariate setting $\boldsymbol{X}_i$. The hierarchical approach allows a convenient way to construct an appropriate model for marginal covariance that takes into account the two sources of variation, among and within individuals.

In contrast, for a nonlinear mean response function, whether one chooses to model the response $\boldsymbol{y}_i$ at the individual level, as is done in the hierarchical nonlinear model framework, or to consider initially population behavior and the marginal moments directly, changes the interpretation of the model and the regression parameter $\boldsymbol{\beta}$. In the first case, where the hierarchical modeling approach is used, the mean response for a given individual is taken to be $\boldsymbol{f}_i(\boldsymbol{\beta}_i)$, and the nature of variation in the population of response vectors $\boldsymbol{y}_i$ follows from the assumptions on $\boldsymbol{\beta}_i$ and $\boldsymbol{b}_i$. The marginal mean response and covariance based on this model may or may not be calculable explicitly, depending on the complexity of the integration required. Alternatively, in the population-based approach, given the apparent response relationship, one might choose to use the same function f to characterize mean response, with a parameter $\boldsymbol{\beta}$ common to all individuals specifying the marginal mean response. One would then model the variation in the population of response vectors $\boldsymbol{y}_i$ directly, taking into account both intra- and inter-individual variation in the form of the model. For a general function f, it is clear from the form of the integral (4.20) that

$$\mathrm{E}\{f(\boldsymbol{x}_{ij}, \boldsymbol{\beta}_i)\} \neq f(\boldsymbol{x}_{ij}, \boldsymbol{\beta})$$

in general. That is, if one uses the same function f to model either individual mean response or population mean response, the two approaches *do not* lead to the same marginal model for the population. From the perspective of the hierarchical nonlinear model, $\boldsymbol{\beta}$ pertains to the population of all parameter values, and hence has the interpretation as the 'typical' fixed effect value. In contrast, from the standpoint of modeling the marginal moments, $\boldsymbol{\beta}$ represents the value producing a 'typical' response $\boldsymbol{y}_i$.

This issue has been discussed by several authors, notably by Zeger *et al.* (1988), who refer to the case of the hierarchical model for an individual as a *subject-specific* model and that of a model for marginal moments as a *population-averaged* model; see also Lindstrom and Bates (1990). Further, detailed discussion is given in Diggle *et al.* (1994, chapter 7).

Our position is that the subject-specific approach, that of modeling at the individual level, seems more natural for the kinds of applications with which we are concerned.

More general covariance structures

(i) *Intra-individual covariance.* The general covariance model (4.3) is predicated on the assumption of a common pattern of intra-individual variation. In many contexts, this is an appropriate model, as discussed in section 4.2.2. In some situations, however, this assumption may be unrealistic.

One could extend the model for intra-individual covariance to allow covariance parameters to vary from individual to individual:

$$\begin{aligned}\mathrm{Cov}(\boldsymbol{e}_i|\boldsymbol{\beta}_i) &= \sigma_i^2 \boldsymbol{G}_i^{1/2}(\boldsymbol{\beta}_i,\boldsymbol{\theta}_i)\boldsymbol{\Gamma}_i(\boldsymbol{\alpha}_i)\boldsymbol{G}_i^{1/2}(\boldsymbol{\beta}_i,\boldsymbol{\theta}_i)\\ &= \boldsymbol{R}_i(\boldsymbol{\beta}_i,\boldsymbol{\xi}_i),\ \ \boldsymbol{\xi}_i = [\sigma_i,\boldsymbol{\theta}_i',\boldsymbol{\alpha}_i']'.\end{aligned}$$

This accommodates the possibility of both variance and correlation parameters that differ among individuals. We do not generally recommend this, however. In our view, it is likely to be impossible to estimate the elements of such a complicated structure reliably. Moreover, the additional complexity necessary to fit such models may degrade inference on first-moment model components. Finally, we remark that it is desirable to strive for a parsimonious representation in modeling variation. A model that assumes a common pattern of variation may well capture enough of the salient features of individual covariance structures, even if the assumption is not exactly correct. For most of the applications covered in this book, we have found that this assumption works well, and we focus for the most part on inferential techniques that are predicated on this assumption in Chapters 5–8.

Caution still should be exercised even under the assumption of a common pattern of intra-individual variation, however. Although we have had good success estimating intra-individual variance parameters when observations on an individual are assumed uncorrelated, estimation of correlation parameters in more complex intra-individual error models has proved to be problematic in some cases, even with data from several individuals. We discuss this issue further in the context of examples in subsequent chapters.

(ii) *Inter-individual covariance.* In the general hierarchical nonlinear model with a parametric specification for $\boldsymbol{\beta}_i$, the form of the inter-individual covariance matrix $\boldsymbol{D}$ describing random individual-to-individual variation is taken to be the same for all individuals. It is possible to relax this assumption; for example, for the soybean data discussed in section 4.2.3, one might specify a separate covariance matrix $\boldsymbol{D}_k$, say, for each of the genotype-weather condition groups indexed by k.

The assumption of a constant inter-individual covariance matrix $\boldsymbol{D}$ is not really a serious restriction. By an appropriate choice of the inter-individual regression model $\boldsymbol{d}$, it is possible to specify a model that allows for nonconstant random variation among the $\boldsymbol{\beta}_i$. An example is given in the context of the data on the population pharmacokinetics of quinidine: in (4.10), for example, the models for clearance, volume of distribution, and absorption rate constant imply that the variance as well as the mean of the distributions of each of these pharmacokinetic parameters depends on individual characteristics such as weight w_i.

Incorporation of intra-individual correlation structures

Throughout our development, we have allowed for the possibility of explicit modeling of within-individual correlation structures. This raises some issues of both interpretation and implementation, which we now discuss.

Historically, one motivation for the hierarchical linear model was the ability, through the assumption of two distinct sources of variation, to induce a special covariance structure at the marginal level that takes into account correlations among observations on a given individual. This was seen as an advantage over an unrestricted multivariate model: although the former involves a more general covariance model, it requires balanced numbers of observations across all individuals (Laird and Ware, 1982). A similar perspective may be taken on the hierarchical nonlinear model: although the form of the marginal distribution may not be written explicitly in most situations, it is intuitively clear that incorporation of individual-specific random effects will serve to induce correlations at the marginal level among measurements on a given individual. In both the linear and nonlinear cases, then, it has often been assumed that, even though there may be a discernible pattern of correlation within individuals, the two-stage framework will yield an adequate model at the marginal level.

In the linear case, Chi and Reinsel (1989) found that incorporation of an AR(1) model for within-subject correlation leads to improved inference on the fixed effects over simply allowing the induced covariance structure to accommodate it, as long as the assumption of AR(1) intra-individual correlation is correct. This is not surprising; it is reasonable to expect that if a systematic intra-individual correlation pattern is a true feature of the data, failure to take it into account will be inefficient. Moreover, it seems likely that such an omission may have a deleterious effect on estimation of the random effects matrix $\boldsymbol{D}$, which will be forced to accommodate both this pattern *and* the variation in the population. Extension of these ideas to the nonlinear case is immediate.

Nonetheless, this extra flexibility in model specification exacts a price. Although, in principle, it seems unquestionable that one would prefer to model the intra-individual correlation pattern explicitly to achieve improved inference, in practice, there may be insufficient information available to allow reliable modeling and estimation. Attempts to fit a within-individual correlation model may be unsuccessful, owing to the paucity of information and computational burden of the estimation procedure itself. Even if estimates are obtained, with limited information, there may be no assurance that the chosen model is appropriate; if it is not, it may well be the case that a model that assumes no correlation within individuals, although also incorrect, may lead to more efficient inferences.

The complicated interplay between intra- and inter-individual covariance components in the hierarchical nonlinear model is difficult to sort out and is not well-understood. Accordingly, our position is that, unless the reasons to do so are compelling and there are a large number of observations per individual, explicit models for within-individual correlation should be approached with caution.

CHAPTER 5

Inference based on individual estimates

5.1 Introduction

Two broad classes of inferential procedures have been proposed for the hierarchical nonlinear model framework with a parametric model specification, described in section 4.3.1. In cases where sufficient measurements are taken on individuals to allow construction of individual-specific regression parameter estimates, such estimates may be used as building blocks for further inference. If this is not the case, the data analyst has recourse to methods based on linearization, discussed in Chapter 6.

In this chapter, we consider the first case, where sufficient data are available for all, or most, of the individuals studied to allow estimation of individual-specific regression parameters. This is often the case in growth studies and for assay data, where sufficient concentration-response data are generated in any specific assay run to determine the regression curve for that run. For pharmacokinetic data, the applicability of the methods in this chapter will depend on the particular study design. For instance, in early pilot studies, such as that for cefamandole described in section 1.1.1, fairly full profile information may be collected on each subject. If pharmacokinetic information is collected in a routine clinical setting, this is less likely to be the case.

In section 5.2, we consider methods of obtaining individual regression parameter estimates. Our development closely parallels that for generalized least squares estimation for data from a single individual, given in section 2.3. When data are available from several individuals, better characterization of the intra-individual covariance structure is possible within the GLS framework; this is discussed in section 5.2.2. The use of individual estimates as building blocks for inference regarding population parameters is considered in section 5.3. Methods in that section are motivated by an assumption of normality for the random effects distribution. Given this assumption, inferential procedures may be derived which closely parallel those for the hierarchical linear model. Computational aspects are discussed in section 5.4, and in section 5.5 the methods are illustrated using data from a pharmacokinetic study of the anti-asthmatic agent theophylline. The chapter concludes with a discussion of some philosophical and practical issues in section 5.6 and brief biographical notes in section 5.7.

5.2 Construction of individual estimates

Methods discussed in this chapter may be classified as *two-stage* methods, as their implementation follows a two-stage scheme: (1) obtain individual estimates $\boldsymbol{\beta}_i^*$; (2) input these $\boldsymbol{\beta}_i^*$ to any procedure that uses individual estimates as building blocks. In this section, we consider stage 1, efficient construction of individual regression parameter estimates $\boldsymbol{\beta}_i^*$.

5.2.1 Generalized least squares

Several methods were discussed in Chapter 2 for estimating $\boldsymbol{\beta}_i$ based on the data from the ith individual only. In particular, we emphasized the use of generalized least squares methods, preferring these techniques to joint normal theory maximum likelihood because of their superior robustness to model misspecification (section 2.5). Another advantage of the GLS approach is that it generalizes readily to the case where data are available from several individuals in a way that joint maximum likelihood does not. Assume that data are available from m individuals and that the model specified by (4.13) in stage 1 of the hierarchy for nonlinear models of section 4.2.4 holds, so that

$$\boldsymbol{y}_i = \boldsymbol{f}_i(\boldsymbol{\beta}_i) + \boldsymbol{e}_i,$$

$$\mathrm{E}(\boldsymbol{e}_i|\boldsymbol{\beta}_i) = \boldsymbol{0} \quad \mathrm{Cov}(\boldsymbol{e}_i|\boldsymbol{\beta}_i) = \boldsymbol{R}_i(\boldsymbol{\beta}_i, \boldsymbol{\xi}),$$

where the functional form of $\boldsymbol{R}_i(\boldsymbol{\beta}_i, \boldsymbol{\xi})$ and the intra-individual covariance parameter $\boldsymbol{\xi}$ are the same across individuals. As before, we will often find it convenient to write

$$\boldsymbol{R}_i(\boldsymbol{\beta}_i, \boldsymbol{\xi}) = \sigma^2 \boldsymbol{S}_i(\boldsymbol{\beta}_i, \gamma),$$

where $\boldsymbol{\xi} = (\sigma, \boldsymbol{\gamma}')'$ and $\boldsymbol{\gamma}$ is the $(v \times 1)$, $v = q + s$, vector of parameters in the functional part of the covariance model.

In a manner completely analogous to the case of data from a single individual, we may define the following GLS estimation scheme:

1. In m separate unweighted regressions (for example, OLS), obtain preliminary estimates $\hat{\boldsymbol{\beta}}_i^{(p)}$ for each individual, $i = 1, \ldots, m$.
2. Use residuals from these preliminary fits to estimate σ and $\boldsymbol{\gamma}$. Form estimated weight matrices based on the estimate $\hat{\boldsymbol{\gamma}}$ obtained from this procedure, along with the preliminary $\hat{\boldsymbol{\beta}}_i^{(p)}$, to form
$$\hat{\boldsymbol{W}}_i = \boldsymbol{S}_i^{-1}(\hat{\boldsymbol{\beta}}_i^{(p)}, \hat{\boldsymbol{\gamma}}).$$
3. Using the estimated weight matrices from step 2, reestimate the $\boldsymbol{\beta}_i$ by m separate minimizations: for individual i, $i = 1, \ldots, m$, minimize in $\boldsymbol{\beta}_i$
$$\{\boldsymbol{y}_i - \boldsymbol{f}_i(\boldsymbol{\beta}_i)\}'\hat{\boldsymbol{W}}_i\{\boldsymbol{y}_i - \boldsymbol{f}_i(\boldsymbol{\beta}_i)\}.$$
Treating the resulting estimators as new preliminary estimators, return to step 2.

The algorithm should be iterated at least once to wash out the effect of potentially inefficient preliminary estimates in step 1; see section 2.3.

5.2.2 Pooled estimation of covariance parameters

Estimation of the within-individual covariance parameters σ and $\boldsymbol{\gamma}$ in step 2 of the GLS algorithm above may be carried out in several ways. An appealing feature of GLS is that inferential methods for $(\sigma, \boldsymbol{\gamma}')'$ based on data from a single individual generalize naturally to the case of several individuals. Essentially, methods based on residuals from a regression fit for a single individual extend by pooling residuals across individuals.

For example, the pseudolikelihood estimators of σ and $\boldsymbol{\gamma}$ for the ith individual minimize in σ and $\boldsymbol{\gamma}$ the expression given in (2.29):

$$PL_i(\hat{\boldsymbol{\beta}}_i^{(p)}, \sigma, \boldsymbol{\gamma}) = \log|\sigma^2 \boldsymbol{S}_i(\hat{\boldsymbol{\beta}}_i^{(p)}, \boldsymbol{\gamma})| + \{\boldsymbol{y}_i - \boldsymbol{f}_i(\hat{\boldsymbol{\beta}}_i^{(p)})\}'\boldsymbol{S}_i^{-1}(\hat{\boldsymbol{\beta}}_i^{(p)}, \boldsymbol{\gamma})\{\boldsymbol{y}_i - \boldsymbol{f}_i(\hat{\boldsymbol{\beta}}_i^{(p)})\}/\sigma^2. \quad (5.1)$$

Generalization to the case of m individuals is trivial; one simply chooses σ and $\boldsymbol{\gamma}$ to minimize $\sum_{i=1}^m PL_i(\hat{\boldsymbol{\beta}}_i^{(p)}, \sigma, \boldsymbol{\gamma})$.

Similarly, a restricted maximum likelihood approach would choose σ and γ to minimize $\sum_{i=1}^{m} REML_i(\hat{\beta}_i^{(p)}, \sigma, \gamma)$, where

$$REML_i(\hat{\beta}_i^{(p)}, \sigma, \gamma) = PL_i(\hat{\beta}_i^{(p)}, \sigma, \gamma)$$
$$-p \log \sigma^2 + \log |X_i'(\hat{\beta}_i^{(p)}) S_i^{-1}(\hat{\beta}_i^{(p)}, \gamma) X_i(\hat{\beta}_i^{(p)})|, \quad (5.2)$$

where $X_i(\beta_i)$ is the the $(n_i \times p)$ matrix with jth row equal to $f'_\beta(x_{ij}, \beta_i)$.

For either objective function, computation of these estimates is no more difficult for the case of m individuals than it is for the case of a single individual. Details for the PL and REML objective functions with a general covariance model may be found in Davidian and Giltinan (1993b).

If the intra-individual covariance matrix is diagonal,

$$R_i(\beta_i, \xi) = \sigma^2 G_i(\beta_i, \gamma),$$

where, adapting the notation of section 2.2.3 to individual i,

$$G_i(\beta_i, \theta) = \text{diag}[g^2(\mu_{i1}, z_{i1}, \theta), \ldots, g^2(\mu_{in_i}, z_{in_i}, \theta)],$$

the methods generalize trivially to the case of m individuals, by summing the PL, REML, or AR objective functions in (2.21)–(2.24) across individuals. The computational strategies outlined in section 2.4.2 generalize similarly; further details may be found in Davidian and Giltinan (1993a).

5.2.3 Confidence regions for covariance parameters

For γ a scalar, an approximate confidence interval may be constructed using standard profile likelihood methods. For the interval based on the PL objective function, using the data for the ith individual only, this is implemented as follows. Note that $PL_i(\beta_i, \sigma, \gamma)$ is minus twice the loglikelihood for the ith individual. Let $\hat{\beta}_i(\gamma)$ and $\hat{\sigma}(\gamma)$ be the estimates from the final GLS iteration where γ is held fixed. Substitute these values in $PL_i(\beta_i, \sigma, \gamma)$, and define the function

$$L_i(\gamma) = -(1/2) PL_i(\hat{\beta}_i(\gamma), \hat{\sigma}(\gamma), \gamma).$$

$L_i(\gamma)$ represents the profile pseudologlikelihood for γ for individual i and is maximized by the pseudolikelihood estimator $\hat{\gamma}$. An approximate $100(1-\alpha)\%$ confidence interval for γ based on the fit for individual i is then

$$\{\gamma : L_i(\gamma) \geq L_i(\hat{\gamma}) - (1/2)\chi^2_{1-\alpha,1}\}, \quad (5.3)$$

where $\chi^2_{1-\alpha,1}$ is the $(1-\alpha)$ quantile of the χ^2 distribution with 1 degree of freedom. Here, as interest is focused on the ith individual only, $\boldsymbol{\beta}_i$ is regarded as a fixed parameter. One usually constructs this confidence interval graphically.

Development of a confidence interval for the pooled estimation case is analogous to (5.3). Define $L(\gamma)$ to be the profile pseudolikelihood for the data from all m individuals, where

$$L(\gamma) = \sum_{i=1}^{m} L_i(\gamma),$$

and now let $\hat{\boldsymbol{\beta}}_i$, $\hat{\sigma}$, and $\hat{\gamma}$ be the final estimates from the pooled GLS algorithm of section 5.2.1. In an obvious extension of (5.3), an approximate $100(1-\alpha)\%$ confidence interval for γ based on the pooled estimates is given by

$$\{\gamma : L(\gamma) \geq L(\hat{\gamma}) - (1/2)\chi^2_{1-\alpha,1}\}. \tag{5.4}$$

For vector-valued $\boldsymbol{\gamma}$, confidence regions may be constructed by extension of these techniques; however, if the dimension of γ is larger than two, simple graphical construction is not possible.

Note that the interval (5.4) is a *conditional* statement; that is, the random individual regression parameters $\boldsymbol{\beta}_i$ are treated as fixed, being replaced by individual estimates $\hat{\boldsymbol{\beta}}_i$. Thus, the interval may be somewhat optimistic, as the additional uncertainty associated with the $\boldsymbol{\beta}_i$ is not taken into account.

5.2.4 Examples

Pharmacokinetics of indomethacin

To illustrate pooled estimation of covariance parameters within the GLS framework, consider the indomethacin data of Table 2.1, discussed in sections 2.1 and 2.3.6. We used the biexponential model with parameterization as in (2.43):

$$f(x, \boldsymbol{\beta}) = e^{\beta_1} \exp(-e^{\beta_2} x) + e^{\beta_3} \exp(-e^{\beta_4} x)$$

modeled the within-subject variance according to the power variance function (2.6), $g(\mu_{ij}, z_{ij}, \boldsymbol{\theta}) = \mu_{ij}^{\theta}$, and assumed that measurements on a given subject are uncorrelated.

The following estimation methods were applied to data for the six subjects:

1. GLS estimation based on each individual's data separately, using each of the three variance function estimation methods PL,

REML, and AR in (2.21)–(2.24). These methods are abbreviated as GLS-PL, GLS-REML, and GLS-AR below.

2. GLS estimation based on pooled estimation of (σ, θ) across subjects, using PL, REML, AR. These methods are abbreviated as GLS-PLPOOL, GLS-REMLPOOL, and GLS-ARPOOL below.
3. OLS estimation, based on individual data, which may be thought of as weighted least squares estimation for a fixed, misspecified value of $\theta = 0$.

The individual data methods GLS-PL, GLS-AR, and GLS-REML gave results virtually identical to one another, as did the pooled methods GLS-PLPOOL, GLS-ARPOOL, and GLS-REMLPOOL. Thus, we present results for OLS, GLS-PL, and GLS-PLPOOL only (Table 5.1). In estimating θ, convergence of the GLS iterative scheme occurred in three to six iterations in all cases. Estimates of θ based on individual data varied considerably from subject to subject. This should be interpreted as reflecting the unreliability of estimating second-moment parameters based on 11 observations per subject, as discussed in section 2.3, rather than true differences in the intrasubject variance patterns. The pooled estimate $\hat{\theta}_{GLS-PLPOOL} = 0.94$ suggests that a constant coefficient of variation model is plausible for these data. With the exception of subject 6, whose estimated value of -0.38 seems suspect, GLS based on individual and pooled data gave generally comparable results in estimating β_i. OLS estimates differed somewhat, with the greatest discrepancies observed in the parameters for the second compartment, β_3 and β_4. Standard errors for the OLS method should not be considered reliable, as they are predicated on an inaccurate assumption about the underlying error structure.

Because the plasma concentration measurements were taken over time, one might expect intra-individual serial correlation of measurements. It is thus reasonable to consider a more general intra-individual covariance structure for these data; specifically, to model both the heterogeneity of variance and correlation structure simultaneously. Examination of lagged residual plots, as in section 2.5, based on GLS-PLPOOL residuals for all subjects, suggested a strong negative association between adjacent observations, with that among observations farther apart falling off rapidly. Thus, we again assumed the power model for within-subject variance, and modeled the correlation by an autoregressive (AR) model of order one, with correlation parameter α, as in section 2.2.3. Because estimation of covariance parameters is problematic when based on

Table 5.1. *Parameter estimates for fits of (2.43) to the data for each subject in the indomethacin study. The four entries for $\hat{\beta}$ are the estimates of the components of β.*

Subject	$\hat{\beta}$ (*SE*)	$\hat{\sigma}, \hat{\theta}$
	OLS	
1	0.71, 0.58, −1.65, −1.79 (0.08, 0.12, 0.58, 0.79)	0.30,–
2	1.04, 0.80, −0.70, −1.64 (0.16, 0.34, 0.69, 0.91)	0.30,–
3	0.82, 0.04, −2.01, −2.58 (0.05, 0.12, 0.95, 2.10)	0.30,–
4	0.79, 0.24, −1.37, −1.60 (0.09, 0.15, 0.88, 0.89)	0.30,–
5	1.27, 1.04, −1.23, −1.51 (0 08, 0.15, 0.49, 0 64)	0 30,–
6	1 10, 1.09, −0 03, −0 87 (0.05, 0.12, 0.14, 0.12)	0 30,–
	GLS-PL	
1	0.74, 0.61, −1.62, −1.76 (0.10, 0.08, 0.09, 0.09)	0.08,1.18
2	1.14, 1.04, −0.41, −1.36 (0 38, 0.34, 0.25, 0.19)	0.19,0.90
3	0.80, −0.03, −2.39, −3.87 (0.06, 0.08, 0.43, 2.93)	0.06,0.75
4	0.79, 0.08, −2.28, −2.96 (0.07, 0.09, 0.44, 1.26)	0.08,0.82
5	1.24, 0.97, −1.43, −1.74 (0.18, 0.14, 0.23, 0.26)	0.13,0.82
6	1.11, 1.26, −0.18, −0.66 (0.04, 0.09, 0.10, 0.10)	0.02, −0.38

sparse data, we did not attempt to fit the model to the individual profiles separately. Rather, we used pooled pseudolikelihood estimation to estimate the covariance parameters σ, θ, and α, denoted as GLS-PLPOOL, AR(1). Because we expected a negative estimate for the correlation parameter α, we chose to approximate the correlation structure for unequally-spaced observations by that in equation (2.10).

Table 5.1. *Continued.*

Subject	$\hat{\beta}$ (*SE*)	$\hat{\sigma}, \hat{\theta}, \hat{\alpha}$
	GLS-PLPOOL	
1	$0.72, 0.60, -1.63, -1.77$	
	$(0.15, 0.14, 0.24, 0.25)$	
2	$1.15, 1.05, -0.41, -1.35$	
	$(0.25, 0.22, 0.15, 0.12)$	
3	$0.79, -0.04, -2.43, -4.18$	Pooled estimates
	$(0.12, 0.15, 0.68, 6.24)$	
4	$0.79, 0.07, -2.31, -3.04$	(0.13,0.94,–)
	$(0.10, 0.13, 0.52, 1.61)$	
5	$1.22, 0.96, -1\ 44, -1.76$	
	$(0.17, 0.12, 0.18, 0.20)$	
6	$0.92, 0.37, -1.12, -1.75$	
	$(0.14, 0.17, 0.30, 0.28)$	
	GLS-PLPOOL, AR(1)	
1	$0.71, 0.59, -1.65, -1.78$	
	$(0.07, 0.07, 0.13, 0.15)$	
2	$1.33, 1.18, -0.37, -1.33$	
	$(0.14, 0.11, 0.07, 0.06)$	
3	$0.79, -0.03, -2.33, -3.42$	Pooled estimates
	$(0.06, 0.08, 0.40, 1.80)$	
4	$0.79, 0.10, -2.15, -2.62$	$(0.07, 0.85, -0.58)$
	$(0.04, 0.06, 0.29, 0.62)$	
5	$1.19, 0.92, -1.50, -1.82$	
	$(0.08, 0.06, 0.10, 0.12)$	
6	$0.91, 0.43, -0.93, -1.57$	
	$(0.07, 0.09, 0.14, 0.13)$	

The results of this fit, which converged in six iterations of the GLS algorithm, are given in Table 5.1. The parameter estimates are similar to those based on the assumption of no intra-individual correlation, with standard errors for the estimates of $\boldsymbol{\beta}$ significantly smaller. The estimate for α is negative, as suggested by preliminary investigation. The estimate for θ agrees roughly with those obtained under the assumption of no correlation, while that for σ is biased downward.

A negative correlation seems counterintuitive. One would expect that observations spaced closely together in time should exhibit positive association. In practice, however, this expectation may be unrealistic. Blood samples from a given individual may not be assayed together, so that inter-assay variation may dominate that within subjects. Moreover, the assay procedure itself may induce unexpected patterns of correlation. For example, the position of a sample on a microtiter plate or the run order of samples may cause associations among observations that are difficult to predict or assess, and such information typically is not available. The negative estimate obtained here may arise in this manner. Thus, although in principle it is possible to specify complex intra-individual covariance structures, in practice, the results may be difficult to interpret.

Some additional cautions are in order. It is likely that there will be some interplay between the mean response function and an assumed correlation structure. Systematic deviations from the assumed form for the mean may be attributable to misspecification of mean function or to autocorrelation; without additional information or assumptions, this may not be possible to determine. Consequences may be computational difficulties or a poor fit of the mean, as the two model components compete to explain the observed pattern of the data. Thus, estimates obtained under such a model should always be critically examined. Furthermore, issues of model identifiability may arise, although this may be difficult to determine. In the hierarchical model, correlation among observations on a given individual at the marginal level will depend on both the assumed inter- and intra-individual covariance matrices, and it may be possible to specify these components so as to produce a nearly unidentifiable marginal model.

Several general messages may be taken from this example. When heterogeneity of intra-individual variance is evident, pooled estimation of variance parameters within the framework of an appropriate intra-individual model usually gives satistfactory results, even when the number of individuals is not large. We strongly recommend this approach when heteroscedasticity is evident. Estimation of within-individual correlation is more difficult to handle. In our view, if such inference proves problematic, the information available does not seem sufficient, or the results are difficult to interpret, it may be preferable to assume an uncorrelated intra-individual error structure. Although this may be a misspecification, the loss in efficiency is likely to be small when compared against the computational difficulties and potential instability associated with a more complicated model. Additional examples are given in Chapter 11.

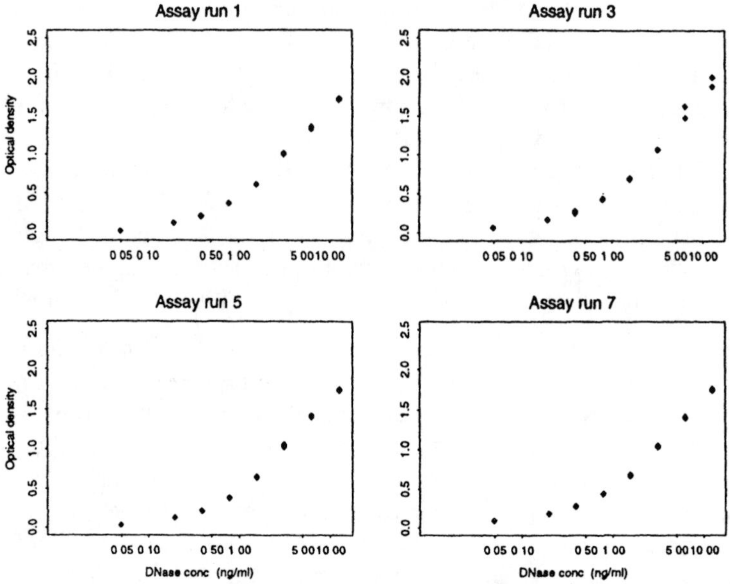

Figure 5.1. *Concentration-response data for four runs of the DNase assay.*

ELISA assay for DNase in rat serum

Consider the data shown in Figure 5.1, generated from four runs during the development of an ELISA assay for the recombinant protein DNase in rat serum. For each of 11 experiments, duplicate absorbance measurements were obtained at each of several known standard concentrations. By convention, responses at zero concentration are plotted at two dilutions below the lowest nonzero concentration.

We fitted the four-parameter logistic model (1.6)

$$y = \beta_1 + \frac{\beta_2 - \beta_1}{1 + \exp\{\beta_4(\log x - \beta_3)\}},$$

and assumed the power variance model with uncorrelated within-assay measurements and power parameter θ. Estimation was carried out by GLS separately for each individual using each of the three variance function estimation methods PL, REML, and AR. GLS estimates based on pooled estimation of (σ, θ) using the same

Table 5.2. *Estimates of (σ, θ), ELISA assay for DNase.*

	Estimates of (σ, θ) by		
Run	*GLS-PL*	*GLS-AR*	*GLS-REML*
1	(0.022,0.549)	(0.022,0.490)	(0.022,0.526)
2	(0.015,0.384)	(0.015,0.359)	(0.015,0.373)
3	(0.038,1.101)	(0.038,1.100)	(0.037,1.059)
4	(0.015,0.427)	(0.014,0.393)	(0.015,0.410)
5	(0.019,1.260)	(0.019,1.261)	(0.017,1.113)
6	(0.016,0.385)	(0.017,0.437)	(0.016,0.380)
7	(0.016,0.776)	(0.015,0.689)	(0.015,0.720)
8	(0.027,0.927)	(0.029,1.014)	(0.026,0 862)
9	(0.026,0.446)	(0.025,0.352)	(0.026,0.416)
10	(0.025,0.182)	(0.024,0.104)	(0.025,0.178)
11	(0.025,0.955)	(0.024,0.904)	(0.024,0.918)
	GLS-PLPOOL	(0.023,0.503)	
	GLS-ARPOOL	(0.024,0.527)	
	GLS-REMLPOOL	(0.023,0.486)	

three methods were also obtained. In this example, estimates of the regression parameters $\boldsymbol{\beta}_i$ were comparable for all methods. As precision is the primary focus in this context, in Table 5.2 we summarize estimates of the variance parameters (σ, θ) only.

In estimating θ, all GLS iterative schemes for both individual and pooled estimation converged in two to five iterations. The pooled variance estimation procedures suggest that a Poisson-like error structure may be suitable for this assay. Individual estimates of (σ, θ) differ considerably from experiment to experiment. This probably reflects imprecision in the individual estimates of θ rather than true underlying differences. Figures 5.2 and 5.3, showing profile pseudolikelihood-based 95% confidence intervals for θ based on data from a single run (#3) using (5.3) and based on the pooled estimation scheme using all runs (5.4), bear this out. The confidence interval for a single run only is very wide, reflecting the high degree of uncertainty in estimating (σ, θ) based on data from only one assay run.

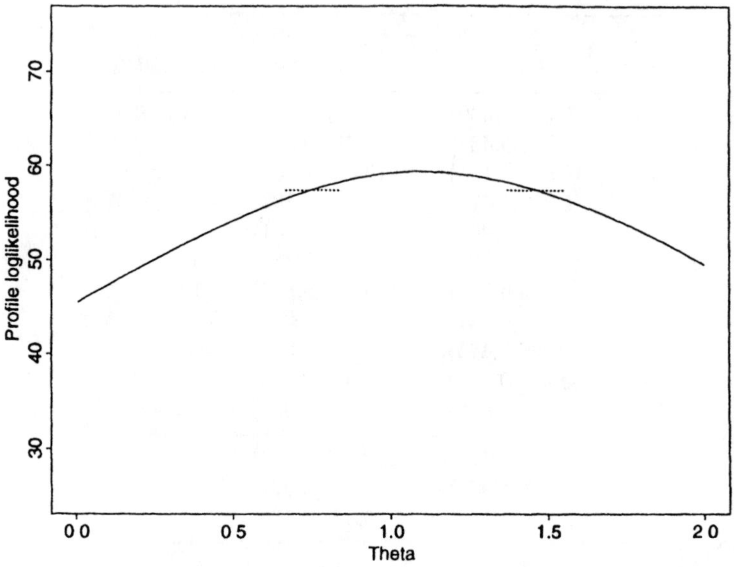

Figure 5.2. *Pseudolikelihood profile confidence interval for θ, DNase assay, based on the individual fit for run #3.*

5.3 Estimation of population parameters

Recall stage 2 of the hierarchy for the fully parametric model specification,

$$\boldsymbol{\beta}_i = \boldsymbol{d}(\boldsymbol{a}_i, \boldsymbol{\beta}, \boldsymbol{b}_i), \quad \boldsymbol{b}_i \sim \mathcal{N}(\boldsymbol{0}, \boldsymbol{D}).$$

Let $\boldsymbol{\beta}_i^*$, $i = 1, \ldots, m$, denote individual estimates obtained from a first stage of estimation. Once the individual estimates $\boldsymbol{\beta}_i^*$ have been obtained, several possibilities exist for implementing stage 2 of a two-stage estimation scheme, that of constructing estimates of $\boldsymbol{\beta}$ and $\boldsymbol{D}$. In our development in this section we assume the specification (4.7),

$$\boldsymbol{\beta}_i = \boldsymbol{A}_i\boldsymbol{\beta} + \boldsymbol{b}_i, \tag{5.5}$$

so that $\boldsymbol{\beta}_i \sim \mathcal{N}(\boldsymbol{A}_i\boldsymbol{\beta}, \boldsymbol{D})$; extension to a nonlinear specification is discussed below. Throughout, we consider only situations where all components of $\boldsymbol{\beta}_i$ are taken to be random; we discuss this issue in section 5.5.

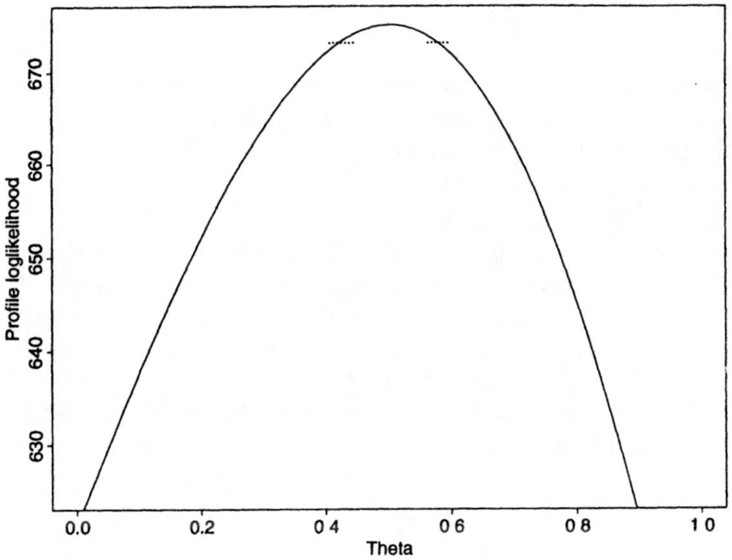

Figure 5.3. *Pseudolikelihood profile confidence interval for θ, DNase assay, based on the pooled fit.*

5.3.1 Standard two-stage method

The standard two-stage (STS) method (Steimer *et al.*, 1984) treats the estimates $\boldsymbol{\beta}_i^*$ as if they were the true $\boldsymbol{\beta}_i$. In the simplest case where $\boldsymbol{A}_i \equiv \boldsymbol{I}_{n_i}$, and the $\boldsymbol{\beta}_i$ are independent and identically distributed $\mathcal{N}(\boldsymbol{\beta}, \boldsymbol{D})$, STS estimates are constructed as the sample mean and covariance of the $\boldsymbol{\beta}_i^*$:

$$\hat{\boldsymbol{\beta}}_{STS} = m^{-1} \sum_{i=1}^{m} \boldsymbol{\beta}_i^*,$$

$$\hat{\boldsymbol{D}}_{STS} = (m-1)^{-1} \sum_{i=1}^{m} (\boldsymbol{\beta}_i^* - \hat{\boldsymbol{\beta}}_{STS})(\boldsymbol{\beta}_i^* - \hat{\boldsymbol{\beta}}_{STS})'.$$

For general $\boldsymbol{A}_i$, the approximation $\boldsymbol{\beta}_i^* \approx \boldsymbol{A}_i\boldsymbol{\beta} + \boldsymbol{b}_i$ suggests modification of these estimates to

$$\hat{\boldsymbol{\beta}}_{STS} = \Big(\sum_{i=1}^{m} \boldsymbol{A}_i'\boldsymbol{A}_i\Big)^{-1} \Big(\sum_{i=1}^{m} \boldsymbol{A}_i'\boldsymbol{\beta}_i^*\Big), \qquad (5.6)$$

$$\hat{\boldsymbol{D}}_{STS} = (m-1)^{-1}\sum_{i=1}^{m}(\boldsymbol{\beta}_i^* - \boldsymbol{A}_i\hat{\boldsymbol{\beta}}_{STS})(\boldsymbol{\beta}_i^* - \boldsymbol{A}_i\hat{\boldsymbol{\beta}}_{STS})'. \quad (5.7)$$

These estimates are appealing on the grounds of computational simplicity, and they do not require the assumption of normality. However, no account is taken of the uncertainty in estimating $\boldsymbol{\beta}_i$, with the result that the estimator for $\boldsymbol{D}$ is upwardly biased, as the following argument illustrates.

For simplicity, suppose that $\boldsymbol{A}_i \equiv \boldsymbol{I}_{n_i}$ and that $\boldsymbol{\beta}_i^*$ is a GLS estimator. Given $\boldsymbol{\beta}_i$, from (2.34), $\boldsymbol{\beta}_i^*$ has asymptotic covariance matrix $\sigma^2\boldsymbol{\Sigma}_i$, where $\boldsymbol{\Sigma}_i^{-1} = \boldsymbol{X}_i'(\boldsymbol{\beta}_i)\boldsymbol{S}_i^{-1}(\boldsymbol{\beta}_i,\gamma)\boldsymbol{X}_i(\boldsymbol{\beta}_i)$. Then $\boldsymbol{\beta}_i^*$ approximately satisfies $\boldsymbol{\beta}_i^* \approx \boldsymbol{\beta}_i + \boldsymbol{e}_i^*$, where $\boldsymbol{e}_i^*$ has covariance matrix $\sigma^2\boldsymbol{\Sigma}_i$. Treating this expression as exact, with $\sigma^2\boldsymbol{\Sigma}_i$ *known* (as it would be for a linear model f with constant variance), it is straightforward to show that $\mathrm{E}(\boldsymbol{\beta}_i^*) = \boldsymbol{\beta}$ and $\mathrm{Cov}(\boldsymbol{\beta}_i^*) = \boldsymbol{D} + \sigma^2\boldsymbol{\Sigma}_i$ from which it follows that

$$\mathrm{E}(\hat{\boldsymbol{D}}_{STS}) = \boldsymbol{D} + \sigma^2 m^{-1}\sum_{i=1}^{m}\boldsymbol{\Sigma}_i. \quad (5.8)$$

Thus, the STS estimate of $\boldsymbol{D}$ is biased upward in a manner corresponding to the average of the asymptotic covariance matrices for the $\boldsymbol{\beta}_i^*$, indicating that the bias of $\hat{\boldsymbol{D}}_{STS}$ is a consequence of the failure of the STS method to take uncertainty of estimation of the $\boldsymbol{\beta}_i$ into account. A further drawback of the STS approach is that no refinement of the individual $\boldsymbol{\beta}_i^*$, such as shrinkage toward the mean, is implemented.

The STS method is not generally recommended, and we do not discuss it further.

5.3.2 Global two-stage method

Estimation

Incorporation of the uncertainty of estimation in $\boldsymbol{\beta}_i^*$ is usually based on the asymptotic theory for $\boldsymbol{\beta}_i^*$ given $\boldsymbol{\beta}_i$. It is assumed that $\boldsymbol{\beta}_i^*|\boldsymbol{\beta}_i$ is approximately $\mathcal{N}(\boldsymbol{\beta}_i, \boldsymbol{C}_i)$, where $\boldsymbol{C}_i$ is the asymptotic covariance matrix for $\boldsymbol{\beta}_i^*$. For GLS as above, $\boldsymbol{C}_i = \sigma^2\boldsymbol{\Sigma}_i$. Under this assumption, the marginal distribution of $\boldsymbol{\beta}_i^*$ is approximately normal with mean $\boldsymbol{A}_i\boldsymbol{\beta}$ and covariance matrix $\boldsymbol{C}_i + \boldsymbol{D}$, so that $\boldsymbol{\beta}_i^*$ may be written as

$$\boldsymbol{\beta}_i^* \approx \boldsymbol{A}_i\boldsymbol{\beta} + \boldsymbol{b}_i + \boldsymbol{e}_i^*, \quad (5.9)$$

where $\boldsymbol{e}_i^*$ has mean $\boldsymbol{0}$ and covariance matrix $\boldsymbol{C}_i$. This approximation suggests estimation of $\boldsymbol{\beta}$ and $\boldsymbol{D}$ by maximum likelihood applied to 'data' $\boldsymbol{\beta}_i^*$; that is, estimate $\boldsymbol{\beta}$ and $\boldsymbol{D}$ by minimizing twice the negative loglikelihood

$$L_{GTS}(\boldsymbol{\beta}, \boldsymbol{D}) = \sum_{i=1}^{m} \log |\boldsymbol{C}_i + \boldsymbol{D}| + \sum_{i=1}^{m} (\boldsymbol{\beta}_i^* - \boldsymbol{A}_i\boldsymbol{\beta})'(\boldsymbol{C}_i + \boldsymbol{D})^{-1}(\boldsymbol{\beta}_i^* - \boldsymbol{A}_i\boldsymbol{\beta}),$$

where $\boldsymbol{C}_i$ is taken to be fixed and known for each i. Standard results on matrix derivatives show that the estimates satisfy

$$\hat{\boldsymbol{\beta}} = \left(\sum_{i=1}^{m} \boldsymbol{A}_i'(\boldsymbol{C}_i + \hat{\boldsymbol{D}})^{-1}\boldsymbol{A}_i\right)^{-1} \sum_{i=1}^{m} \boldsymbol{A}_i'(\boldsymbol{C}_i + \hat{\boldsymbol{D}})^{-1}\boldsymbol{\beta}_i^* \qquad (5.10)$$

and

$$\hat{\boldsymbol{D}} = m^{-1}\sum_{i=1}^{m} \boldsymbol{c}_i^*\boldsymbol{c}_i^{*\prime} + m^{-1}\sum_{i=1}^{m}(\boldsymbol{C}_i^{-1} + \hat{\boldsymbol{D}}^{-1})^{-1}, \qquad (5.11)$$

where

$$\boldsymbol{c}_i^* = (\boldsymbol{C}_i^{-1} + \hat{\boldsymbol{D}}^{-1})^{-1}\boldsymbol{C}_i^{-1}(\boldsymbol{\beta}_i^* - \boldsymbol{A}_i\hat{\boldsymbol{\beta}}).$$

In implementing this approach, an estimate of the asymptotic covariance matrix $\boldsymbol{C}_i$ for individual i would be used. For example, in the case where the $\boldsymbol{\beta}_i^*$ were obtained using the pooled GLS methods of section 5.2, the estimate of $\boldsymbol{C}_i$ would be constructed by substituting the pooled estimates for σ and γ and the final individual estimate $\boldsymbol{\beta}_i^*$ into the expression for the asymptotic covariance. The resulting estimate would then be treated as fixed in minimization of $L_{GTS}(\boldsymbol{\beta}, \boldsymbol{D})$.

The estimates satisfying (5.10) and (5.11) may be found by directly maximizing the likelihood, using, for example, a Newton–Raphson approach. This yields estimates for $\boldsymbol{\beta}$ and $\boldsymbol{D}$ equivalent to applying the methods of section 3.3.2 to the hierarchical linear model (5.9). Estimates of the $\boldsymbol{b}_i$ may also be obtained as described in that section.

Alternatively, an iterative EM algorithm (see section 3.4.1) to compute the normal theory maximum likelihood estimates, referred to as the global two-stage method (GTS), is described by Steimer *et al.* (1984), modified here to apply to the model (5.5). The 'Expectation step' of the algorithm consists of computing the conditional expectation of the $\boldsymbol{\beta}_i$, given the 'data' $\boldsymbol{\beta}_i^*$, at current estimates of $\boldsymbol{\beta}$ and $\boldsymbol{D}$; new estimates of $\boldsymbol{\beta}$ and $\boldsymbol{D}$ are then computed in the 'Maximization step.' At iteration $(c+1)$:

1. '*E-step.*' Produce refined estimates of the $\boldsymbol{\beta}_i$, $i = 1, \ldots, m$:

$$\hat{\boldsymbol{\beta}}_{i,(c+1)} = (\boldsymbol{C}_i^{-1} + \hat{\boldsymbol{D}}_{(c)}^{-1})^{-1}(\boldsymbol{C}_i^{-1}\boldsymbol{\beta}_i^* + \hat{\boldsymbol{D}}_{(c)}^{-1}\boldsymbol{A}_i\hat{\boldsymbol{\beta}}_{(c)}) \qquad (5.12)$$

2. '*M-step.*' Obtain updated estimates of the population parameters:

$$\hat{\boldsymbol{\beta}}_{(c+1)} = \sum_{i=1}^{m} \boldsymbol{W}_{i,(c)}\hat{\boldsymbol{\beta}}_{i,(c+1)}, \qquad (5.13)$$

where

$$\boldsymbol{W}_{i,(c)} = \Big(\sum_{i=1}^{m} \boldsymbol{A}_i'\hat{\boldsymbol{D}}_{(c)}^{-1}\boldsymbol{A}_i\Big)^{-1}\boldsymbol{A}_i'\hat{\boldsymbol{D}}_{(c)}^{-1},$$

$$\hat{\boldsymbol{D}}_{(c+1)} = m^{-1}\sum_{i=1}^{m}(\boldsymbol{C}_i^{-1} + \hat{\boldsymbol{D}}_{(c)}^{-1})^{-1} \qquad (5.14)$$

$$+ m^{-1}\sum_{i=1}^{m}(\hat{\boldsymbol{\beta}}_{i,(c+1)} - \boldsymbol{A}_i\hat{\boldsymbol{\beta}}_{(c+1)})(\hat{\boldsymbol{\beta}}_{i,(c+1)} - \boldsymbol{A}_i\hat{\boldsymbol{\beta}}_{(c+1)})'.$$

Denote the final estimates as $\hat{\boldsymbol{\beta}}_{GTS}$, $\hat{\boldsymbol{D}}_{GTS}$, and $\hat{\boldsymbol{\beta}}_{i,GTS}$. The algorithm is iterated until the M-step converges; that is, until the difference between successive estimates of $\boldsymbol{\beta}$ and $\boldsymbol{D}$ is sufficiently small. Suitable starting values for the algorithm are $\hat{\boldsymbol{\beta}}_{STS}$ and $\hat{\boldsymbol{D}}_{STS}$ given in (5.6) and (5.7).

It is not immediately obvious that the the algorithm does indeed yield the estimators minimizing $L_{GTS}(\boldsymbol{\beta}, \boldsymbol{D})$. To show this, note that, at convergence, successive iterates of estimated values for $\boldsymbol{\beta}$, $\boldsymbol{D}$, and $\boldsymbol{\beta}_i$ are identical. Denote the estimators at convergence by $\hat{\boldsymbol{\beta}}$, $\hat{\boldsymbol{D}}$, and $\hat{\boldsymbol{\beta}}_i$, suppressing the subscript GTS. It is straightforward to show that

$$(\hat{\boldsymbol{\beta}}_i - \boldsymbol{A}_i\hat{\boldsymbol{\beta}}) = (\boldsymbol{C}_i^{-1} + \hat{\boldsymbol{D}}^{-1})^{-1}\boldsymbol{C}_i(\boldsymbol{\beta}_i^* - \boldsymbol{A}_i\hat{\boldsymbol{\beta}}) = \boldsymbol{c}_i^*.$$

Substituting this into (5.14) gives the estimating equation (5.11) for $\hat{\boldsymbol{D}}$. $\hat{\boldsymbol{\beta}}$ satisfies, from (5.13),

$$\hat{\boldsymbol{\beta}} = \Big(\sum_{i=1}^{m} \boldsymbol{A}_i'(\hat{\boldsymbol{D}})^{-1}\boldsymbol{A}_i\Big)^{-1}\sum_{i=1}^{m} \boldsymbol{A}_i'\hat{\boldsymbol{D}}^{-1}\hat{\boldsymbol{\beta}}_i.$$

Replacing $\hat{\beta}_i$ by

$$\hat{\beta}_i = (C_i^{-1} + \hat{D}^{-1})^{-1}(C_i^{-1}\beta_i^* + \hat{D}^{-1}A_i\hat{\beta}), \qquad (5.15)$$

using the identity $(B_1^{-1} + B_2^{-1})^{-1} = B_1 - B_1(B_1 + B_2)^{-1}B_1$, and rearranging yields (5.10), as required.

In addition to estimates of the fixed parameters, the EM algorithm to minimize $L_{GTS}(\beta, D)$ produces estimates of the individual regression parameters β_i. From (5.15), these estimates have an empirical Bayes interpretation, as they are Bayes estimates evaluated at $\hat{\beta}_{GTS}$ and $\hat{D}_{GTS}$. The estimate $\hat{\beta}_{i,GTS}$ represents a compromise between the individual estimate β_i^* and the estimated population value $\hat{\beta}_{GTS}$, where β_i^* is 'shrunk' toward the population value in the usual fashion. These estimates may be preferred to the β_i^* when individual-specific estimates are desired, as they 'borrow' information across all individuals. In particular, $\hat{\beta}_{i,GTS}$ might be preferred if n_i is small relative to the amount of information on other individuals. Note that estimates of the individual random effects b_i may be obtained as $\hat{b}_{i,GTS} = \hat{\beta}_{i,GTS} - \hat{\beta}_{GTS}$ and have a similar empirical Bayes interpretation.

Although the GTS procedure seems predicated on the approximate normality of the distribution of $\beta_i^*|\beta_i$, the method has general appeal. The salient feature is that the first two marginal moments of β_i^* may be written as $A_i\beta$ and $C_i + D$, respectively; this follows if the representation (5.9), which is linear in b_i and e_i^*, is used. Under this approximate linear marginal model, the estimator for β in (5.10) has the form of a GLS estimator, and thus would be a natural choice even if the distributions of b_i and e_i^* were not exactly normal. This is not unexpected; the marginal covariance of β_i^* does not depend on β, so that the ML and GLS estimators for β in (5.9) coincide. Consequently, the GTS method is likely to perform well even if the assumptions of normality of the b_i and e_i^* are not exactly correct. Thus, the important aspect of the approximation is the accuracy of the first two moments of the asymptotic distribution of $\beta_i^*|\beta_i$ rather than the normality assumption.

Because no correction is made for loss of degrees of freedom for estimation of β, the GTS estimator for D is likely to be biased downward, analogous to results for the hierarchical linear model.

Confidence intervals and hypothesis testing

Under the assumption that the β_i^* are exactly marginally normally distributed, inference may be based on an appeal to standard asymptotic theory for maximum likelihood, allowing the number of individuals m to become large. Specifically, one may derive the asymptotic covariance matrix for $\hat{\boldsymbol{\beta}}_{GTS}$ and the elements of $\hat{\boldsymbol{D}}_{GTS}$ as the inverse of the information matrix associated with the normal likelihood $L_{GTS}(\boldsymbol{\beta}, \boldsymbol{D})$. This matrix may be estimated by evaluation at the estimates $\hat{\boldsymbol{\beta}}_{GTS}$ and $\hat{\boldsymbol{D}}_{GTS}$, and estimated standard errors obtained as the square roots of the diagonal elements.

The estimate of the asymptotic covariance matrix for the final estimates $\hat{\boldsymbol{\beta}}_{GTS}$ based on this approach is given by

$$\hat{\boldsymbol{\Sigma}}_{GTS} = \left(\sum_{i=1}^{m} \boldsymbol{A}_i'(\boldsymbol{C}_i + \hat{\boldsymbol{D}}_{GTS})^{-1}\boldsymbol{A}_i\right)^{-1}. \qquad (5.16)$$

The form of the asymptotic covariance matrix for $\hat{\boldsymbol{\beta}}_{GTS}$ is unchanged even away from normality, so that inference for $\boldsymbol{\beta}$ based on (5.16) is still valid.

Using (5.16), one may construct confidence intervals and hypothesis tests for functions of $\boldsymbol{\beta}$ in a manner analogous to the development in section 2.3.5.

5.3.3 Bayesian method

Racine-Poon (1985) proposes a Bayesian approach to estimation of $\boldsymbol{\beta}$, $\boldsymbol{D}$, and the $\boldsymbol{\beta}_i$, which is implemented by an EM algorithm similar to the GTS procedure in section 5.3.2. As with GTS, the procedure is based on the assumption that $\boldsymbol{\beta}_i^*|\boldsymbol{\beta}_i$ is approximately $\mathcal{N}(\boldsymbol{\beta}_i, \boldsymbol{C}_i)$, where, again, $\boldsymbol{C}_i$ is the asymptotic covariance matrix for $\boldsymbol{\beta}_i^*$. Analogous to the Bayesian approach to the hierarchical linear model, given in section 3.2.3, prior distributions are specified for $\boldsymbol{\beta}$ and $\boldsymbol{D}^{-1}$; vague prior information is assumed on $\boldsymbol{\beta}$, and a Wishart distribution with ρ degrees of freedom and matrix $\boldsymbol{P}$ is taken for $\boldsymbol{D}^{-1}$. The matrix $\rho^{-1}\boldsymbol{P}$ may be considered as an initial estimate for $\boldsymbol{D}$; Racine-Poon (1985) notes that the choice of $\boldsymbol{P}$ has little effect on the results, except in cases where m is small. The reader is referred to Racine *et al.* (1986) for discussion and a detailed application of this approach to pharmacokinetic data.

5.3.4 General inter-individual models

The linear specification (5.5) will be sufficiently general in many applications. As discussed in section 4.2.3, such a linear model allows incorporation of features such as several treatment groups;

moreover, by judicious parameterization of the function f, more complex specifications may also be expressed in linear form. In cases where a nonlinear model for $\boldsymbol{\beta}_i$ is required, estimation of population parameters using the methods of section 5.3.2 may be extended to certain nonlinear specifications. Consider the model

$$\boldsymbol{\beta}_i = \boldsymbol{d}(\boldsymbol{a}_i, \boldsymbol{\beta}) + \boldsymbol{b}_i, \tag{5.17}$$

where $\boldsymbol{d}$ is a p-dimensional function; (5.17) allows the first two moments of the marginal distribution of the $\boldsymbol{\beta}_i^*$ to be specified readily. An approximate nonlinear hierarchical model for $\boldsymbol{\beta}_i^*$ may then be written as

$$\boldsymbol{\beta}_i^* \approx \boldsymbol{d}(\boldsymbol{a}_i, \boldsymbol{\beta}) + \boldsymbol{b}_i + \boldsymbol{e}_i^*. \tag{5.18}$$

One may write down the marginal normal likelihood for this model as before and in principle maximize the likelihood directly. Because the approximate marginal nonlinear model (5.18) has the form of the approximate linearization models derived in Chapter 6, this may be accomplished using methods in that chapter; see section 6.6.

Alternatively, it is possible to modify the GTS EM algorithm to accommodate (5.18) in several ways. One simple approach is based on noting that, for $\boldsymbol{\beta}$ close to some $\hat{\boldsymbol{\beta}}$, say, by a Taylor series in $\boldsymbol{\beta}$ about $\hat{\boldsymbol{\beta}}$,

$$\boldsymbol{\beta}_i^* - \boldsymbol{d}(\boldsymbol{a}_i, \boldsymbol{\beta}) \approx \boldsymbol{\beta}_i^* - \{\boldsymbol{d}(\boldsymbol{a}_i, \hat{\boldsymbol{\beta}}) - \tilde{\boldsymbol{A}}_i(\hat{\boldsymbol{\beta}})\hat{\boldsymbol{\beta}})\} - \tilde{\boldsymbol{A}}_i(\hat{\boldsymbol{\beta}})\boldsymbol{\beta}$$

where $\boldsymbol{A}_i(\hat{\boldsymbol{\beta}})$ is the $(p \times r)$ matrix of derivatives of the components of $\boldsymbol{d}(\boldsymbol{a}_i, \hat{\boldsymbol{\beta}})$ with respect to $\boldsymbol{\beta}$, evaluated at $\hat{\boldsymbol{\beta}}$. Inserting (5.19) in the objective function based on (5.18) yields an approximate criterion linear in $\boldsymbol{\beta}$, which may be optimized by a suitable adaptation of the GTS algorithm given in section 5.3.2.

Often, a nonlinear inter-individual model of the form (5.17) may be replaced by a linear specification by suitable reparameterization of the nonlinear regression function at Stage 1 of the hierarchy. For example, it is common to represent pharmacokinetic models in terms of the logarithms of the parameters clearance, Cl, and volume of distribution, V, and adopt a linear model relating $\log Cl$ and $\log V$ for the ith individual to individual-specific covariates, with an additive random effect. Not only does this enforce positivity of individual estimates, it implies constant coefficient of variation on the original Cl and V scales, which is likely to represent the true behavior of these parameters. Moreover, this approach allows straightforward application of inferential methods, such as the GTS algorithm, that are most easily implemented for linear specifications. We make use of parameterizations of this type often in this book (see section 5.5 and the examples in Chapters 6 and 9). We favor this approach both for the reasons noted above and for improved stability of estimation likely to be achieved.

5.3.5 Choice of individual estimates

It is intuitively clear that the reliability of the population estimates obtained by two-stage methods is predicated on the quality of the estimates $\boldsymbol{\beta}_i^*$ input to the second stage. The pooled GLS estimation strategy described in section 5.2 is one way to obtain individual estimates of the $\boldsymbol{\beta}_i$. Of course, any estimators $\boldsymbol{\beta}_i^*$, such as those obtained from GLS or ML fits to each individual's data separately, may also be used. When the assumption of a common intra-individual covariance structure is reasonable, however, individual estimates $\boldsymbol{\beta}_i^*$ based on pooling information across individuals ought to be more precise. Thus, improving estimation of $\boldsymbol{\beta}_i$ by better characterization of intra-individual variation may be expected to increase precision of inference on $\boldsymbol{\beta}$ and $\boldsymbol{D}$. Davidian and Giltinan (1993a) offer simulation evidence to support this claim.

5.4 Computational aspects

Commercial software that directly implements the two stage procedures discussed in this chapter is not available. However, the necessary computations may be accomplished by taking advantage of the similarity to those for the hierarchical linear model, individual nonlinear regression, and methods discussed in Chapter 6.

In the first stage, the pooled GLS estimation methods discussed in section 5.2 may be implemented using standard nonlinear regression software for both estimation of the $\boldsymbol{\beta}_i$ and pooled estimation of the intra-individual covariance parameters; see Davidian and Giltinan (1993a,b) for details. For the second stage, the approximate model (5.9) may be viewed as a hierarchical linear model, so that the software discussed in section 3.4.3 may be used to estimate $\boldsymbol{\beta}$ and $\boldsymbol{D}$ as well as to obtain empirical Bayes estimates of the $\boldsymbol{b}_i$, as long as the software allows specification of a known within-individual covariance matrix. The estimates $\hat{\boldsymbol{b}}_i$ and $\hat{\boldsymbol{\beta}}$, say, so obtained may be used to construct empirical Bayes estimates of the $\boldsymbol{\beta}_i$ in the obvious way, by substitution in (5.5): $\hat{\boldsymbol{\beta}}_i = \boldsymbol{A}_i\hat{\boldsymbol{\beta}} + \hat{\boldsymbol{b}}_i$. From the discussion in section 5.3.4, for the case of nonlinear specifications $\boldsymbol{d}$, software for the linearization methods considered in Chapter 6 may be used for inference; this is discussed in section 6.6.

Alternatively, it is straightforward to program the GTS algorithm, the EM algorithm for the Bayesian approach of section 5.3.3, and the modified GTS algorithm in section 5.3.4 using a matrix programming language such as GAUSS, SAS PROC IML, or Splus. We have used GAUSS and PROC IML to implement the GTS EM algorithm; in our experience, convergence to sensible estimates occurs for most problems, but may be slow in some cases.

5.5 Example

To illustrate the methods discussed in this chapter, we apply them to concentration-time profiles obtained from a study of the kinetics of the anti-asthmatic agent theophylline reported by Boeckmann, Sheiner and Beal (1992). In this experiment, the drug was administered orally to twelve subjects, and serum concentrations were measured at ten time points per subject over the subsequent 25 hours. Figure 5.4 shows the resulting concentration-time profiles for four subjects. The original data set also contains concentration values at time zero; all but three of these are zero, as might be expected in subjects naive to theophylline who have received the drug orally. The presence of these structural zeroes complicates modeling of the within-subject mean variance relationship, and we have excluded the zero time point from the analyses reported here.

A common model for the kinetics of theophylline following oral administration is the one-compartment open model with first-order absorption and elimination. The usual parameterization of this model is in terms of absorption rate constant, k_a, clearance, Cl, and volume of distribution, V. Denoting serum concentration for subject i at time t following dose D_i by $C_i(t)$, the model is

$$C_i(t) = \frac{FD_ik_{ai}}{V_i(k_{ai} - Cl_i/V_i)}\left\{\exp\left(-\frac{Cl_i}{V_i}t\right) - \exp(-k_{ai}t)\right\}. \quad (5.19)$$

In (5.19), (k_{ai}, Cl_i, V_i) are the subject-specific pharmacokinetic parameters to be estimated; we set the fraction of drug available $F \equiv 1$ in the analyses given here. For the theophylline study, dose D_i (mg/kg) was given on a per-weight basis, so that (k_{ai}, Cl_i, V_i) have units 1/hr, L/hr/kg, and L/kg, respectively. Because each of these parameters is necessarily positive to be meaningful, we enforced this constraint in fitting (5.19) by reparameterizing in terms

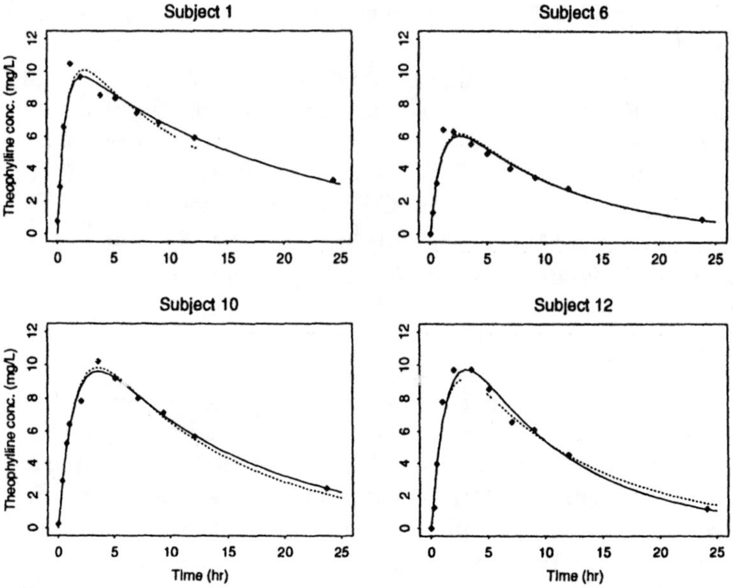

Figure 5.4. *Concentration-time data from 12 patients in a pharmacokinetic study of theophylline. Solid lines, GLS fits based on pooled PL variance parameter estimation; dashed lines, empirical Bayes fits from GTS algorithm.*

of $\boldsymbol{\beta}_i = (\beta_{1i}, \beta_{2i}, \beta_{3i})' = (\log k_{ai}, \log Cl_i, \log V_i)'$. This parameterization ensures non-negativity and, under the model

$$\boldsymbol{\beta}_i = \boldsymbol{\beta} + \boldsymbol{b}_i, \tag{5.20}$$

the assumption of normally distributed random effects $\boldsymbol{b}_i$ is more likely to be fulfilled.

Individual estimates were obtained using the GLS method of section 5.2, fitting the power-of-the-mean variance model by minimizing the pooled pseudolikelihood objective function under the assumption of uncorrelated within-subject measurements. For these data, the pooled GLS algorithm converged in ten iterations to the pooled PL estimate of $(\sigma, \theta) = (0.29, 0.59)$. Predicted values obtained from the final individual GLS fits $\boldsymbol{\beta}_i^*$, based on these variance parameter values, are depicted by the solid curves in the individual panels of Figure 5.4.

Table 5.3. *GTS fit to the theophylline data.*

Parameter	*Estimate of Population Mean (SE)*
$\log k_a$	0.265 (0.158)
$\log Cl$	−3.207 (0.071)
$\log V$	−0.795 (0.045)
D_{11}	0.2433
D_{12}	0.0026
D_{13}	0.0012
D_{22}	0.0537
D_{23}	0.0296
D_{22}	0.0164

The subject-specific pooled GLS estimates, β_i^*, were input to the GTS scheme outlined in section 5.3.2 assuming the inter-subject model (5.20). These computations were conducted using GAUSS and the PROC IML matrix language in the SAS system; both implementations yielded comparable results. Table 5.3 and Figure 5.4 summarize the results of this analysis. Estimates of the population average, $\boldsymbol{\beta} = (\log k_a, \log Cl, \log V)'$, and of the inter-subject covariance matrix, $\boldsymbol{D}$, are shown in the table. The estimate of $\boldsymbol{D}$ indicates an extremely high degree of correlation ($\rho = 0.998$) between $\log Cl_i$ and $\log V_i$ for these subjects. This may possibly be related to an association of both to a common underlying covariate. The only available potential covariate is weight; however, because dose was given on a per-weight basis, the effect of weight is already taken into account in Cl and V.

Predicted response profiles based on the empirical Bayes estimates, $\hat{\boldsymbol{\beta}}_{i,GTS}$, are represented by the dashed line fits shown in Figure 5.4. For these data, the difference between the individual GLS estimates, $\boldsymbol{\beta}_i^*$, input to the GTS algorithm and the final empirical Bayes estimates is not dramatic. This presumably reflects the high degree of balance in the design.

5.6 Discussion

Validity of asymptotic theory approximation

It is important to note that the success of the GTS and Bayesian approaches of sections 5.3.2–5.3.4, which take uncertainty in the first-stage estimation of $\boldsymbol{\beta}_i$ into account, depends critically on the relevance of the asymptotic theory for nonlinear regression. The form of the first two approximate moments of the distribution of $\boldsymbol{\beta}_i^*|\boldsymbol{\beta}_i$ is more critical than whether the asymptotic distribution is precisely normal for the GTS procedure; this is also true for the Bayesian method. Such large-sample approximation is likely to be good when the n_i are reasonably large, or when the variation about the nonlinear function f is small for each individual (the 'small-σ' situation described in section 2.5); exactly how 'large' the n_i must be is problem-dependent. Otherwise, the validity of this approximation may be suspect. In this case, better results may be obtained from the linearization methods of Chapter 6. In our experience, for growth, assay, and experimental pharmacokinetic data, where the within-individual variation is often small relative to the range of the mean response and a moderate number of responses is available on each individual, the approximation seems to work well; this observation may not be true in other situations, however.

Flexibility

In the development in section 5.2, estimation of the $\boldsymbol{\beta}_i^*$ is by generalized least squares. Because second-stage estimation of $\boldsymbol{\beta}$ and $\boldsymbol{D}$ is unaltered by the method used to obtain the $\boldsymbol{\beta}_i^*$, alternative techniques, such as those based on robust regression criteria, could be incorporated easily into the first-stage estimation.

Randomness of the components of $\boldsymbol{\beta}_i$

As in any model containing both fixed and random effects, designation of a particular component of $\boldsymbol{\beta}_i$ as fixed is often subject to debate. In the absence of cogent theoretical justification, treating certain components of $\boldsymbol{\beta}_i$ as fixed does not seem entirely defensible, at least for the kinds of applications that are the focus of this book. In our view, it is unlikely that biological phenomena such as growth or drug absorption and disposition will be such that certain features vary in the population while others are fixed across individuals. Rather, a more realistic perspective is that the variation in some characteristics in the population of individuals may be small enough to be regarded as negligible for practical purposes. In this situation, computational difficulties may be encountered that favor setting certain components of $\boldsymbol{\beta}_i$ as fixed in order to achieve stability in model fitting. A conservative strategy, then, is to adopt initially the convention that all elements of β_i are random. Should model fitting prove difficult, or if the magnitudes of population covariance estimates suggest that variation in a certain component is small, the analyst may be led to reexamine this assumption. If an element is then designated as fixed, this should not be taken to reflect the belief that the fixed component is truly constant across individuals; a more appropriate interpretation is that, because variation in this component is small, the true underlying process may be approximated reasonably by regarding the component as fixed. In some cases it may be necessary to regard certain components as fixed to avoid model misspecification; an example is given in section 11.4. See Lindstrom and Bates (1990), Vonesh (1992), Vonesh and Carter (1992), and Pinheiro and Bates (1995) for other perspectives on this issue.

The use of two-stage estimation methods does not seem entirely sensible when some components of β_i are taken as fixed. In the first stage, all components of the β_i are estimated for each individual, thus, the tacit assumption that all components differ across individuals is made. Consequently, our development in this chapter is based on the implicit assumption that there is a random component corresponding to each of the p elements of β_i. It is possible to use two-stage procedures when some components are fixed. This may be incorporated in the model (5.5), for example, by setting the relevant rows of A_i to zero in stage 2. On philosophical grounds, this approach still seems somewhat difficult to defend, as the remaining (random) elements of the β_i^* are obtained in stage 1 by treating the components assumed to be fixed as random. When some elements of β_i are taken as fixed, the linearization methods of Chapter 6, which do not summarize the data by means of individual estimation, may be more appropriate.

Time-dependent covariates

Two-stage estimation of population parameters does not generalize readily to the situation where the values of individual-specific covariates change over the course of observation, discussed in section 4.2.5. The first stage requires that there be a single regression parameter for each individual, which may be estimated using observations on that individual via standard regression methods. Consquently, the model given in (4.15), (4.16), which assumes a possibly different value for the regression parameter at each condition of measurement, is not accommodated. In cases where time-dependent covariates are involved, the analyst may appeal to the inferential methods in Chapters 6–8, which extend to handle this situation.

5.7 Bibliographic notes

The theme of two-stage estimation, in which estimates of individual regression parameters are used as 'data' in a subsequent population analysis, is common in the context of linear and nonlinear models. Swamy (1971) and Gumpertz and Pantula (1989) consider linear random coefficient regression models. The idea for the EM algorithm used in the global two-stage method for pharmacokinetic analysis is generally attributed to unpublished work of Prevost (1977). In addition to the GTS approach, Steimer *et al.* (1984) also discuss a more computationally intensive scheme they refer to as the iterative two-stage (ITS) algorithm, where the β_i^* and estimated covariance matrices C_i are updated at each iteration. Racine-Poon (1985) reports that the improvement attained by use of this modification over GTS or her Bayesian procedure is marginal; see also Steimer *et al.* (1984).

CHAPTER 6

Inference based on linearization

6.1 Introduction

The two-stage estimation methods considered in Chapter 5 comprise one main class of approaches to the analysis of nonlinear repeated measurement data. A requirement for the use of these methods is the availability of sufficient data to obtain regression parameter estimates for each individual. In many instances, this requirement is not a restriction.

In other situations, however, individual-specific model fitting may not be possible for some or all individuals. For example, it is not uncommon in agricultural field studies for some plots to be damaged by weather or parasitic infestation during a growing season, resulting in incomplete time-growth profiles for these units. If the information is too sparse, these plots may have to be excluded from a two-stage analysis. This issue is even more critical for the analysis of pharmacokinetic data from a clinical study, such as the quinidine data described in section 1.1.2. For the quinidine data, the postulated concentration-time model given in (1.2)–(1.4) is a function of four pharmacokinetic parameters (Cl, V, k_a, F), but four or more observations are available on fewer than 20% of the subjects; only 66% of the subjects have two or more measurements. Thus, reliable individual regression parameter estimates cannot be obtained on most subjects, even for those with more than four measurements. It is clear that two-stage estimation is infeasible for these data. Even in a situation where only a few individuals lack sufficient data for individual fitting, the need to eliminate these individuals under a two-stage approach is troubling.

Sparseness of individual data is one motivation for a second main class of estimation techniques for nonlinear repeated measurement data, methods based on *linearization* of the hierarchical nonlinear model summarized in section 4.2.4. Linearization methods approximate the hierarchical nonlinear model with a form additive in random effects and individual errors. Given the validity of the linear approximation, estimation procedures may then be developed that parallel those for the linear case and that may be used with sparse data. A second motivation for the linearization approach is thus the ability to adapt existing inferential techniques for the hierarchical linear model.

In this chapter, we discuss several linear approximations to the hierarchical nonlinear model. We begin in section 6.2 with the first-order linearization advocated by Beal and Sheiner (1982). In section 6.3, we discuss a refinement of the first-order approximation suggested by Lindstrom and Bates (1990). Inferential techniques based on linearization are relevant to the fully parametric model specification (section 4.3.1); methods based on each approximation are reviewed in each section. Section 6.4 gives a brief overview of available software, and the methods are applied to two examples in section 6.5. Section 6.6 discusses some philosophical and computational issues, as well as other methods and extensions. A list of related references is provided in section 6.7.

6.2 First-order linearization

The linearization approach to inference for the hierarchical nonlinear model was first suggested in the pharmacokinetics literature by Sheiner, Rosenberg and Melmon (1972) and Beal and Sheiner (1982). In section 6.2.1, we derive the approximation to the hierarchical nonlinear model that forms the basis for inference. Assuming normality of both the random effects and within-individual errors in this approximate model, one can derive the marginal likelihood explicitly. Beal and Sheiner proposed joint maximum likelihood inference within this framework; this is discussed in section 6.2.2. Alternative, staged procedures based on GLS ideas, are described in section 6.2.3. Connections to the literature on generalized estimating equations (GEE), first mentioned in section 2.6.2, are noted in section 6.2.4.

6.2.1 Approximate model specification

Recall the general form of the hierarchical nonlinear model given in (4.13) and (4.14) of section 4.2.4:

$$\boldsymbol{y}_i = \boldsymbol{f}_i(\boldsymbol{\beta}_i) + \boldsymbol{e}_i, \quad \boldsymbol{e}_i|\boldsymbol{b}_i \sim \Big(\boldsymbol{0}, \boldsymbol{R}_i(\boldsymbol{\beta}_i, \boldsymbol{\xi})\Big),$$

$$\boldsymbol{\beta}_i = \boldsymbol{d}(\boldsymbol{a}_i, \boldsymbol{\beta}, \boldsymbol{b}_i).$$

Throughout this chapter, the assumption $\boldsymbol{b}_i \sim (\mathbf{0}, \boldsymbol{D})$ is made, and we impose no structure on the matrix $\boldsymbol{D}$ in our development. As discussed in section 4.4, it is not possible in general to derive the marginal distribution of $\boldsymbol{y}_i$ explicitly and thereby employ classical statistical techniques such as maximum likelihood in a straightforward manner.

In the case of the hierarchical *linear model*, the ability to write down the marginal distribution of $\boldsymbol{y}_i$ is predicated upon two key features: the random effects $\boldsymbol{b}_i$ and the individual error vectors $\boldsymbol{e}_i$ enter the model in an additive, linear fashion and $\boldsymbol{b}_i$ and $\boldsymbol{e}_i$ are independent; under the assumption of normality of the distributions of $\boldsymbol{e}_i$ and $\boldsymbol{b}_i$, the marginal distribution of $\boldsymbol{y}_i$ is also normal. The approximation suggested by Beal and Sheiner (1982) exploits this fact. Write

$$\boldsymbol{e}_i = \boldsymbol{R}_i^{1/2}(\boldsymbol{\beta}_i, \boldsymbol{\xi})\boldsymbol{\epsilon}_i,$$

where $\boldsymbol{\epsilon}_i$ has mean zero, covariance matrix $\boldsymbol{I}_{n_i}$, and is independent of $\boldsymbol{b}_i$; and $\boldsymbol{R}_i^{1/2}(\boldsymbol{\beta}_i, \boldsymbol{\xi})$ is the Cholesky decomposition of $\boldsymbol{R}_i(\boldsymbol{\beta}_i, \boldsymbol{\xi})$. The first-stage model may then be rewritten as

$$\boldsymbol{y}_i = \boldsymbol{f}_i\{\boldsymbol{d}(\boldsymbol{a}_i, \boldsymbol{\beta}, \boldsymbol{b}_i)\} + \boldsymbol{R}_i^{1/2}\{\boldsymbol{d}(\boldsymbol{a}_i, \boldsymbol{\beta}, \boldsymbol{b}_i), \boldsymbol{\xi}\}\boldsymbol{\epsilon}_i, \tag{6.1}$$

where $\boldsymbol{\beta}_i$ has been replaced by $\boldsymbol{d}$ to show explicitly the dependence on the random effect $\boldsymbol{b}_i$. A Taylor series expansion of (6.1) in $\boldsymbol{b}_i$ about the mean value $\mathrm{E}(\boldsymbol{b}_i) = \mathbf{0}$, retaining the first two terms in in the expansion for $\boldsymbol{f}_i\{\boldsymbol{d}(\boldsymbol{a}_i, \boldsymbol{\beta}, \boldsymbol{b}_i)\}$ and the leading term of the expansion of $\boldsymbol{R}_i^{1/2}\{\boldsymbol{d}(\boldsymbol{a}_i, \boldsymbol{\beta}, \boldsymbol{b}_i), \boldsymbol{\xi}\}\boldsymbol{\epsilon}_i$, yields

$$\begin{aligned} \boldsymbol{y}_i &\approx \boldsymbol{f}_i\{\boldsymbol{d}(\boldsymbol{a}_i, \boldsymbol{\beta}, \mathbf{0})\} + \boldsymbol{F}_i(\boldsymbol{\beta}, \mathbf{0})\boldsymbol{\Delta}_{bi}(\boldsymbol{\beta}, \mathbf{0})\boldsymbol{b}_i \\ &+ \boldsymbol{R}_i^{1/2}\{\boldsymbol{d}(\boldsymbol{a}_i, \boldsymbol{\beta}, \mathbf{0}), \boldsymbol{\xi}\}\boldsymbol{\epsilon}_i, \end{aligned} \tag{6.2}$$

where $\boldsymbol{F}_i(\boldsymbol{\beta}, \mathbf{0})$ is the $(n_i \times p)$ matrix of derivatives of $\boldsymbol{f}_i(\boldsymbol{\beta}_i)$ with respect to $\boldsymbol{\beta}_i$ evaluated at $\boldsymbol{\beta}_i = \boldsymbol{d}(\boldsymbol{a}_i, \boldsymbol{\beta}, \mathbf{0})$, and $\boldsymbol{\Delta}_{bi}(\boldsymbol{\beta}, \mathbf{0})$ is the $(p \times k)$ matrix of derivatives of $\boldsymbol{d}(\boldsymbol{a}_i, \boldsymbol{\beta}, \boldsymbol{b}_i)$ with respect to $\boldsymbol{b}_i$ evaluated at $\boldsymbol{b}_i = \mathbf{0}$. This notation highlights the fact that the expansion has been taken about $\boldsymbol{b}_i = \mathbf{0}$.

In (6.2), the approximation to the mean function $\boldsymbol{f}_i\{\boldsymbol{d}(\boldsymbol{a}_i, \boldsymbol{\beta}, \boldsymbol{b}_i)\}$ is based on two terms of a Taylor series expansion, whereas the approximation to the within-subject error $\boldsymbol{R}_i^{1/2}\{\boldsymbol{d}(\boldsymbol{a}_i, \boldsymbol{\beta}, \boldsymbol{b}_i), \boldsymbol{\xi}\}\boldsymbol{\epsilon}_i$ is based only on the leading term. This seeming inconsistency is defensible, because misspecification of first-moment properties is more serious than that of second moments.

Defining the $(n_i \times k)$ matrix $\boldsymbol{Z}_i(\boldsymbol{\beta}, \boldsymbol{0}) = \boldsymbol{F}_i(\boldsymbol{\beta}, \boldsymbol{0})\boldsymbol{\Delta}_{bi}(\boldsymbol{\beta}, \boldsymbol{0})$ and $\boldsymbol{e}_i^* = \boldsymbol{R}_i^{1/2}\{\boldsymbol{d}(\boldsymbol{a}_i, \boldsymbol{\beta}, \boldsymbol{0}), \boldsymbol{\xi}\}\boldsymbol{\epsilon}_i$, (6.2) may be written as

$$\boldsymbol{y}_i \approx \boldsymbol{f}_i\{\boldsymbol{d}(\boldsymbol{a}_i, \boldsymbol{\beta}, \boldsymbol{0})\} + \boldsymbol{Z}_i(\boldsymbol{\beta}, \boldsymbol{0})\boldsymbol{b}_i + \boldsymbol{e}_i^*. \qquad (6.3)$$

Comparison with the first stage of the linear hierarchical specification (3.5) shows that the approximate model (6.3) has a similar form. The random effects $\boldsymbol{b}_i$ and 'within-individual' errors $\boldsymbol{e}_i^*$ enter the model in the same linear, additive fashion. Two differences from the linear case are worth noting. The matrix $\boldsymbol{Z}_i(\boldsymbol{\beta}, \boldsymbol{0})$ is not a fixed design matrix, but rather depends on the parameter $\boldsymbol{\beta}$. The fixed part of the model, $\boldsymbol{f}_i\{\boldsymbol{d}(\boldsymbol{a}_i, \boldsymbol{\beta}, \boldsymbol{0})\}$, is a nonlinear function of $\boldsymbol{\beta}$. An additional departure from the linear case is that the covariance matrix of the error $\boldsymbol{e}_i^*$ in (6.3) no longer depends on i only through its dimension, but is a function of individual-specific covariates $\boldsymbol{a}_i$, fixed effects $\boldsymbol{\beta}$, and covariance parameters $\boldsymbol{\xi}$.

The important consequence of (6.3) is that the marginal mean and covariance of $\boldsymbol{y}_i$ may be specified readily:

$$\begin{aligned} \mathrm{E}(\boldsymbol{y}_i) &\approx \boldsymbol{f}_i\{\boldsymbol{d}(\boldsymbol{a}_i, \boldsymbol{\beta}, \boldsymbol{0})\}, \\ \mathrm{Cov}(\boldsymbol{y}_i) &\approx \boldsymbol{R}_i\{\boldsymbol{d}(\boldsymbol{a}_i, \boldsymbol{\beta}, \boldsymbol{0}), \boldsymbol{\xi}\} + \boldsymbol{Z}_i(\boldsymbol{\beta}, \boldsymbol{0})\boldsymbol{D}\boldsymbol{Z}_i'(\boldsymbol{\beta}, \boldsymbol{0}) \\ &\equiv \boldsymbol{V}_i(\boldsymbol{\beta}, \boldsymbol{0}, \boldsymbol{\omega}), \end{aligned} \qquad (6.4)$$

where, as in section 4.2.4, $\boldsymbol{\omega}$ is the w-dimensional vector of parameters consisting of the intra-individual covariance parameter $\boldsymbol{\xi}$ and the distinct elements of $\boldsymbol{D}$. As in the linear case, if $\boldsymbol{b}_i$ and $\boldsymbol{e}_i^*$ are assumed to be normally distributed, it follows from (6.3) that the marginal distribution may be taken as approximately normal with moments given by (6.4). An important feature of (6.4), which is different from the linear case, is that $\mathrm{Cov}(\boldsymbol{y}_i)$ depends on the fixed effects $\boldsymbol{\beta}$; this is discussed further in section 6.2.3.

In model (6.3), the parameters to be estimated are thus the r-dimensional vector of fixed effects $\boldsymbol{\beta}$ and the w-dimensional vector of covariance parameters $\boldsymbol{\omega}$, for a total of $P = r + w$ parameters. The approaches to inference outlined in the next two sections are based on taking the approximate model (6.3) and the ensuing marginal moments (6.4) as exact.

6.2.2 Maximum likelihood

Estimation

Beal and Sheiner (1982) suggest assuming that the approximation (6.3) is exact and that $\boldsymbol{b}_i$ and $\boldsymbol{e}_i^*$ are normally distributed. Under these assumptions, they propose joint maximum likelihood estimation of $\boldsymbol{\beta}$ and $\boldsymbol{\omega}$ ($\boldsymbol{\xi}$ and $\boldsymbol{D}$). The estimates $\hat{\boldsymbol{\beta}}_{FO}$ and $\hat{\boldsymbol{\omega}}_{FO}$ obtained by this procedure, which is known as the first-order method in the pharmacokinetics literature, thus minimize the objective function

$$L_{FO}(\boldsymbol{\beta},\boldsymbol{\omega}) = \sum_{i=1}^{m}\Big(\log|\boldsymbol{V}_i(\boldsymbol{\beta},\boldsymbol{0},\boldsymbol{\omega})| + [\boldsymbol{y}_i - \boldsymbol{f}_i\{\boldsymbol{d}(\boldsymbol{a}_i,\boldsymbol{\beta},\boldsymbol{0})\}]'\boldsymbol{V}_i^{-1}(\boldsymbol{\beta},\boldsymbol{0},\boldsymbol{\omega})[\boldsymbol{y}_i - \boldsymbol{f}_i\{\boldsymbol{d}(\boldsymbol{a}_i,\boldsymbol{\beta},\boldsymbol{0})\}]\Big), \tag{6.5}$$

which is twice the negative marginal normal loglikelihood under (6.3).

Numerical techniques to implement this method, such as the Newton–Raphson algorithm and its variants (section 2.4.1), require that suitable starting values be specified. One suggestion for a starting value for $\boldsymbol{\beta}$ is the ordinary least squares estimator $\hat{\boldsymbol{\beta}}_{OLS}$ obtained by treating the data from all individuals as an independent sample of size N, which minimizes in $\boldsymbol{\beta}$

$$\sum_{i=1}^{m}[\boldsymbol{y}_i - \boldsymbol{f}_i\{\boldsymbol{d}(\boldsymbol{a}_i,\boldsymbol{\beta},\boldsymbol{0})\}]'[\boldsymbol{y}_i - \boldsymbol{f}_i\{\boldsymbol{d}(\boldsymbol{a}_i,\boldsymbol{\beta},\boldsymbol{0})\}]. \tag{6.6}$$

$\hat{\boldsymbol{\beta}}_{OLS}$ is known as the 'naive pooled data' estimator in the pharmacokinetics literature. A possible starting value for $\boldsymbol{D}$ is the $(k \times k)$ identity matrix. Specification of sensible starting values for other covariance parameters will depend both on context and the complexity of the intra-individual covariance model. See Boeckmann, Sheiner and Beal (1992) for a more extensive discussion.

Confidence intervals and hypothesis testing

Under the approximation (6.3) and the assumption that the random effects and intra-individual errors are normally distributed, inference may be based on standard asymptotic theory for maximum likelihood, allowing the number of individuals m to become large (Beal, 1984b; Ramos, 1993). The asymptotic covariance matrix for $\hat{\boldsymbol{\beta}}_{FO}$ and $\hat{\boldsymbol{\omega}}_{FO}$ may be estimated by the inverse of the information matrix evaluated at the estimates, and standard errors for the estimates may be obtained as the square roots of the diagonal elements of this matrix.

Implementation of likelihood ratio test procedures for inference is common in the pharmacokinetics literature (Boeckmann *et al.*,

1992), particularly for purposes of selection of an appropriate model $\boldsymbol{d}$ within a series of nested models. For comparing two nonnested models, A and B, say, with parameter dimensions P_A and P_B, respectively, Boeckmann *et al.* (1992) suggest use of a penalized likelihood criterion such as the Akaike Information Criterion (AIC). Because a model with a greater number of parameters will always produce a smaller value for $\hat{L}_{FO}$, the objective function evaluated at the parameter estimates, a penalty related to the dimension of a model is added to L_{FO} to offset this advantage. For model A, in obvious notation,

$$AIC_A = \hat{L}_{FO,A} + 2P_A,$$

and similarly for model B; one would compare AIC_A and AIC_B and prefer the model with the smaller value. The use of information criteria for model selection is discussed further in section 7.3.

Alternatively, one may construct confidence intervals and hypothesis tests for functions of the parameters in a manner analogous to the development in section 2.3.5. For a d-dimensional vector-valued function $\boldsymbol{a}(\boldsymbol{\beta})$ of $\boldsymbol{\beta}$, let $\boldsymbol{A}$ be the $(d \times p)$ matrix of derivatives of $\boldsymbol{a}(\boldsymbol{\beta})$. For the d-dimensional nonlinear hypotheses

$$H_0 : \boldsymbol{a}(\boldsymbol{\beta}) = \mathbf{0} \text{ vs. } H_1 : \boldsymbol{a}(\boldsymbol{\beta}) \neq \mathbf{0}, \tag{6.7}$$

the test statistic is

$$T^2 = \boldsymbol{a}'(\hat{\boldsymbol{\beta}}_{FO})(\hat{\boldsymbol{A}}\hat{\boldsymbol{\Sigma}}_{FO}\hat{\boldsymbol{A}}')^{-1}\boldsymbol{a}(\hat{\boldsymbol{\beta}}_{FO}), \tag{6.8}$$

where $\hat{\boldsymbol{A}}$ is the matrix $\boldsymbol{A}$ with $\boldsymbol{\beta}$ replaced by $\hat{\boldsymbol{\beta}}_{FO}$ and $\hat{\boldsymbol{\Sigma}}_{FO}$ is the estimated asymptotic covariance matrix for $\hat{\boldsymbol{\beta}}_{FO}$. The test is conducted by comparing T^2 to the appropriate χ^2_d critical value. As discussed in section 2.3.5, such Wald-type inference is not likely to be as reliable as inference based on likelihood principles.

It is important to note that these inferential procedures are approximate, as they are based on the assumption that the model (6.3) obtained by linearization is exact. Moreover, even if the linearization model were exact, they depend strongly on the validity of the normality assumption. If the test procedure (6.8) is used rather than the likelihood ratio approach, this is still an issue, as the form of the asymptotic covariance matrix is derived under the assumption of normality.

6.2.3 *Generalized least squares*

Estimation

For nonlinear modeling of individual data, we observed in section 2.7 that normal theory maximum likelihood estimation can be sensitive to nonnormality of the response and misspecification of the individual covariance structure, in particular when the covariance depends on the regression parameter. GLS methods provide an appealing alternative in this situation. The mean-covariance specification for the first order linearization (6.4), if taken as exact, is a multivariate analog to the models discussed in Chapter 2, where the responses are now vectors of length n_i from a sample of size m. The covariance matrix $\boldsymbol{V}_i$ for this model depends on the fixed effects $\boldsymbol{\beta}$, and estimation of covariance parameters $\boldsymbol{\omega}$ is required. Consequently, the same issues regarding the contrast between ML and GLS estimation, discussed in section 2.6, arise in this context. In the situation where the first order linearization (6.3) is an adequate approximation, then, Vonesh and Carter (1992) advocate the use of generalized least squares methods for inference as an alternative to maximum likelihood.

Because the error structure of the underlying model (6.3) is more complicated than that of the individual models considered in Chapter 2, a number of possibilities exist for implementing a GLS approach. Here, we consider three such procedures.

GLS algorithm. One strategy is to adapt the GLS algorithm for univariate response, given in sections 2.3.2–2.3.4, to the multivariate setting under the mean-covariance specification (6.4). Let $\boldsymbol{W}_i(\boldsymbol{\beta}, \boldsymbol{0}, \boldsymbol{\omega}) = \boldsymbol{V}_i^{-1}(\boldsymbol{\beta}, \boldsymbol{0}, \boldsymbol{\omega})$. A natural extension of the algorithm to the moments in (6.4) is as follows:

1. Estimate $\boldsymbol{\beta}$ by a preliminary estimator $\hat{\boldsymbol{\beta}}^{(p)}$. A reasonable choice would be the ordinary least squares estimator $\hat{\boldsymbol{\beta}}_{OLS}$ minimizing (6.6).

2. Estimate the covariance parameters $\boldsymbol{\omega}$ by $\hat{\boldsymbol{\omega}}$, and form estimated weight matrices $\boldsymbol{W}_i(\hat{\boldsymbol{\beta}}^{(p)}, \boldsymbol{0}, \hat{\boldsymbol{\omega}})$, $i = 1, \ldots, m$.

3. Using the estimates $\boldsymbol{W}_i(\hat{\boldsymbol{\beta}}^{(p)}, \boldsymbol{0}, \hat{\boldsymbol{\omega}})$ from step 2, reestimate $\boldsymbol{\beta}$ by minimizing in $\boldsymbol{\beta}$

$$\sum_{i=1}^{m}[\boldsymbol{y}_i - \boldsymbol{f}_i\{\boldsymbol{d}(\boldsymbol{a}_i, \boldsymbol{\beta}, \boldsymbol{0})\}]'\boldsymbol{W}_i(\hat{\boldsymbol{\beta}}^{(p)}, \boldsymbol{0}, \hat{\boldsymbol{\omega}})[\boldsymbol{y}_i - \boldsymbol{f}_i\{\boldsymbol{d}(\boldsymbol{a}_i, \boldsymbol{\beta}, \boldsymbol{0})\}],$$

or, equivalently, solving the set of estimating equations

$$\sum_{i=1}^{m} \boldsymbol{X}_i'(\boldsymbol{\beta},\boldsymbol{0})\boldsymbol{W}_i(\hat{\boldsymbol{\beta}}^{(p)},\boldsymbol{0},\hat{\boldsymbol{\omega}})[\boldsymbol{y}_i - \boldsymbol{f}_i\{\boldsymbol{d}(\boldsymbol{a}_i,\boldsymbol{\beta},\boldsymbol{0})\}] = \boldsymbol{0}, \quad (6.9)$$

where $\boldsymbol{X}_i(\boldsymbol{\beta},\boldsymbol{0}) = \boldsymbol{F}_i(\boldsymbol{\beta},\boldsymbol{0})\boldsymbol{\Delta}_{\beta i}(\boldsymbol{\beta},\boldsymbol{0})$ and $\boldsymbol{\Delta}_{\beta i}(\boldsymbol{\beta},\boldsymbol{0})$ is the $(p \times r)$ matrix of derivatives of $\boldsymbol{d}(\boldsymbol{a}_i,\boldsymbol{\beta},\boldsymbol{b}_i)$ with respect to $\boldsymbol{\beta}$ evaluated at $\boldsymbol{b}_i = \boldsymbol{0}$.

Treating this estimate as a new preliminary estimate $\hat{\boldsymbol{\beta}}^{(p)}$, return to step 2.

As in the univariate case, this scheme may be iterated a fixed number of times or to convergence, with at least one iteration recommended; convergence is not guaranteed. By analogy with the development in Chapter 2, we use the subscript GLS to refer to estimators obtained by this approach.

Estimation of $\boldsymbol{\omega}$ in step 2 may be accomplished by extension of the techniques in section 2.3. A pseudolikelihood approach would minimize in $\boldsymbol{\omega}$ $\sum_{i=1}^{m} PL_i(\hat{\boldsymbol{\beta}}^{(p)},\boldsymbol{\omega})$, where

$$PL_i(\hat{\boldsymbol{\beta}}^{(p)},\boldsymbol{\omega}) = [\boldsymbol{y}_i - \boldsymbol{f}_i\{\boldsymbol{d}(\boldsymbol{a}_i,\hat{\boldsymbol{\beta}}^{(p)},\boldsymbol{0})\}]'\boldsymbol{V}_i^{-1}(\hat{\boldsymbol{\beta}}^{(p)},\boldsymbol{0},\boldsymbol{\omega}) \times[\boldsymbol{y}_i - \boldsymbol{f}_i\{\boldsymbol{d}(\boldsymbol{a}_i,\hat{\boldsymbol{\beta}}^{(p)},\boldsymbol{0})\}] + \log|\boldsymbol{V}_i(\hat{\boldsymbol{\beta}}^{(p)},\boldsymbol{0},\boldsymbol{\omega})| \quad (6.10)$$

To account for preliminary estimation of $\boldsymbol{\beta}$, a restricted maximum likelihood approach would minimize $\sum_{i=1}^{m} REML_i(\hat{\boldsymbol{\beta}}^{(p)},\boldsymbol{\omega})$, where

$$REML_i(\hat{\boldsymbol{\beta}}^{(p)},\boldsymbol{\omega}) = PL_i(\hat{\boldsymbol{\beta}}^{(p)},\boldsymbol{\omega}) + \log|\boldsymbol{X}_i'(\hat{\boldsymbol{\beta}}^{(p)},\boldsymbol{0})\boldsymbol{V}_i^{-1}(\hat{\boldsymbol{\beta}}^{(p)},\boldsymbol{0},\boldsymbol{\omega})\boldsymbol{X}_i(\hat{\boldsymbol{\beta}}^{(p)},\boldsymbol{0})|, \quad (6.11)$$

where $\boldsymbol{X}_i(\boldsymbol{\beta},\boldsymbol{0})$ is defined as above.

The preliminary estimator for $\boldsymbol{\beta}$ in step 1 may be viewed as a starting value. Choice of starting values for step 2 may be undertaken as for the maximum likelihood method of the previous section.

As in the univariate case, PL or REML estimation at the second step of the GLS algorithm does not depend on normality, requiring only moment assumptions, and thus has omnibus appeal; as in section 2.3.3, the estimating equations associated with (6.10) may be interpreted as multivariate 'weighted least squares.' We comment on this issue further in section 6.2.4.

Vonesh and Carter algorithm. An alternative GLS algorithm based on assuming that (6.4) gives the true marginal moments has been proposed by Vonesh and Carter (1992). Their approach is similar in spirit to that given above; the fundamental difference is in the method used to compute the estimate of ω used to form the estimated inverse covariance matrix $W_i(\hat{\beta}^{(p)}, 0, \hat{\omega})$ used in step 2. Rather than employ methods for covariance parameter estimation that are multivariate analogs of the PL and REML procedures, Vonesh and Carter acknowledge explicitly the two-component error structure in (6.3) and base covariance parameter estimation on estimates of the random effects b_i. In their approach, the two components of ω, ξ and D, are estimated separately in two stages.

The original proposal of Vonesh and Carter is specific to the case where $R_i(\beta_i, \xi) \equiv \sigma^2 I_{n_i}$; here, we present a modification given by Davidian and Giltinan (1993b) suitable for use with general $R_i(\beta_i, \xi)$. As we did in section 2.3.4, it proves useful to separate out the scale parameter σ, writing $R_i(\beta_i, \xi) = \sigma^2 S_i(\beta_i, \gamma)$, $\xi = [\sigma, \gamma']'$.

The initial and final steps of the Vonesh and Carter procedure are identical to steps 1 and 3 given above. Step 2 is replaced by the following:

2. (a) Form residuals $\hat{s}_i^* = y_i - f_i\{d(a_i, \hat{\beta}^{(p)}, 0)\}$ from the preliminary fit, and iterate the following steps:

   (i) For each individual, obtain an initial estimate of b_i,

$$\hat{b}_i^0 = \{Z_i'(\hat{\beta}^{(p)}, 0) Z_i(\hat{\beta}^{(p)}, 0)\}^{-1} Z_i'(\hat{\beta}^{(p)}, 0)\hat{s}_i^*.$$

   (ii) Form residuals $\hat{r}_i = \hat{s}_i^* - Z_i'(\hat{\beta}^{(p)}, 0)\hat{b}_i^0$ and estimate ξ by $\hat{\xi}$ minimizing in ξ $\sum_{i=1}^m O_i(\hat{\beta}^{(p)}, \xi)$, where O_i is one of the objective functions given below.

   (iii) Form $\hat{S}_i = S_i\{d(a_i, \hat{\beta}^{(p)}, 0), \hat{\gamma}\}$ for each individual and update estimates of b_i by the weighted least squares estimates

$$\hat{b}_i = [Z_i'(\hat{\beta}^{(p)}, 0)\hat{S}_i^{-1} Z_i(\hat{\beta}^{(p)}, 0)]^{-1} Z_i'(\hat{\beta}^{(p)}, 0)\hat{S}_i^{-1}\hat{s}_i^*.$$

   Treating this estimate as a new initial estimate $\hat{b}_i^0$, return to step (ii).

(b) Using the final estimates $\hat{b}_i$ and $\hat{\xi} = [\hat{\sigma}, \hat{\gamma}']'$ from step 2(a), let

$$S_{bb} = (m-1)^{-1} \sum_{i=1}^m (\hat{b}_i - \bar{b})(\hat{b}_i - \bar{b})',$$

where $\bar{\boldsymbol{b}}$ is the mean of the $\hat{\boldsymbol{b}}_i$. Estimate $\boldsymbol{D}$ by

$$\hat{\boldsymbol{D}} = \begin{cases} \boldsymbol{S}_{bb} - \hat{\sigma}^2 m^{-1} \sum_{i=1}^m [\boldsymbol{Z}_i'(\hat{\boldsymbol{\beta}}^{(p)}, \boldsymbol{0}) \hat{\boldsymbol{S}}_i^{-1} \boldsymbol{Z}_i(\hat{\boldsymbol{\beta}}^{(p)}, \boldsymbol{0})]^{-1}, \\ \quad \hat{\lambda} > \hat{\sigma}^2 \\ \boldsymbol{S}_{bb} - \hat{\lambda} m^{-1} \sum_{i=1}^m [\boldsymbol{Z}_i'(\hat{\boldsymbol{\beta}}^{(p)}, \boldsymbol{0}) \hat{\boldsymbol{S}}_i^{-1} \boldsymbol{Z}_i(\hat{\boldsymbol{\beta}}^{(p)}, \boldsymbol{0})]^{-1}, \\ \quad \hat{\lambda} \leq \hat{\sigma}^2, \end{cases}$$

where $\hat{\boldsymbol{S}}_i = \boldsymbol{S}_i\{\boldsymbol{d}(\boldsymbol{a}_i, \hat{\boldsymbol{\beta}}^{(p)}, \boldsymbol{0}), \hat{\boldsymbol{\gamma}}\}$ and $\hat{\lambda}$ is the smallest root of

$$|\boldsymbol{S}_{bb} - \lambda m^{-1} \sum_{i=1}^m [\boldsymbol{Z}_i'(\hat{\boldsymbol{\beta}}^{(p)}, \boldsymbol{0}) \hat{\boldsymbol{S}}_i^{-1} \boldsymbol{Z}_i(\hat{\boldsymbol{\beta}}^{(p)}, \boldsymbol{0})]^{-1}| = 0.$$

At the end of steps 2(a) and 2(b), collect $\hat{\boldsymbol{\xi}}$ and the distinct elements of $\hat{\boldsymbol{D}}$ into $\hat{\boldsymbol{\omega}}$ and form estimates $\boldsymbol{W}_i(\hat{\boldsymbol{\beta}}^{(p)}, \boldsymbol{0}, \hat{\boldsymbol{\omega}})$, $i = 1, \ldots, m$.

As before, it is recommended that steps 1–3 of the Vonesh–Carter algorithm be iterated; results often stabilize after a few iterations. Note that if $\boldsymbol{R}_i(\boldsymbol{\beta}_i, \boldsymbol{\xi}) \equiv \sigma^2 \boldsymbol{I}_{n_i}$, no iteration is needed in step 2(a), as $\hat{\sigma}^2$ is calculated readily as the mean square of the residuals in (ii). Call the final estimates of the fixed parameters from the algorithm $\hat{\boldsymbol{\beta}}_{VC}$ and $\hat{\boldsymbol{\omega}}_{VC}$. Note that final estimates of the random effects, denoted $\hat{\boldsymbol{b}}_{i,VC}$, are also generated.

Step 2(a) is based on the observation that, if $\boldsymbol{\beta}$ were known, (6.3) may be rearranged as

$$[\boldsymbol{y}_i - \boldsymbol{f}_i\{\boldsymbol{d}(\boldsymbol{a}_i, \boldsymbol{\beta}, \boldsymbol{0})\}] = \boldsymbol{Z}_i(\boldsymbol{\beta}, \boldsymbol{0})\boldsymbol{b}_i + \boldsymbol{e}_i^*,$$

which has the form of a 'linear regression' for individual i with 'response' $[\boldsymbol{y}_i - \boldsymbol{f}_i\{\boldsymbol{d}(\boldsymbol{a}_i, \boldsymbol{\beta}, \boldsymbol{0})\}]$, 'regression parameter' $\boldsymbol{b}_i$, and general covariance structure $\boldsymbol{R}_i\{\boldsymbol{d}(\boldsymbol{a}_i, \boldsymbol{\beta}, \boldsymbol{0}), \boldsymbol{\xi}\}$. Suitable objective functions O_i for estimation of $\boldsymbol{\xi}$ in (ii) include the pseudolikelihood function

$$PL_i^*(\hat{\boldsymbol{\beta}}^{(p)}, \boldsymbol{\xi}) = \\ \log|\sigma^2 \boldsymbol{S}_i\{\boldsymbol{d}(\boldsymbol{a}_i, \hat{\boldsymbol{\beta}}^{(p)}, \boldsymbol{0}), \hat{\boldsymbol{\gamma}}\}| + \hat{\boldsymbol{r}}_i' \boldsymbol{S}_i^{-1}\{\boldsymbol{d}(\boldsymbol{a}_i, \hat{\boldsymbol{\beta}}^{(p)}, \boldsymbol{0}), \hat{\boldsymbol{\gamma}}\}\hat{\boldsymbol{r}}_i / \sigma^2 \tag{6.12}$$

and the restricted maximum likelihood function

$$REML_i^*(\hat{\boldsymbol{\beta}}^{(p)}, \boldsymbol{\xi}) = PL_i^*(\hat{\boldsymbol{\beta}}^{(p)}, \boldsymbol{\xi}) \\ + \log|\sigma^{-2} \boldsymbol{Z}_i'(\hat{\boldsymbol{\beta}}^{(p)}, \boldsymbol{0}) \boldsymbol{S}_i^{-1}\{\boldsymbol{d}(\boldsymbol{a}_i, \hat{\boldsymbol{\beta}}^{(p)}, \boldsymbol{0}), \hat{\boldsymbol{\gamma}}\} \boldsymbol{Z}_i(\hat{\boldsymbol{\beta}}^{(p)}, \boldsymbol{0})|. \tag{6.13}$$

In the case where the matrix $\boldsymbol{S}_i$ is diagonal, these reduce to the objective functions (2.21) and (2.23) given in section 2.3.3; an analogous version of the absolute residuals objective function (2.24) could also then be used.

The estimator for $\boldsymbol{D}$ in step 2(b) is an approximation to a similar estimator proposed for use with linear random coefficient models. See Carter and Yang (1986) and Vonesh and Carter (1987); justification is based on an argument similar to that used to exhibit the bias of the standard two-stage (STS) estimator of $\boldsymbol{D}$, see (5.8) in section 5.3.1. Conditional on the current estimate for $\boldsymbol{\beta}$, $\hat{\boldsymbol{D}}_{VC}$ is an unbiased method of moments estimator, where the adjustment based on the estimated smallest root $\hat{\lambda}$ is undertaken to guarantee a positive-semidefinite estimate. Vonesh and Carter (1992) discuss examination of $\hat{\lambda}$ as a diagnostic tool for model evaluation.

This method differs from the previous GLS method as well as the maximum likelihood approach in that a natural estimator for the individual regression parameters $\boldsymbol{\beta}_i$ is easily identified: $\hat{\boldsymbol{\beta}}_{i,VC} = \boldsymbol{d}(\boldsymbol{a}_i, \hat{\boldsymbol{\beta}}_{VC}, \hat{\boldsymbol{b}}_{i,VC})$. Although superficially it may appear more complicated, the staged nature of the Vonesh and Carter procedure often makes it easier to calculate than either the normal theory likelihood or the previous GLS method. In the simplest case where $\boldsymbol{R}_i(\boldsymbol{\beta}_i, \boldsymbol{\xi}) = \sigma^2 \boldsymbol{I}_{n_i}$, execution of step 2 is particularly straightforward. Even for more complicated intra-individual covariance models, step 2(a) is not too intensive, as the 'regression fits' are linear; if $\boldsymbol{R}_i(\boldsymbol{\beta}_i, \boldsymbol{\xi})$ is diagonal, the straightforward computational methods of section 2.4.2 are used in (ii).

The estimators for the covariance components in the Vonesh and Carter algorithm are simple, moment-type estimators. Hence, it is likely that they will be inefficient relative to PL or REML estimators, which, although also moment-based in the spirit of generalized least squares, as discussed above, are based on appropriately weighted functions of residuals. However, the simpler estimators offer a computational advantage; because estimation of $\boldsymbol{\omega}$ is separated into two stages, no starting value for $\boldsymbol{D}$ is required. Consequently, this method may be valuable as a means of obtaining suitable starting values for use with a more computationally demanding procedure.

The Vonesh and Carter method may be used with sparse data as long as $n_i \geq k$ for all i; individuals with fewer observations than the number of random effects must be excluded from the analysis.

Generalized estimating equation approaches. A third GLS strategy under the mean-covariance model (6.4) may be based on the GEE approach taken by Zeger *et al.* (1988); A number of variations on this idea are possible; here, we describe briefly a version of a proposal given by Schabenberger (1994) for the simple case where $\boldsymbol{R}_i(\boldsymbol{\beta}_i, \boldsymbol{\xi}) = \sigma^2 \boldsymbol{I}_{n_i}$.

As with the Vonesh and Carter scheme, this approach differs from the GLS algorithm outlined at the beginning of this section in the method used to estimate the covariance parameters $\boldsymbol{\omega}$ at step 2; the appropriately weighted PL or REML estimating equation is again replaced by the use of simpler moment estimators. In particular, the initial and final steps of the GLS algorithm are identical to steps 1 and 3 above, as before; step 2 is replaced by the following:

2. Let $\hat{\boldsymbol{Z}}_i = \boldsymbol{Z}_i(\hat{\boldsymbol{\beta}}^{(p)}, \boldsymbol{0})$, and define

$$\hat{\boldsymbol{U}}_i = [\boldsymbol{y}_i - \boldsymbol{f}_i\{\boldsymbol{d}(\boldsymbol{a}_i, \hat{\boldsymbol{\beta}}^{(p)}, \boldsymbol{0})\}][\boldsymbol{y}_i - \boldsymbol{f}_i\{\boldsymbol{d}(\boldsymbol{a}_i, \hat{\boldsymbol{\beta}}^{(p)}, \boldsymbol{0})\}]'.$$

The updated estimate of $\boldsymbol{D}$ is then given by

$$\hat{\boldsymbol{D}} = m^{-1} \sum_{i=1}^{m} (\hat{\boldsymbol{Z}}_i' \hat{\boldsymbol{Z}}_i)^{-1} \hat{\boldsymbol{Z}}_i' (\hat{\boldsymbol{U}}_i - \hat{\sigma}^2 \boldsymbol{I}_{n_i}) \hat{\boldsymbol{Z}}_i (\hat{\boldsymbol{Z}}_i' \hat{\boldsymbol{Z}}_i)^{-1},$$

where

$$\hat{\sigma}^2 = m^{-1} \sum_{i=1}^{m} [\boldsymbol{y}_i - \boldsymbol{f}_i\{\boldsymbol{d}(\boldsymbol{a}_i, \hat{\boldsymbol{\beta}}^{(p)}, \boldsymbol{0})\}]' \hat{\boldsymbol{W}}_i [\boldsymbol{y}_i - \boldsymbol{f}_i\{\boldsymbol{d}(\boldsymbol{a}_i, \hat{\boldsymbol{\beta}}^{(p)}, \boldsymbol{0})\}],$$

$\hat{\boldsymbol{W}}_i = \boldsymbol{W}_i(\hat{\boldsymbol{\beta}}^{(p)}, \boldsymbol{0}, \hat{\boldsymbol{\omega}})$, and $\hat{\boldsymbol{\omega}}$ is the estimate from the previous iteration.

Again, estimation of the covariance components will likely be inefficient relative to estimation based on PL or REML, but offers the advantage of relative ease of implementation. Schabenberger (1994) suggests use of these estimates as sensible starting values for more computationally burdensome approaches. Denote the final estimates from this procedure as $\hat{\boldsymbol{\beta}}_{ZLA}$ and $\hat{\boldsymbol{\omega}}_{ZLA}$. Estimates of the random effects may then be obtained at convergence as

$$\hat{\boldsymbol{b}}_{i,ZLA} = \hat{\boldsymbol{D}}_{ZLA} \boldsymbol{Z}_i'(\hat{\boldsymbol{\beta}}_{ZLA}, \boldsymbol{0}) \boldsymbol{W}_i(\hat{\boldsymbol{\beta}}_{ZLA}, \boldsymbol{0}, \hat{\boldsymbol{\omega}}_{ZLA})$$

$$\times [\boldsymbol{y}_i - \boldsymbol{f}_i\{\boldsymbol{d}(\boldsymbol{a}_i, \hat{\boldsymbol{\beta}}_{ZLA}, \boldsymbol{0})\}],$$

mimicking the form of such estimates in the linear case, see (3.14). These may be used to construct final estimates of the individual regression parameters in the obvious fashion.

Confidence intervals and hypothesis testing

If the linearization model (6.3) is taken to hold exactly, standard asymptotic theory may be invoked to show that as m grows large, $\hat{\boldsymbol{\beta}}_{GLS}$ is asymptotically normal with mean $\boldsymbol{\beta}$ and covariance matrix

$$\boldsymbol{\Sigma}_{GLS} = \Big(\sum_{i=1}^{m} \boldsymbol{X}_i'(\boldsymbol{\beta},\boldsymbol{0})\boldsymbol{V}_i^{-1}(\boldsymbol{\beta},\boldsymbol{0},\boldsymbol{\omega})\boldsymbol{X}_i(\boldsymbol{\beta},\boldsymbol{0})\Big)^{-1}. \tag{6.14}$$

Under additional regularity conditions, it may be shown (Vonesh and Carter, 1992; Ramos, 1993) that $\hat{\boldsymbol{\beta}}_{VC}$ is asymptotically normal with mean $\boldsymbol{\beta}$ and covariance matrix $\boldsymbol{\Sigma}_{VC} = \boldsymbol{\Sigma}_{GLS}$. As in the univariate case, the assumption that the marginal distribution of $\boldsymbol{y}_i$ is normal is not required. A similar result may be invoked for $\hat{\boldsymbol{\beta}}_{ZLA}$.

For all GLS approaches, an estimate $\hat{\boldsymbol{\Sigma}}_{GLS}$ may be obtained by evaluation of (6.14) at the final estimates of $\boldsymbol{\beta}$ and $\boldsymbol{\omega}$ with estimated standard errors calculated as the square roots of the diagonal elements. Hypothesis tests may be conducted in a manner analogous to that presented in (6.7) and (6.8).

As before, inference based on the GLS methods should be interpreted with caution, as it is based on the assumption that the linearization is exact.

6.2.4 Connection with generalized estimating equations

As in section 2.6.2 in the case of univariate response, the GLS procedures for repeated measurement data discussed in section 6.2.3 are moment methods, based on the assumption that the model for the marginal moments (6.4) is exact. We now note the direct correspondence between the methods for the first order linearization model discussed in this section, and the methods based on generalized estimating equations (GEEs).

The first-order linearization yields an approximate model for the first two marginal moments of the multivariate response $\boldsymbol{y}_i$:

$$\begin{aligned} \mathrm{E}(\boldsymbol{y}_i) &\approx \boldsymbol{f}_i\{\boldsymbol{d}(\boldsymbol{a}_i,\boldsymbol{\beta},\boldsymbol{0})\}, \\ \mathrm{Cov}(\boldsymbol{y}_i) &\approx \boldsymbol{R}_i\{\boldsymbol{d}(\boldsymbol{a}_i,\boldsymbol{\beta},\boldsymbol{0}),\boldsymbol{\xi}\} + \boldsymbol{Z}_i(\boldsymbol{\beta},\boldsymbol{0})\boldsymbol{D}\boldsymbol{Z}_i'(\boldsymbol{\beta},\boldsymbol{0}) \\ &\equiv \boldsymbol{V}_i(\boldsymbol{\beta},\boldsymbol{0},\boldsymbol{\omega}). \end{aligned} \tag{6.15}$$

Liang and Zeger (1986) consider a multivariate regression problem of this type, where the first two marginal moments of a multivariate response are modeled directly; this has been termed a 'population-averaged' model. For estimation of $\boldsymbol{\beta}$, they advocate solving an estimating equation of the form (6.9) in step 3 of the GLS algorithm in section 6.2.3 using a suitable estimate for the covariance components $\boldsymbol{\omega}$ to form the estimated weight matrices. Their original suggestion was to use simple functions of residuals to obtain this estimate. The methods of Vonesh and Carter (1992) and Schabenberger (1994) are in this spirit.

Prentice (1988), Zhao and Prentice (1990), Prentice and Zhao (1991) and Liang *et al.* (1992) have discussed the use of a second set of estimating equations along with those for $\boldsymbol{\beta}$ to obtain more efficient estimates of $\boldsymbol{\omega}$. As discussed in section 2.6.2, these are referred to as quadratic estimating equations to emphasize their dependence on quadratic functions of residuals. For the marginal moments given by (6.15), this set of estimating equations has the form of those corresponding to the PL criterion (6.10); analogous to the univariate case, differentiation of (6.10) shows that the quadratic estimating equations given in equation (2) of Prentice and Zhao (1991) have the same form as the PL equations. Thus, the GEE approach given in section 2 of Prentice and Zhao (1991) and called 'GEE1' by Liang *et al.* (1992), applied to the linearized marginal model (6.4), is essentially equivalent to the multivariate GLS algorithm with PL estimation of the covariance parameters at step 2, iterated to convergence.

Prentice and Zhao (1991, section 3) modify the estimating equations in an attempt to achieve more efficient inference. Specifically, they replace (6.9), which is linear in $\boldsymbol{y}_i$, by an equation quadratic in $\boldsymbol{y}_i$ and propose joint solution of this equation with that for the covariance parameters. This approach, referred to as 'GEE2' by Liang *et al.* (1992), with the Gaussian 'working matrix' assumption, is essentially equivalent to full normal theory maximum likelihood for the marginal moment model (6.15), outlined in section 6.2.2.

6.3 Conditional first-order linearization

A key aspect of the approximate linearization model (6.3) discussed in section 6.2 is that the random variation among individuals is incorporated only through the additive, linear term $\boldsymbol{Z}_i(\boldsymbol{\beta}, \mathbf{0})\boldsymbol{b}_i$; the implied marginal mean function given in (6.4) does not reflect inter-individual variation. If this component of variability is substantial, the model (6.3) may be a poor approximation, leading to biased and imprecise estimation of the fixed parameters. To alleviate this difficulty, Lindstrom and Bates (1990) advocate basing inference on a refinement of the linearization that may be expected to provide a better approximation. Under the refined approximation, they propose an iterative GLS estimation scheme, which we consider in section 6.3.2. Our exposition differs somewhat from that of Lindstrom and Bates (1990) in order to more closely parallel that of the last section. Alternative derivations and methods are possible; these are discussed in section 6.3.3.

6.3.1 Approximate model specification

Consider again the general hierarchical nonlinear model and the development in the first two paragraphs of section 6.2.1, so that the first-stage model may be written as in (6.1):

$$\boldsymbol{y}_i = \boldsymbol{f}_i\{\boldsymbol{d}(\boldsymbol{a}_i, \boldsymbol{\beta}, \boldsymbol{b}_i)\} + \boldsymbol{R}_i^{1/2}\{\boldsymbol{d}(\boldsymbol{a}_i, \boldsymbol{\beta}, \boldsymbol{b}_i), \boldsymbol{\xi}\}\boldsymbol{\epsilon}_i. \qquad (6.16)$$

Lindstrom and Bates (1990) argue that the approximation obtained by expansion of (6.16) about the expectation of the random effects $\boldsymbol{b}_i = \boldsymbol{0}$ may be poor. Instead, they consider linearization of (6.16) in the random effects about some value $\boldsymbol{b}_i^*$, say, that is closer to $\boldsymbol{b}_i$ than its expectation, $\boldsymbol{0}$. Specifically, retaining the first two terms of the Taylor series expansion about $\boldsymbol{b}_i = \boldsymbol{b}_i^*$ of $\boldsymbol{f}_i\{\boldsymbol{d}(\boldsymbol{a}_i, \boldsymbol{\beta}, \boldsymbol{b}_i)\}$ and the leading term of the Taylor expansion of $\boldsymbol{R}_i^{1/2}\{\boldsymbol{d}(\boldsymbol{a}_i, \boldsymbol{\beta}, \boldsymbol{b}_i), \boldsymbol{\xi}\}\boldsymbol{\epsilon}_i$ yields

$$\begin{aligned} \boldsymbol{y}_i &\approx \boldsymbol{f}_i\{\boldsymbol{d}(\boldsymbol{a}_i, \boldsymbol{\beta}, \boldsymbol{b}_i^*)\} + \boldsymbol{F}_i(\boldsymbol{\beta}, \boldsymbol{b}_i^*)\boldsymbol{\Delta}_{bi}(\boldsymbol{\beta}, \boldsymbol{b}_i^*)(\boldsymbol{b}_i - \boldsymbol{b}_i^*) \\ &+ \boldsymbol{R}_i^{1/2}\{\boldsymbol{d}(\boldsymbol{a}_i, \boldsymbol{\beta}, \boldsymbol{b}_i^*), \boldsymbol{\xi}\}\boldsymbol{\epsilon}_i, \end{aligned} \qquad (6.17)$$

where $\boldsymbol{F}_i(\boldsymbol{\beta}, \boldsymbol{b}_i^*)$ is the $(n_i \times p)$ matrix of derivatives of $\boldsymbol{f}_i(\boldsymbol{\beta}_i)$ with respect to $\boldsymbol{\beta}_i$ evaluated at $\boldsymbol{\beta}_i = \boldsymbol{d}(\boldsymbol{a}_i, \boldsymbol{\beta}, \boldsymbol{b}_i^*)$, and $\boldsymbol{\Delta}_{bi}(\boldsymbol{\beta}, \boldsymbol{b}_i^*)$ is the $(p \times k)$ matrix of derivatives of $\boldsymbol{d}(\boldsymbol{a}_i, \boldsymbol{\beta}, \boldsymbol{b}_i)$ with respect to $\boldsymbol{b}_i$ evaluated at $\boldsymbol{b}_i = \boldsymbol{b}_i^*$. This notation makes explicit the fact that the expansion in $\boldsymbol{b}_i$ has been taken about $\boldsymbol{b}_i^*$.

Defining the $(n_i \times k)$ matrix $\boldsymbol{Z}_i(\boldsymbol{\beta}, \boldsymbol{b}_i) = \boldsymbol{F}_i(\boldsymbol{\beta}, \boldsymbol{b}_i)\boldsymbol{\Delta}_{bi}(\boldsymbol{\beta}, \boldsymbol{b}_i)$ and $\boldsymbol{e}_i^* = \boldsymbol{R}_i\{\boldsymbol{d}(\boldsymbol{a}_i, \boldsymbol{\beta}, \boldsymbol{b}_i^*), \boldsymbol{\xi}\}\boldsymbol{\epsilon}_i$, (6.17) may be written as

$$\boldsymbol{y}_i = \boldsymbol{f}_i\{\boldsymbol{d}(\boldsymbol{a}_i, \boldsymbol{\beta}, \boldsymbol{b}_i^*)\} - \boldsymbol{Z}_i(\boldsymbol{\beta}, \boldsymbol{b}_i^*)\boldsymbol{b}_i^* + \boldsymbol{Z}_i(\boldsymbol{\beta}, \boldsymbol{b}_i^*)\boldsymbol{b}_i + \boldsymbol{e}_i^*. \qquad (6.18)$$

This suggests an approximation to the distribution of $\boldsymbol{y}_i$ given $\boldsymbol{b}_i$ for $\boldsymbol{b}_i$ near $\boldsymbol{b}_i^*$. From (6.18), paralleling the development in section 6.2.1, it follows that, for $\boldsymbol{b}_i$ close to $\boldsymbol{b}_i^*$, the approximate marginal mean and covariance of $\boldsymbol{y}_i$ is

$$\begin{aligned}
\mathrm{E}(\boldsymbol{y}_i) &\approx \boldsymbol{f}_i\{\boldsymbol{d}(\boldsymbol{a}_i,\boldsymbol{\beta},\boldsymbol{b}_i^*)\} - \boldsymbol{Z}_i(\boldsymbol{\beta},\boldsymbol{b}_i^*)\boldsymbol{b}_i^*, \\
\mathrm{Cov}(\boldsymbol{y}_i) &\approx \boldsymbol{R}_i\{\boldsymbol{d}(\boldsymbol{a}_i,\boldsymbol{\beta},\boldsymbol{b}_i^*),\boldsymbol{\xi}\} + \boldsymbol{Z}_i(\boldsymbol{\beta},\boldsymbol{b}_i^*)\boldsymbol{D}\boldsymbol{Z}_i'(\boldsymbol{\beta},\boldsymbol{b}_i^*) \\
&\equiv \boldsymbol{V}_i(\boldsymbol{\beta},\boldsymbol{b}_i^*,\boldsymbol{\omega}), \qquad (6.19)
\end{aligned}$$

As in section 6.2.1, $\boldsymbol{\omega}$ is the vector of distinct covariance parameters, and, if $\boldsymbol{b}_i$ and $\boldsymbol{e}_i^*$ are assumed to be normally distributed, the approximate marginal distribution will be normal as well.

In order to construct an estimation procedure for $\boldsymbol{\beta}$ and $\boldsymbol{\omega}$ based on this expansion, one must have available a reasonable choice for the value $\boldsymbol{b}_i^*$. A natural strategy is to obtain a suitable estimate of $\boldsymbol{b}_i$, insert this estimate as the value $\boldsymbol{b}_i^*$ in (6.19), and, treating the estimate as fixed, use either a GLS or ML scheme to estimate $\boldsymbol{\beta}$ and $\boldsymbol{\omega}$. This process can be iterated: the estimates from the GLS or ML step can be used to obtain an updated estimate for $\boldsymbol{b}_i$, and the process repeated. This is the fundamental idea of Lindstrom and Bates (1990) and is the basis for the inferential techniques discussed in the next two sections.

6.3.2 *Generalized least squares*

Estimation

Lindstrom and Bates (1990) propose their iterative GLS scheme under a more restrictive version of the hierarchical nonlinear model of section 4.2.4. They assume that (i) the inter-individual regression function $\boldsymbol{d}$ is linear in $\boldsymbol{\beta}$ and $\boldsymbol{b}_i$, as in (4.9); and (ii) that the intra-individual covariance matrix $\boldsymbol{R}_i(\boldsymbol{\beta}_i,\boldsymbol{\xi})$ does not depend on $\boldsymbol{\beta}_i$ (and hence on $\boldsymbol{b}_i$), but rather depends on i only through its dimension. Here, we extend their technique to accommodate the more general model of section 4.2.4. Generalization of (i) to nonlinear dependence is straightforward, and is incorporated in the development below. In contrast, accounting for (ii) requires some care. Accordingly, we first describe the Lindstrom and Bates procedure under the restriction (ii), and then show how it may be modified to account for more general intra-individual covariance structures. Initially, we write the intra-individual covariance matrix as $\boldsymbol{R}_i(\boldsymbol{\xi})$ to emphasize that it is taken not to depend on $\boldsymbol{b}_i$ or $\boldsymbol{\beta}$.

Lindstrom and Bates discuss a natural extension to the methods for the hierarchical linear model discussed in section 3.3.2. By analogy to (3.11), for *known* $\boldsymbol{\omega}$ (that is, known $\boldsymbol{D}$ and $\boldsymbol{\xi}$), the values of $\boldsymbol{\beta}$ and $\boldsymbol{b}_i$ that jointly minimize

$$\sum_{i=1}^{m}\Big(\log|\boldsymbol{D}| + \boldsymbol{b}_i'\boldsymbol{D}^{-1}\boldsymbol{b}_i + \log|\boldsymbol{R}_i(\boldsymbol{\xi})| \tag{6.20}$$
$$+[\boldsymbol{y}_i - \boldsymbol{f}_i\{\boldsymbol{d}(\boldsymbol{a}_i,\boldsymbol{\beta},\boldsymbol{b}_i)\}]'\boldsymbol{R}_i^{-1}(\boldsymbol{\xi})[\boldsymbol{y}_i - \boldsymbol{f}_i\{\boldsymbol{d}(\boldsymbol{a}_i,\boldsymbol{\beta},\boldsymbol{b}_i)\}]\Big)$$

have the same interpretations as in the linear case; (6.20) is twice the negative log of the posterior density of $\boldsymbol{b}_i$ for fixed $\boldsymbol{\beta}$ and twice the negative loglikelihood for $\boldsymbol{\beta}$ for fixed $\boldsymbol{b}_i$. Thus, an obvious strategy for estimation of $\boldsymbol{b}_i$ is to minimize (6.20) evaluated at a suitable estimate of $\boldsymbol{\omega}$. Estimation of $\boldsymbol{\omega}$ may be accomplished by the appropriate generalization of ML and REML techniques for the linear model to the nonlinear case.

Lindstrom and Bates (1990) thus propose the following two-step estimation scheme:

1. Given the current estimate of $\boldsymbol{\omega}$, $\hat{\boldsymbol{\omega}}$ (and thus $\hat{\boldsymbol{\xi}}$ and $\hat{\boldsymbol{D}}$), minimize in $\boldsymbol{\beta}$ and $\boldsymbol{b}_i$, $i = 1,\ldots,m$,

$$\sum_{i=1}^{m}\Big(\log|\hat{\boldsymbol{D}}| + \boldsymbol{b}_i'\hat{\boldsymbol{D}}^{-1}\boldsymbol{b}_i + \log|\boldsymbol{R}_i(\hat{\boldsymbol{\xi}})|$$
$$+[\boldsymbol{y}_i - \boldsymbol{f}_i\{\boldsymbol{d}(\boldsymbol{a}_i,\boldsymbol{\beta},\boldsymbol{b}_i)\}]'\boldsymbol{R}_i^{-1}(\hat{\boldsymbol{\xi}})[\boldsymbol{y}_i - \boldsymbol{f}_i\{\boldsymbol{d}(\boldsymbol{a}_i,\boldsymbol{\beta},\boldsymbol{b}_i)\}]\Big).$$

   Denote the resulting estimates as $\hat{\boldsymbol{b}}_i$ and $\hat{\boldsymbol{\beta}}_0$.

2. Estimate $\boldsymbol{\beta}$ and $\boldsymbol{\omega}$ as the values $\hat{\boldsymbol{\beta}}$ and $\hat{\boldsymbol{\omega}}$ that minimize

$$L_{LB}(\boldsymbol{\beta},\boldsymbol{\omega}) = \sum_{i=1}^{m}\Big(\log|\boldsymbol{V}_i(\hat{\boldsymbol{\beta}}_0,\hat{\boldsymbol{b}}_i,\boldsymbol{\omega})|$$
$$+\boldsymbol{r}_i^{*\prime}(\boldsymbol{\beta},\hat{\boldsymbol{b}}_i,\hat{\boldsymbol{\beta}}_0)\boldsymbol{V}_i^{-1}(\hat{\boldsymbol{\beta}}_0,\hat{\boldsymbol{b}}_i,\boldsymbol{\omega})\boldsymbol{r}_i^{*}(\boldsymbol{\beta},\hat{\boldsymbol{b}}_i,\hat{\boldsymbol{\beta}}_0)\Big), \tag{6.21}$$

   where $\boldsymbol{r}_i^{*}(\boldsymbol{\beta},\hat{\boldsymbol{b}}_i,\hat{\boldsymbol{\beta}}_0) = \boldsymbol{y}_i - \boldsymbol{f}_i\{\boldsymbol{d}(\boldsymbol{a}_i,\boldsymbol{\beta},\hat{\boldsymbol{b}}_i)\} + \boldsymbol{Z}_i(\hat{\boldsymbol{\beta}}_0,\hat{\boldsymbol{b}}_i)\hat{\boldsymbol{b}}_i$. (6.21) is twice the negative approximate marginal normal loglikelihood. Alternatively, take a restricted maximum likelihood approach, and minimize

$$L_{LB,REML}(\boldsymbol{\beta},\boldsymbol{\omega}) = L_{LB}(\boldsymbol{\beta},\boldsymbol{\omega})$$
$$+\log|\boldsymbol{X}_i'(\hat{\boldsymbol{\beta}}_0,\hat{\boldsymbol{b}}_i)\boldsymbol{V}_i(\hat{\boldsymbol{\beta}}_0,\hat{\boldsymbol{b}}_i,\boldsymbol{\omega})\boldsymbol{X}_i(\hat{\boldsymbol{\beta}}_0,\hat{\boldsymbol{b}}_i)|; \tag{6.22}$$

   in (6.22), $\boldsymbol{X}_i(\boldsymbol{\beta},\hat{\boldsymbol{b}}_i) = \boldsymbol{F}_i(\boldsymbol{\beta},\hat{\boldsymbol{b}}_i)\boldsymbol{\Delta}_{\beta i}(\boldsymbol{\beta},\hat{\boldsymbol{b}}_i)$, where $\boldsymbol{\Delta}_{\beta i}(\boldsymbol{\beta},\hat{\boldsymbol{b}}_i)$ is the $(p \times r)$ matrix of derivatives of $\boldsymbol{d}(\boldsymbol{a}_i,\boldsymbol{\beta},\boldsymbol{b}_i)$ with respect to $\boldsymbol{\beta}$ evaluated at $\boldsymbol{b}_i = \hat{\boldsymbol{b}}_i$.

Iterate this process to convergence, denoting the final estimates as $\hat{\beta}_{LB}$, $\hat{\omega}_{LB}$, and $\hat{b}_{i,LB}$.

Lindstrom and Bates call step 1 the 'pseudo-data' step, because joint estimation of β and b_i by minimizing (6.20) may be accomplished simultaneously by specifying an augmented *nonlinear* least squares problem, analogous to the linear case. Defining y, b, X, Z, R, and $\tilde{D}$ as in section 3.3.2, letting $f(\beta, b) = [f_1'\{d(a_1, \beta, b_1)\}, \ldots, f_m'\{d(a_m, \beta, b_m)\}]'$, $\hat{R}_i = R_i(\hat{\xi})$, and defining $\hat{R} = \text{diag}(\hat{R}_1, \ldots, \hat{R}_m)$ $(N \times N)$ and $\hat{\tilde{D}} = \text{diag}(\hat{D}, \ldots, \hat{D})$ $(km \times km)$, step 1 may be achieved by regressing the $(N + km \times 1)$ augmented data vector

$$\begin{bmatrix} \hat{R}^{-1/2} y \\ 0_{km \times 1} \end{bmatrix}$$

on the nonlinear regression function

$$\begin{bmatrix} \hat{R}^{-1/2} f(\beta, b) \\ \hat{\tilde{D}}^{-1/2} b \end{bmatrix}.$$

Wolfinger (1993) has shown that this nonlinear regression may be further approximated by a linear problem, which may be expressed in a form analogous to the mixed model equations (3.12) in the linear case. Thus, this step is the nonlinear version of the standard approach to estimation of β and b_i in the linear case, and the resulting estimator for β has the form of a GLS estimator.

The second step may be motivated as maximum likelihood or restricted maximum likelihood using an approximation to the marginal distribution of y_i based on (6.19), specifically, the normal distribution with mean and covariance matrix

$$\begin{aligned} \text{E}(y_i) &= f_i\{d(a_i, \beta, \hat{b}_i)\} - Z_i(\hat{\beta}_0, \hat{b}_i)\hat{b}_i, \\ \text{Cov}(y_i) &= R_i(\xi) + Z_i(\hat{\beta}_0, \hat{b}_i) D Z_i'(\hat{\beta}_0, \hat{b}_i) \\ &\equiv V_i(\hat{\beta}_0, \hat{b}_i, \omega). \end{aligned} \tag{6.23}$$

In (6.23), $\hat{b}_i$ and $\hat{\beta}_0$ are estimates from the first step. Note that the matrix $Z_i(\beta, b_i^*)$, which is evaluated at these estimates in both the mean and covariance, is taken as fixed for the minimization of step 2. This matrix corresponds to the fixed design matrix Z_i in the hierarchical linear model, so regarding it as fixed in (6.23) maintains the correspondence with the linear case. Thus, $\text{Cov}(y_i)$

is constant with respect to $\boldsymbol{\beta}$ in (6.23). Analogous to results for the hierarchical linear model, then, minimizing (6.21) or (6.22) in $\boldsymbol{\beta}$ and $\boldsymbol{\omega}$ also yields the (GLS) estimator for $\boldsymbol{\beta}$ minimizing (6.20) evaluated at the the estimate for $\boldsymbol{\omega}$. As in the linear case, because the estimator for $\boldsymbol{\beta}$ has the same form in both steps, convergence of the two step iterative algorithm is expected; see Wolfinger (1993) for further discussion.

Lindstrom and Bates refer to $\hat{\boldsymbol{\beta}}$ and $\hat{\boldsymbol{\omega}}$ minimizing $L_{LB}(\boldsymbol{\beta}, \boldsymbol{\omega})$ in step 2 as maximum likelihood estimators; however, a more appropriate characterization of these estimators might be as GLS and pseudolikelihood estimators, respectively. In particular, because $\text{Cov}(\boldsymbol{y}_i)$ does not depend on $\boldsymbol{\beta}$, it is straightforward to show by differentiation of the objective function (6.21) that these estimators are identical to those that would be obtained by iterating to convergence the following GLS algorithm:

(i) Estimate $\boldsymbol{\beta}$ by a preliminary estimator $\hat{\boldsymbol{\beta}}^{(p)}$. A reasonable choice would be the ordinary least squares estimator minimizing

$$\sum_{i=1}^{m} \boldsymbol{r}_i^{*\prime}(\boldsymbol{\beta}, \hat{\boldsymbol{b}}_i, \hat{\boldsymbol{\beta}}_0)\boldsymbol{r}_i^{*}(\boldsymbol{\beta}, \hat{\boldsymbol{b}}_i, \hat{\boldsymbol{\beta}}_0).$$

(ii) Estimate $\boldsymbol{\omega}$ by the value $\hat{\boldsymbol{\omega}}$ minimizing $L_{LB}(\hat{\boldsymbol{\beta}}^{(p)}, \boldsymbol{\omega})$. Form estimates $\boldsymbol{W}_i(\hat{\boldsymbol{\beta}}_0, \hat{\boldsymbol{b}}_i, \hat{\boldsymbol{\omega}}) = \boldsymbol{V}_i^{-1}(\hat{\boldsymbol{\beta}}_0, \hat{\boldsymbol{b}}_i, \hat{\boldsymbol{\omega}})$, $i = 1, \ldots, m$.

(iii) Using the estimates $\boldsymbol{W}_i(\hat{\boldsymbol{\beta}}_0, \hat{\boldsymbol{b}}_i, \hat{\boldsymbol{\omega}})$ from step 2, reestimate $\boldsymbol{\beta}$ by minimizing in $\boldsymbol{\beta}$

$$\sum_{i=1}^{m} \boldsymbol{r}_i^{*\prime}(\boldsymbol{\beta}, \hat{\boldsymbol{b}}_i, \hat{\boldsymbol{\beta}}_0)\boldsymbol{W}_i(\hat{\boldsymbol{\beta}}_0, \hat{\boldsymbol{b}}_i, \hat{\boldsymbol{\omega}})\boldsymbol{r}_i^{*}(\boldsymbol{\beta}, \hat{\boldsymbol{b}}_i, \hat{\boldsymbol{\beta}}_0).$$

Treating the estimate as a new preliminary estimate $\hat{\boldsymbol{\beta}}^{(p)}$, return to step (ii).

The estimator for $\boldsymbol{\omega}$ is the pseudolikelihood estimator, as it maximizes the normal likelihood evaluated at a preliminary estimate for $\boldsymbol{\beta}$. The restricted maximum likelihood approach in step 2 of the two-step algorithm is also equivalent to iterating this GLS algorithm to convergence, replacing the pseudolikelihood criterion with the REML objective function $L_{LB,REML}(\hat{\boldsymbol{\beta}}^{(p)}, \boldsymbol{\omega})$ in step (ii).

Step 2 of the algorithm is referred to as the 'linear mixed effects' step for the following reason. A further approximation allows the minimization in step 2 to be expressed as a linear estimation problem. Specifically, approximate $\boldsymbol{f}_i\{\boldsymbol{d}(\boldsymbol{a}_i, \boldsymbol{\beta}, \hat{\boldsymbol{b}}_i)\}$ by a linear function of $\boldsymbol{\beta}$ by expansion about the current estimate $\hat{\boldsymbol{\beta}}_0$ from step 1:

$$\begin{aligned} \boldsymbol{r}_i^*(\boldsymbol{\beta}, \hat{\boldsymbol{b}}_i, \hat{\boldsymbol{\beta}}_0) &= \boldsymbol{y}_i - \boldsymbol{f}_i\{\boldsymbol{d}(\boldsymbol{a}_i, \boldsymbol{\beta}, \hat{\boldsymbol{b}}_i)\} + \boldsymbol{Z}_i(\hat{\boldsymbol{\beta}}_0, \hat{\boldsymbol{b}}_i)\hat{\boldsymbol{b}}_i \\ &\approx \boldsymbol{y}_i - [\boldsymbol{f}_i\{\boldsymbol{d}(\boldsymbol{a}_i, \hat{\boldsymbol{\beta}}_0, \hat{\boldsymbol{b}}_i)\} + \boldsymbol{X}_i(\hat{\boldsymbol{\beta}}_0, \hat{\boldsymbol{b}}_i)(\boldsymbol{\beta} - \hat{\boldsymbol{\beta}}_0)] \\ &\quad + \boldsymbol{Z}_i(\hat{\boldsymbol{\beta}}_0, \hat{\boldsymbol{b}}_i)\hat{\boldsymbol{b}}_i \\ &= \boldsymbol{y}_i^* - \boldsymbol{X}_i(\hat{\boldsymbol{\beta}}_0, \hat{\boldsymbol{b}}_i)\boldsymbol{\beta} \end{aligned} \quad (6.24)$$

where $\boldsymbol{y}_i^* = \boldsymbol{y}_i - \boldsymbol{f}_i\{\boldsymbol{d}(\boldsymbol{a}_i, \hat{\boldsymbol{\beta}}_0, \hat{\boldsymbol{b}}_i)\} + \boldsymbol{X}_i(\hat{\boldsymbol{\beta}}_0, \hat{\boldsymbol{b}}_i)\hat{\boldsymbol{\beta}}_0 + \boldsymbol{Z}_i(\hat{\boldsymbol{\beta}}_0, \hat{\boldsymbol{b}}_i)\hat{\boldsymbol{b}}_i$. Substituting $\boldsymbol{y}_i^* - \boldsymbol{X}_i(\hat{\boldsymbol{\beta}}_0, \hat{\boldsymbol{b}}_i)\boldsymbol{\beta}$ for $\boldsymbol{r}_i^*(\boldsymbol{\beta}, \hat{\boldsymbol{b}}_i, \hat{\boldsymbol{\beta}}_0)$ in L_{LB} (6.21) or $L_{LB,REML}$ (6.22) yields a problem similar in form to that in the linear case; see (3.17) or (3.18), respectively. Computational methods for the hierarchical linear model may then be used to execute this step; see section 3.4.

The $\hat{\boldsymbol{b}}_{i,LB}$ may be regarded as empirical Bayes estimates of the random effects; these estimates are approximate posterior modes for $\boldsymbol{b}_i$, where the fixed parameters $\boldsymbol{\beta}$ and $\boldsymbol{\omega}$ are replaced by estimates. Based on the final estimates $\hat{\boldsymbol{\beta}}_{LB}$ and $\hat{\boldsymbol{b}}_{i,LB}$, the individual regression parameters $\boldsymbol{\beta}_i$ may then be estimated by

$$\hat{\boldsymbol{\beta}}_{i,LB} = \boldsymbol{d}(\boldsymbol{a}_i, \hat{\boldsymbol{\beta}}_{LB}, \hat{\boldsymbol{b}}_{i,LB}). \quad (6.25)$$

The $\hat{\boldsymbol{\beta}}_{i,LB}$ also may be regarded as empirical Bayes estimates.

Lindstrom and Bates (1990, section 6) discuss a number of computational issues regarding implementation of the algorithm, including recommendations for selecting starting values, which are required for all parameters at step 1 of the initial iterate. As with the procedures we already have discussed, they suggest the OLS estimator minimizing (6.6) for $\boldsymbol{\beta}$ and $\boldsymbol{b}_i = \boldsymbol{0}$, $i = 1, \ldots, m$. For $\boldsymbol{R}_i(\boldsymbol{\xi})$ of the form $\sigma^2 \boldsymbol{S}_i(\boldsymbol{\gamma})$, say, where $\boldsymbol{S}_i(\boldsymbol{\gamma})$ depends on i only through its dimension, they recommend obtaining a starting value for $\boldsymbol{\gamma}$ by setting $\boldsymbol{S}_i(\boldsymbol{\gamma}) = \boldsymbol{I}_{n_i}$. Starting values for σ^2 and $\boldsymbol{D}$ may be obtained by appealing to the recommendation of Laird *et al.* (1987) for the linear case, given in (3.29) and (3.30), evaluating the corresponding versions of these formulæ for the approximate linear model (6.24) using the starting values for $\boldsymbol{\beta}$ and $\boldsymbol{b}_i$.

Now consider the case where the intra-individual covariance matrix $\boldsymbol{R}_i(\boldsymbol{\beta}_i, \boldsymbol{\xi}) = \boldsymbol{R}_i\{\boldsymbol{d}(\boldsymbol{a}_i, \boldsymbol{\beta}, \boldsymbol{b}_i)\}$ depends on $\boldsymbol{\beta}_i$ (and hence on $\boldsymbol{b}_i$

and β). By analogy to (6.20), consider the objective function

$$\sum_{i=1}^{m}\Big(\log|D| + b_i' D^{-1} b_i + \log|R_i\{d(a_i,\beta,b_i),\xi\}| \\ +[y_i - f_i\{d(a_i,\beta,b_i)\}]' \\ \times R_i^{-1}\{d(a_i,\beta,b_i),\xi\}|[y_i - f_i\{d(a_i,\beta,b_i)\}]\Big), \quad (6.26)$$

with ω known. For fixed β, (6.26) is twice the negative log of the posterior density for the b_i, as before, and for fixed b_i, minimization of (6.26) should yield the maximum likelihood estimator for β. In (6.26), both β and the random effects b_i appear in a more complicated fashion than in (6.20); in particular, $R_i(\beta_i, \xi)$ may depend on these parameters nonlinearly. Consequently, minimization of this criterion by, for example, construction of a 'pseudo-data' problem, is not straightforward. Furthermore, because the fixed effects parameter β appears in the intra-individual covariance matrix, the resulting estimator for β, which maximizes a normal likelihood, will not have the form of a GLS estimator (see section 2.6).

One possible computational strategy for this situation, which modifies the basic two-step approach of Lindstrom and Bates (1990) sensibly and maintains the GLS spirit of their procedure, is the following. Given estimates $\hat{b}_i$ and $\hat{\beta}_0$, take the approximate marginal distribution for y_i to be the normal distribution with mean and covariance

$$\begin{aligned} \mathrm{E}(y_i) &= f_i\{d(a_i,\beta,\hat{b}_i)\} - Z_i(\hat{\beta}_0,\hat{b}_i)\hat{b}_i, \\ \mathrm{Cov}(y_i) &= R_i\{d(a_i,\hat{\beta}_0,\hat{b}_i),\xi\} + Z_i(\hat{\beta}_0,\hat{b}_i) D Z_i'(\hat{\beta}_0,\hat{b}_i) \\ &\equiv V_i(\hat{\beta}_0,\hat{b}_i,\omega). \end{aligned} \quad (6.27)$$

In (6.27), $R_i\{d(a_i,\beta,b_i),\xi\}$ is approximated by its value at $\hat{\beta}_0$ and $\hat{b}_i$, so that the expression for $\mathrm{Cov}(y_i)$ under (6.27) does not depend on unknown β, as in the simpler specification (6.23). Similarly, to maintain the correspondence with the form of the estimators for β and b_i based on (6.20), approximate $R_i\{d(a_i,\beta,b_i),\xi\}$ in (6.26) $R_i\{d(a_i,\hat{\beta}_{00},\hat{b}_{i,0}),\xi\}$, where $\hat{\beta}_{00}$ and $\hat{b}_{i,0}$ are some previous estimates of β and b_i. Under these conditions, one may define a two-step algorithm that, although based on further approximation, ought to have behavior comparable to that in the simpler case and may be implemented similarly:

1. Given the current estimate of ω, $\hat{\omega}$ (and thus $\hat{\xi}$ and $\hat{D}$), and previous estimates $\hat{\beta}_{00}$ and $\hat{b}_{i,0}$, $i = 1,\ldots,m$, maximize in β

and b_i, $i = 1, \ldots, m$,

$$\sum_{i=1}^{m} \Big(\log|\hat{D}| + b_i'\hat{D}^{-1} b_i + \log|R_i\{d(a_i, \hat{\beta}_{00}, \hat{b}_{i,0}), \hat{\xi}\}|$$
$$+ [y_i - f_i\{d(a_i, \beta, b_i)\}]'$$
$$\times R_i^{-1}\{d(a_i, \hat{\beta}_{00}, \hat{b}_{i,0}), \hat{\xi}\}[y_i - f_i\{d(a_i, \beta, b_i)\}] \Big).$$

Denote the resulting estimates as $\hat{b}_i$ and $\hat{\beta}_0$.

2. Estimate β and ω as the values $\hat{\beta}$ and $\hat{\omega}$ that minimize (6.21) or (6.22), where $V_i(\hat{\beta}_0, \hat{b}_i, \omega)$ is defined as in (6.27). Update $\hat{\beta}_{00}$ to equal $\hat{\beta}$ and $\hat{b}_{i,0}$ to equal $\hat{b}_i$, $i = 1, \ldots, m$.

Iterate this process to convergence, denoting the final estimates as $\hat{\beta}_{LB}$, $\hat{\omega}_{LB}$, and $\hat{b}_{i,LB}$. As before, the $\hat{b}_{i,LB}$ have an empirical Bayes interpretation, and individual regression parameter estimates may be obtained as in (6.25).

Note that because β and b_i are replaced in $R_i(\beta_i, \xi)$ by estimates, step 1 may be implemented as a 'pseudo-data' regression problem as before. Moreover, because these parameters are held fixed in $R_i(\beta_i, \xi)$ in this step as well as in step 2, the estimator for β will have the form of a GLS estimator.

To select suitable starting values for this modified approach, one may adapt in a straightforward manner the proposal of Lindstrom and Bates outlined above. Because the form of $R_i\{d(a_i, \beta, b_i), \xi\}$ as a function of β is likely to have been derived based on knowledge of the area of application, suitable starting values for ξ may be identified accordingly.

Confidence intervals and hypothesis testing

Lindstrom and Bates propose using the final estimated inverse Hessian matrix from a Newton–Raphson implementation of step 2 to estimate the covariance matrix of the parameter estimates for β and ω. This corresponds roughly to treating the approximate marginal moments (6.19) evaluated at $\hat{b}_{i,LB}$ as exact and deriving the usual asymptotic results for GLS, as in section 6.2.3. Estimated standard errors may be obtained and hypothesis tests conducted as previously indicated, based on $\hat{\beta}_{LB}$ and $\hat{\Sigma}_{LB}$, the estimated covariance matrix for $\hat{\beta}_{LB}$, in an obvious fashion. Lindstrom and Bates also advocate the use of the Akaike Information Criterion, discussed in section 6.2.2, as a way of choosing between competing models, where the computation of AIC is based on the approximate likelihood (6.21) evaluated at the converged values. As before, inference based on these approaches is approximate; moreover, no adjustment for estimation of the random effects is incorporated.

6.3.3 Alternative derivations and methods

An alternative derivation of the Lindstrom and Bates procedure of the previous section is given by Wolfinger (1993). He notes that the algorithm may be derived by applying Laplace's approximation to the exact marginal likelihood specified by the hierarchical nonlinear model, assuming normality of the random effects and the conditional distribution of the within-individual errors $\boldsymbol{e}_i$, and treating $\boldsymbol{\omega}$ as known. A number of authors have invoked this approximation to derive inferential strategies for the nonlinear hierarchy and related models; see Beal and Sheiner (1992), Breslow and Clayton (1993), and Pinheiro and Bates (1995); here, we sketch the basic idea. Throughout, assume that $\boldsymbol{\omega}$ is fixed and known.

From (4.20), the marginal likelihood is given by

$$\prod_{i=1}^{m} \int p_{y|b}(\boldsymbol{y}_i|\boldsymbol{x}_{i1},\ldots,\boldsymbol{x}_{in_i},\boldsymbol{a}_i,\boldsymbol{\beta},\boldsymbol{\xi},\boldsymbol{b}_i)\, p_b(\boldsymbol{b}_i|\boldsymbol{D})\, d\boldsymbol{b}_i; \qquad (6.28)$$

here, $p_{y|b}$ and p_b are normal densities. It is straightforward to show that (6.28) may be written, up to a constant, as the product of integrals of the form

$$\int \exp\{-\ell_i(\boldsymbol{b}_i)\}\, d\boldsymbol{b}_i, \qquad (6.29)$$

where

$$\begin{aligned}\ell_i(\boldsymbol{b}_i) &= \log|\boldsymbol{D}| + \boldsymbol{b}_i'\boldsymbol{D}^{-1}\boldsymbol{b}_i + \log|\boldsymbol{R}_i(\boldsymbol{\xi})| \\ &\quad +[\boldsymbol{y}_i - \boldsymbol{f}_i\{\boldsymbol{d}(\boldsymbol{a}_i,\boldsymbol{\beta},\boldsymbol{b}_i)\}]'\boldsymbol{R}_i^{-1}(\boldsymbol{\xi})[\boldsymbol{y}_i - \boldsymbol{f}_i\{\boldsymbol{d}(\boldsymbol{a}_i,\boldsymbol{\beta},\boldsymbol{b}_i)\}];\end{aligned}$$

thus, the empirical Bayes estimates $\hat{\boldsymbol{b}}_i$ satisfy $\partial/\partial\boldsymbol{b}_i\ell(\hat{\boldsymbol{b}}_i) = 0$, $i = 1,\ldots,m$. Laplace's approximation to an integral of the form (6.29) states that if $\ell_i(\boldsymbol{b}_i)$ has a unique minimum at $\hat{\boldsymbol{b}}_i$,

$$\int \exp\{-\ell_i(\boldsymbol{b}_i)\}\, d\boldsymbol{b}_i \approx (2\pi)^{k/2} \exp\{\ell(\hat{\boldsymbol{b}}_i)\}|\ell''(\hat{\boldsymbol{b}}_i)|^{-1/2}, \qquad (6.30)$$

where $\ell''(\hat{\boldsymbol{b}}_i) = \partial^2/\partial\boldsymbol{b}_i\boldsymbol{b}_i'\{\ell_i(\boldsymbol{b}_i)\}$ evaluated at $\hat{\boldsymbol{b}}_i$.

Given suitable estimates for the $\boldsymbol{b}_i$, the approximation to the likelihood (6.30) may then be used as a basis for estimation of $\boldsymbol{\beta}$ and

ω. Under a further approximation of (6.30), Wolfinger (1993) derives the two-stage algorithm of Lindstrom and Bates (1990). Beal and Sheiner (1992) construct several full maximum likelihood approaches based on this approximation; their first-order conditional estimation method, which may be thought of as a full normal theory maximum likelihood version of the method of Lindstrom and Bates (1990) for the approximate marginal model (6.19), is motivated using the Laplacian approximation. All of these procedures involve updated estimation of the random effects $\boldsymbol{b}_i$, so that final estimates are produced at convergence.

Another approach to inference for the conditional linearization model is given by Burnett, Ross and Krewski (1995), who propose methods based on the GEE approach of Zeger *et al.* (1988). These authors separate estimation of $\boldsymbol{D}$ and $\boldsymbol{\xi}$, using a simpler moment-type estimator for the former.

6.4 Software implementation

Software offering the capacity to fit hierarchical nonlinear models is not readily available at this time from commercial sources such as SAS or BMDP. However, a number of programs that implement inferential methods based on linearization have been developed independently.

The software package NONMEM (Beal and Sheiner, 1992), which performs maximum likelihood estimation based on the first-order and conditional first-order linearization models (sections 6.2.2 and 6.3.3), has been used widely by practitioners in the area of pharmacokinetic and pharmacodynamic analysis. Routines to fit standard pharmacokinetic and pharmacodynamic models are built into the program, allowing immediate analysis of data based on these models. Because pharmacokinetic and pharmacodynamic data typically exhibit within-individual variability that is dependent on the level of individual mean response, the program supports quite general modeling of intra-individual covariance structures. A number of additional, specialized features are available. The program and supporting documentation are available from the authors.

Other available software requires some effort by the user to program certain aspects of a specific model, for example, the mean function f and possibly its derivatives. NLME, a collection of Splus functions written by J. Pinheiro, D. Bates, and M. Lindstrom, is an extension of the nonlinear modeling facilities in release 3 of S and Splus that implements the two-step algorithm described in section 6.3.2 for the approximate model (6.23) in the special case where $\boldsymbol{R}_i(\boldsymbol{\beta}_i, \boldsymbol{\xi}) = \sigma^2 \boldsymbol{I}_{n_i}$. A general description of the capabilities of NLME is given by Pinheiro, Bates and Lindstrom (1993); the code is available in the Statlib collection maintained through Carnegie-Mellon University. At this time, the code does not support estimation of more general within-individual covariance structures. A SAS macro written by E. Vonesh, MIXNLIN, implements the GLS algorithm of Vonesh and Carter given in section 6.2.3, including extension to covariance structures that are a function of the random effects. Details are given in a user's guide (Vonesh, 1993), available, with the code, from the author.

It is possible to use existing commercial software for the hierarchical linear model (section 3.4.3) to implement some of the strategies based on linearization outlined in this chapter. Wolfinger and O'Connell (1993) describe how to use procedures such as SAS PROC MIXED or Module 5V of BMDP to fit hierarchical models that have the particular structure imposed by the generalized linear model framework; see section 6.6. Details on a SAS macro, GLIMMIX, implementing their proposal, may be obtained from R. Wolfinger. A SAS macro written by R. Wolfinger, NLINMIX, which implements the Lindstrom and Bates GLS algorithm as described in Wolfinger (1993) and Schabenberger (1994) is also available.

6.5 Graphical approaches to model selection

A useful graphical device for investigating possible relationships between parameter values and individual attributes is to plot empirical Bayes estimates of random effects or random parameters against potential covariates. Examination of plots of the estimated components of $\boldsymbol{b}_i$ or $\boldsymbol{\beta}_i$ plotted against candidate elements of the vector $\boldsymbol{a}_i$ provides a visual basis for specification of the component functions of the inter-individual regression model $\boldsymbol{d}(\boldsymbol{a}_i, \boldsymbol{\beta}, \boldsymbol{b}_i)$.

This approach was first suggested by Maitre *et al.* (1991), who propose a three-step strategy for model building. At the first step, one fits the hierarchical nonlinear model with no covariates; that is, the model is fitted with a second-stage specification for $\boldsymbol{\beta}_i$ containing no covariates, for example, $\boldsymbol{\beta}_i = \boldsymbol{\beta} + \boldsymbol{b}_i$. At the second step, one computes individual empirical Bayes estimates of the $\boldsymbol{\beta}_i$ based on this fit. The individual components of these estimates are then plotted against potential covariates, one at a time, and the plots are examined for possible systematic relationships.

Alternatively, empirical Bayes estimates of the $\boldsymbol{b}_i$ may be used. Based on this graphical evidence, a model $\boldsymbol{\beta}_i = \boldsymbol{d}(\boldsymbol{a}_i, \boldsymbol{\beta}, \boldsymbol{b}_i)$ is proposed, chosen based on the nature of the apparent relationship between each component and the individual candidate covariates. The third step consists of fitting the updated model. Further refinements to the model may be considered by returning to step 2, plotting updated empirical Bayes estimates against the remaining covariates, and adjusting the model. Maitre *et al.* suggest using formal hypothesis tests, as described in sections 6.2.2 and 6.3.2, to assess formally whether inclusion of an additional covariate in the model yields an improved fit.

The original proposal by Maitre *et al.* (1991) was in connection with linearization methods; however, it may be regarded as a general strategy. Davidian and Gallant (1992) propose examination of such plots in the context of semiparametric inference; see section 7.6. Wakefield (1995) uses graphics for model-building with the Bayesian methods discussed in Chapter 8.

A related idea is suggested by Mandema, Verotta and Sheiner (1992), who argue that one-at-a-time *ad hoc* examination of covariates in the graphical approach may be inefficient. They replace step 2 with a formal technique that allows consideration of several covariates at once. Specifically, for each component of $\boldsymbol{\beta}_i$, Mandema *et al.* (1992) fit a series of generalized additive models (e.g. Chambers and Hastie, 1992) involving all potential covariates and use a formal stepwise addition/deletion procedure to deduce the covariates to be retained. The form of the final fitted generalized additive model is used to specify the updated model in step 3.

6.6 Examples

Pharmacokinetics of theophylline

Recall the data from the study of the pharmacokinetics of theophylline, introduced in section 5.5. As for two-stage estimation, the zero time point has been excluded in the analyses reported here.

We again considered the one-compartment open model with first-order absorption and elimination given in (5.19), parameterized in terms of $\boldsymbol{\beta}_i = [\beta_{1i}, \beta_{2i}, \beta_{3i}]' = [\log k_{ai}, \log Cl_i, \log V_i]'$, and the population model $\boldsymbol{\beta}_i = \boldsymbol{\beta} + \boldsymbol{b}_i$. Estimates were obtained using the Vonesh and Carter GLS algorithm given in section 6.2.3 and the Lindstrom and Bates GLS approach described in section 6.3.2 assuming uncorrelated within-subject responses.

Table 6.1. *Vonesh and Carter GLS fit to the theophylline data.*

Parameter	*Estimate of Population Mean (SE)*
$\log k_a$	0.453 (0.233)
$\log Cl$	−3.264 (0.093)
$\log V$	−0.748 (0.039)
D_{11}	0.6116
D_{12}	−0.0552
D_{13}	0.0033
D_{22}	0.0971
D_{23}	0.0304
D_{33}	0.0109

Table 6.2. *Lindstrom and Bates GLS fit to the theophylline data.*

Parameter	*Estimate of Population Mean (SE)*
$\log k_a$	0.329 (0.197)
$\log Cl$	−3.214 (0.075)
$\log V$	−0.789 (0.043)
D_{11}	0.4141
D_{12}	−0.0186
D_{13}	−0.0097
D_{22}	0.0595
D_{23}	0.0291
D_{33}	0.0143

For the Vonesh and Carter procedure, we assumed the power variance function and estimated the power parameter θ in step 2(a)(ii) by minimizing the pseudolikelihood criterion. These computations were programmed in GAUSS. The algorithm converged in eight iterations, yielding the estimate of $(\sigma, \theta) = (0.27,0.63)$; estimates of $\boldsymbol{\beta}$ and $\boldsymbol{D}$ are given in Table 6.1. Predicted values based on the final individual estimates $\hat{\boldsymbol{\beta}}_{i,VC}$ were indistinguishable from those obtained from the GTS algorithm for most subjects.

We used the NLME collection of Splus functions to obtain the GLS fit under the conditional first-order linearization. Because this implementation does not support specification of nonconstant within-individual variance nor estimation of within-individual variance function parameters, we accommodated apparent intra-subject heteroscedasticity using the TBS approach outlined in section 2.7. Specifically, for comparison with the two-stage and Vonesh and Carter analyses, we applied the Box–Cox transformation with $\lambda = 0.4$, which corresponds to $\theta = 0.6$ in the power variance model, to both the pharmacokinetic model (5.19) and the theophylline concentrations and assumed constant within-subject variance on this scale. Note that selection of the transformation parameter requires that within-subject variance has been evaluated by some other means. The results of this fit are displayed in Table 6.2. As with the Vonesh and Carter estimates, predicted values obtained using $\hat{\boldsymbol{\beta}}_{i,LB}$ were quite similar to those obtained using the two-stage approach.

Comparison of the results in section 5.5 to those here shows that the GTS, Vonesh and Carter, and Lindstrom and Bates methods give remarkably consistent estimates for the population means of $\log Cl$ and $\log V$. Moreover, the Vonesh and Carter and pooled estimates of (σ, θ) are quite similar. The estimates for population mean $\log k_a$ do not agree well; this parameter is often difficult to estimate because of the paucity of information on absorption relative to that on clearance and volume under typical sampling protocols. The information on absorption seems adequate for many of the subjects in this study; however, Figure 5.4 shows that several subjects exhibit a rapid absorption phase, so that only one or two observations are available in this portion of the profile. The disparity of the $\log k_a$ estimates across methods thus may reflect this imbalance. The estimates of the population covariance matrix $\boldsymbol{D}$ also differ across methods. With only 12 individuals, it is not surprising that the off-diagonal elements of $\boldsymbol{D}$ disagree; these parameters are inherently difficult to estimate. The diagonal element corresponding to the variance of the population random effect for $\log k_a$ differs across fits; again, this may reflect the quality of information on absorption. Note that the estimated lower (2×2) submatrices of $\boldsymbol{D}$ corresponding to $(\log Cl, \log V)$ for the GTS and Lindstrom and Bates fits are in fairly good agreement and different from that obtained from the Vonesh and Carter procedure. These data represent full concentration-time profiles, so they contain good information on individual mean response. Both the GTS and Lindstrom and Bates procedures exploit this information through individual estimates and estimation of individual random effects, respectively, in estimation of the mean response, while the Vonesh and Carter method, based on the first order linearization, does not. Thus, the latter approach must attribute variation that may be due to differences in individual mean response to variation in the population, leading to a different estimate.

Pharmacokinetics of phenobarbital

The data in Table 6.3 are concentration-time-dose history information for two patients in a study of the neonatal pharmacokinetics of phenobarbital reported by Grasela and Donn (1985). Data were collected on $m = 59$ preterm infants given phenobarbital for prevention of seizures during the first 16 days after birth. Each individual received an initial dose followed by one or more sustaining doses by intravenous administration. A total of anywhere from $n_i = 1$ to 6 concentration measurements were obtained from each individual at times other than dose times as part of routine monitoring, for a total of $N = 155$ measurements. In addition, birth weight and 5-minute Apgar score were recorded for each subject. The data are described in detail by Grasela and Donn (1985) and analyzed by these authors and Boeckmann *et al.* (1992).

The pharmacokinetics of phenobarbital may be described by a one-compartment open model with intravenous administration and first-order elimination (Grasela and Donn, 1985). The model states that mean concentration at time t for subject i following a single dose D_{id} administered at time t_{id} is given by

$$\frac{D_{id}}{V_i} \exp\left\{-\frac{Cl_i}{V_i}(t - t_{id})\right\},$$

$t > t_{id}$, where Cl_i and V_i are clearance and volume of distribution, respectively, for subject i. Because individuals received several doses over the study period, concentration at time t, $C_i(t)$, is a sum of such terms:

$$C_i(t) = \sum_{d:t_{id}<t} \frac{D_{id}}{V_i} \exp\left\{-\frac{Cl_i}{V_i}(t - t_{id})\right\}. \tag{6.31}$$

This model may also be expressed in recursive form; see Grasela and Donn (1985). Figure 6.1 depicts a typical mean concentration pattern given by this model for the data from subject #9.

Several hierarchical nonlinear models were fitted to the phenobarbital data using the first-order (ML) method described in section 6.2.2 as implemented in the software package NONMEM. A thorough discussion of model-building strategies using this package applied to these data may be found in Boeckmann *et al.* (1992). Here, we summarize only a sequence of models and results; these

Table 6 3. *Data for two subjects, pharmacokinetic study of phenobarbital. Birth weight and Apgar score are constant across the observation period.*

	Subject 9			*Subject 50*	
time (hrs)	*dose* (μg/kg)	*conc.* (μg/L)	*time* (hrs)	*dose* (μg/kg)	*conc.* (μg/L)
0.0	27.0	–	0.0	20.0	–
1.1	–	22.1	3.0	–	22.2
11.1	3.2	–	12.5	2.5	–
22.3	3.2	–	24.5	2.5	–
34.6	3.2	–	36.5	2.5	–
46.6	3.2	–	48.0	2.5	–
58.7	3.2	–	60.5	2.5	–
70.9	3.2	–	72.5	2.5	–
82.7	–	29.2	81.0	–	30.5
83.2	3.2	–	84.5	2.5	–
94.6	3.2	–	88.0	30.0	–
106.6	3.2	–	89.0	–	67.9
118.6	3.2	–	96.5	2.5	–
130.6	3.2	–	108.5	3.5	–
142.1	–	34.2	120.5	3.5	–
142.6	3.2	–	132.5	3.5	–
312.6	–	19.6	144.5	3.5	–
			157.0	3.5	–
			162.0	–	58.7
Apgar	8		Apgar	6	
Weight	1.4 kg		Weight	1.1 kg	

are not meant to constitute a full analysis, but rather to illustrate the range of modeling possibilities and provide representative fits. The reader is referred to Boeckmann *et al.* (1992) and Grasela and Donn (1985) for further details and explanation.

In all cases, in the following sequence of models, the intra-individual structural model for subject i relating phenobarbital concentration to time, dose history, and the pharmacokinetic parameters Cl_i and V_i is the one-compartment open model with intravenous administration and first-order elimination given in (6.31). The differences among the hierarchical models lie in the specification of both the systematic and random components of inter-individual variation and in the model for intra-subject error. In all cases, $\boldsymbol{b}_i$ is

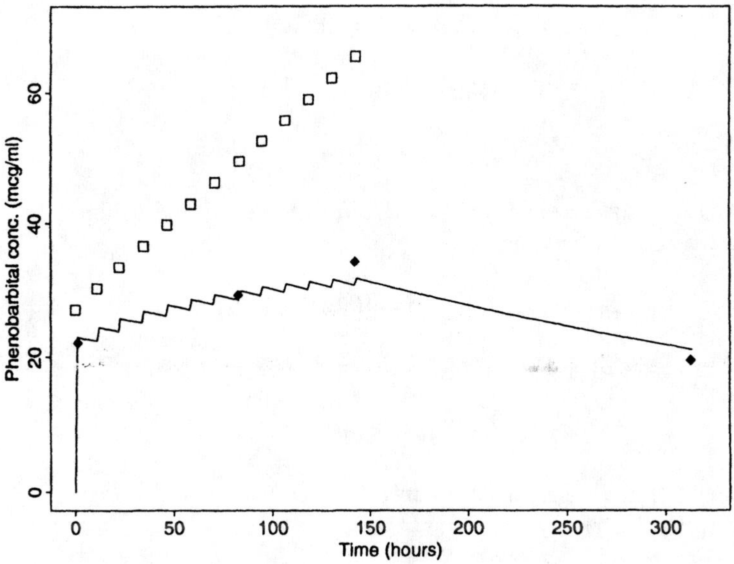

Figure 6.1. *Phenobarbital data, subject #9. The diamonds are phenobarbital concentration measurements, the open squares represent cumulative dose, in μg/kg, and the solid line is a fit of the one-compartment model (6.31).*

taken to have mean $\mathbf{0}$ and covariance $\boldsymbol{D}$, and the components of the intra-individual error term $\boldsymbol{e}_i$ are taken to be uncorrelated, so that $\boldsymbol{R}_i(\boldsymbol{\beta}_i, \boldsymbol{\xi})$ is a diagonal matrix with diagonal elements $\text{Var}(e_{ij}|\boldsymbol{\beta}_i)$. Let w_i denote the birth weight of the ith subject.

The following models were fit to the phenobarbital data:

Model 1. (No covariates, additive error structures)

$$\boldsymbol{\beta}_i = [Cl_i, V_i]', \quad \boldsymbol{\beta} = [\beta_1, \beta_2]'$$

$$\beta_{1i} = \beta_1 + b_{1i}, \quad \beta_{2i} = \beta_2 + b_{2i},$$

$$\text{Var}(e_{ij}|\boldsymbol{\beta}_i) = \sigma^2$$

Model 2. (No covariates, multiplicative error structures)

$$\boldsymbol{\beta}_i = [Cl_i, V_i]', \quad \boldsymbol{\beta} = [\beta_1, \beta_2]'$$

$$\beta_{1i} = \beta_1(1 + b_{1i}), \quad \beta_{2i} = \beta_2(1 + b_{2i}),$$

$$\text{Var}(e_{ij}|\boldsymbol{\beta}_i) = \sigma^2 f^2(\boldsymbol{x}_{ij}, \boldsymbol{\beta}_i)$$

Model 3. (Birth weight as covariate for clearance and volume, additive error structures)

$$\boldsymbol{\beta}_i = [Cl_i, V_i]', \quad \boldsymbol{\beta} = [\beta_1, \beta_2, \beta_3, \beta_4]'$$

$$\beta_{1i} = \beta_1 + \beta_3 w_i + b_{1i}, \quad \beta_{2i} = \beta_2 + \beta_4 w_i + b_{2i},$$

$$\text{Var}(e_{ij}|\boldsymbol{\beta}_i) = \sigma^2$$

Model 4. (Birth weight as covariate for clearance and volume, multiplicative error structures)

$$\boldsymbol{\beta}_i = [Cl_i, V_i]', \quad \boldsymbol{\beta} = [\beta_1, \beta_2, \beta_3, \beta_4]'$$

$$\beta_{1i} = (\beta_1 + \beta_3 w_i)(1 + b_{1i}), \quad \beta_{2i} = (\beta_2 + \beta_4 w_i)(1 + b_{2i}),$$

$$\text{Var}(e_{ij}|\boldsymbol{\beta}_i) = \sigma^2 f^2(\boldsymbol{x}_{ij}, \boldsymbol{\beta}_i)$$

Model 5. (Birth weight as covariate for clearance, birth weight and dichotomized Apgar score as covariates for volume, multiplicative error structures)

$$\boldsymbol{\beta}_i = [Cl_i, V_i]', \quad \boldsymbol{\beta} = [\beta_1, \beta_2, \beta_3, \beta_4, \beta_5]'$$

$$\beta_{1i} = (\beta_1 + \beta_3 w_i)(1 + b_{1i}),$$

$$\beta_{2i} = (\beta_2 + \beta_4 w_i)(1 + \beta_5 I_{[\text{Apgar}<5]})(1 + b_{2i}),$$

$$\text{Var}(e_{ij}|\boldsymbol{\beta}_i) = \sigma^2 f^2(\boldsymbol{x}_{ij}, \boldsymbol{\beta}_i)$$

Model 6. (Birth weight as covariate for clearance, birth weight and dichotomized Apgar score as covariates for volume, multiplicative error structures)

$$\boldsymbol{\beta}_i = [Cl_i, V_i]', \quad \boldsymbol{\beta} = [\beta_3, \beta_4, \beta_5]'$$

$$\beta_{1i} = \beta_3 w_i \exp(b_{1i}),$$

$$\beta_{2i} = \beta_4 w_i (1 + \beta_5 I_{[\text{Apgar}<5]}) \exp(b_{2i}),$$

$$\text{Var}(e_{ij}|\boldsymbol{\beta}_i) = \sigma^2 f^2(\boldsymbol{x}_{ij}, \boldsymbol{\beta}_i),$$

where $I_{[A]}$ is the indicator function for the event A.

Results for fitting this model sequence, constraining the covariance matrix $\boldsymbol{D}$ to be diagonal in all cases, are summarized in Table 6.4. It is evident from the table that, whether one adopts an additive or multiplicative error structure in both stages of the model, birth weight is an important explanatory variable for systematic variation in both clearance and volume. Apgar score may be important (Models 5 and 6); however, there is relatively little information on this issue. Only five of the 59 infants had Apgar scores less than 3 and only ten had Apgar scores less than 5 in these data. Note that estimates of the covariance components are not comparable across additive and multiplicative error structures.

Model 6 is the final specification fit to these data in the original analysis given by Grasela and Donn (1985). These authors incorporated multiplicative inter-individual error in a different form from that given in Models 2, 4, and 5. Moreover, they observed that the intercept terms in an analogous version of Model 4 were negligible, and thus eliminated such terms in the final fit.

For comparison, we fitted a similar sequence of models by the conditional first-order GLS procedure of Lindstrom and Bates using the NLME collection of Splus functions. In all cases, we constrained $\boldsymbol{D}$ to be diagonal as above. Models 1 and 3 were fitted exactly as described. This implementation does not support nonlinear specifications $\boldsymbol{\beta}_i = \boldsymbol{d}(\boldsymbol{a}_i, \boldsymbol{\beta}, \boldsymbol{b}_i)$ or nonconstant within-individual variance models. Thus, to obtain fits comparable to those above for Models 2, 4, and 6 (Model 5 was not fitted), we reparameterized the pharmacokinetic model in terms of $\boldsymbol{\beta}_i = (\log Cl_i, \log V_i)'$ to represent multiplicative errors on the original scales of these parameters. In addition, we applied a log transformation to both phenobarbital concentrations and the pharmacokinetic model (6.31) and assumed constant within-subject variance on this scale, so that the estimated intra-individual standard deviation would correspond roughly to the coefficient of variation in the first-order fits above. The inter-individual model specifications were thus as follows:

Table 6.4. *Fits of Models 1–6 by the first-order method using the software NONMEM. Note that the parameter σ corresponds to the assumed constant intra-individual standard deviation for Models 1 and 3, and to within-individual CV in Models 2, 4, 5, and 6.*

	Model 1	2	3
β_1	5.48×10^{-3}	5.55×10^{-3}	9.57×10^{-11}
(SE)	(4.86×10^{-4})	(3.49×10^{-4})	(6.35×10^{-10})
β_2	1.40	1.34	1.21×10^{-1}
(SE)	(0.078)	(0.080)	(1.46×10^{-1})
β_3	–	–	4.77×10^{-3}
(SE)	–	–	(2.24×10^{-4})
β_4	–	–	9.18×10^{-1}
(SE)	–	–	(1.13×10^{-1})
β_5	–	–	–
(SE)	–	–	–
D_{11}	6.85×10^{-6}	0.247	1.36×10^{-6}
D_{22}	2.86×10^{-1}	0.142	7.51×10^{-2}
σ	2.83	0.13	2.95

Table 6.4. *Continued.*

	Model 4	5	6
β_1	9.55×10^{-5}	2.21×10^{-5}	–
(SE)	(1.25×10^{-3})	(5.26×10^{-4})	–
β_2	6.21×10^{-2}	4.11×10^{-2}	–
(SE)	(1.05×10^{-1})	(9.39×10^{-2})	–
β_3	4.58×10^{-3}	4.66×10^{-3}	4.68×10^{-3}
(SE)	(9.41×10^{-4})	(6.33×10^{-4})	(2.03×10^{-4})
β_4	9.40×10^{-1}	9.34×10^{-1}	9.65×10^{-1}
(SE)	(8.21×10^{-2})	(7.70×10^{-2})	(2.48×10^{-2})
β_5	–	0.15	0.15
(SE)	–	(0.07)	(0.07)
D_{11}	5.01×10^{-2}	4.48×10^{-2}	4.30×10^{-2}
D_{22}	2.77×10^{-2}	2.61×10^{-2}	2.71×10^{-2}
σ	0.10	0.10	0.10

Model 2. (No covariates, multiplicative error structures)*

$$\beta_{1i} = \beta_1 + b_{1i}, \quad \beta_{2i} = \beta_2 + b_{2i}, \quad \boldsymbol{\beta} = [\beta_1, \beta_2]'$$

Model 4. (Birth weight as covariate for log clearance and log volume, multiplicative error structures)*

$$\beta_{1i} = \beta_1 + \beta_3 w_i + b_{1i}, \quad \beta_{2i} = \beta_2 + \beta_4 w_i + b_{2i},$$

$$\boldsymbol{\beta} = [\beta_1, \beta_2, \beta_3, \beta_4]'$$

Model 6. (Birth weight as covariate for log clearance, birth weight and dichotomized Apgar score as covariates for log volume, multiplicative error structures)*

$$\beta_{1i} = \beta_1 + \beta_3 w_i + b_{1i}, \quad \beta_{2i} = \beta_2 + \beta_4 w_i + \beta_5 I_{[\mathrm{Apgar}<5]} + b_{2i},$$

$$\boldsymbol{\beta} = [\beta_1, \beta_2, \beta_3, \beta_4, \beta_5]'.$$

The results of these fits are given in Table 6.5. Again, a strong association between birth weight and both clearance and volume is evident, regardless of error specifications. The conditional GLS fits to Models 1 and 3 are directly comparable to the ML fits given in Table 6.4; there is considerable divergence in the estimates from each method for both models. This may reflect the contrast between the first-order and conditional first order-linearizations, the use of normal theory ML techniques versus GLS, or both.

Figure 6.2 shows estimated random effects from the fit of Model 2* plotted against birth weight and Apgar category for each subject. The plots suggest a strong relationship between birth weight and both pharmacokinetic parameters; the visual evidence for relationships with Apgar score is not striking. Figure 6.3 displays the same plots for the fit of Model 4*, where birth weight has been added as a covariate in the specifications for both $\log Cl_i$ and $\log V_i$. The plots indicate that the association between birth weight and the parameters has been taken into account in the fit. Moreover, there is little graphical evidence suggesting a relationship between either parameter and Apgar score. This agrees with the fit of Model 6*, for which the magnitude of the estimate of β_5 relative to its estimated standard error is negligible. Because weight and Apgar score are highly correlated, once birth weight has been included in the model, Apgar score has little additional explanatory power.

Table 6.5. *Fits of Models 1, 2*, 3, 4*, and 6* by the conditional first-order GLS method using the software NLME. Note that the parameter σ corresponds to the assumed constant intra-individual standard deviation for Models 1 and 3, and to within-individual CV in Models 2*, 4*, and 6*. Entries marked with an asterisk (*) should be multiplied by 10^{-4}.*

	Model				
	1	*2**	*3*	*4**	*6**
β_1	27.9*	−5.847	−2.30*	−6.472	−6.747
(SE)	(2.19*)	0.088	(5.27*)	(0.167)	(0.168)
β_2	1.592	0.363	−0.093	−0.490	−0.499
(SE)	(0.123)	(0.062)	(0.014)	(0.069)	(0.070)
β_3	–	–	23.4*	0.654	0.657
(SE)	–	–	(3.96*)	(0.102)	(0.102)
β_4	–	–	1.082	0.551	0.552
(SE)	–	–	(0.095)	(0.042)	(0.042)
β_5	–	–	–	–	0.049
(SE)	–	–	–	–	(0.075)
D_{11}	1.18×10^{-6}	0.216	$4\ 70 \times 10^{-7}$	0.115	0.115
D_{22}	0.828	0.205	0.121	0.033	0.032
σ	3.38	0.14	3.65	0.13	0.13

6.7 Discussion

Comparison among linearization methods

Extensive empirical or theoretical comparison of the different linearization methods has not been undertaken; simulation work that is available (Vonesh, 1992; Ramos, 1993) is limited to a few specific models and methods, and is not conclusive. Thus, no general recommendations on the choice among linearization methods can be made. The major issue, choosing between methods based on first-order or conditional first-order linearization, is unresolved; one would expect the conditional approach to reduce bias in parameter estimates because the approximate marginal moments may be closer to the true moments. There is conflicting evidence whether or not the extra computational burden required to implement the conditional approach is worthwhile in this regard. As pointed out by Beal and Sheiner (1992), it is likely that the need for methods based on the conditional linearization will be greatest in models that are highly nonlinear in the $\boldsymbol{b}_i$. The extent to which results

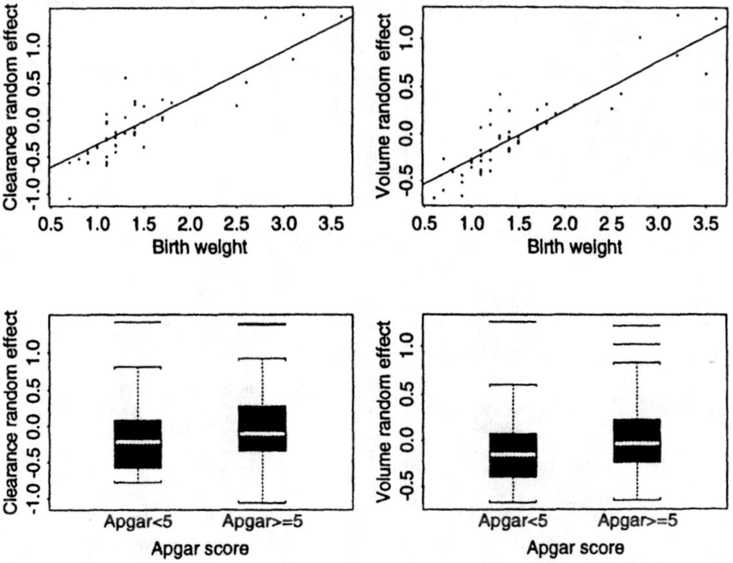

Figure 6.2. *Estimated random effects for log Cl and log V plotted against birth weight and Apgar score for the Lindstrom and Bates fit of Model 2*, phenobarbital data. The plots for birth weight have an ordinary least squares line superimposed; those for Apgar score are boxplots with the median represented by the white bar.*

differ between first-order and conditional first-order analyses will decrease as the number of observations per individual decreases, because the empirical Bayes estimate of $\boldsymbol{b}_i$ is 'shrunk' toward its mean, $\mathbf{0}$, and the shrinkage is greater the smaller the amount of data available on individual i.

Time-dependent covariates

It is straightforward to extend linearization methods to accommodate the situation where the vector of individual attributes for each individual may vary with time, given in (4.15), (4.16):

$$y_{ij} = f(\boldsymbol{x}_{ij}, \boldsymbol{\beta}_{ij}) + e_{ij}, \tag{6.32}$$

$$\boldsymbol{\beta}_{ij} = \boldsymbol{d}(\boldsymbol{a}_{ij}, \boldsymbol{\beta}, \boldsymbol{b}_i). \tag{6.33}$$

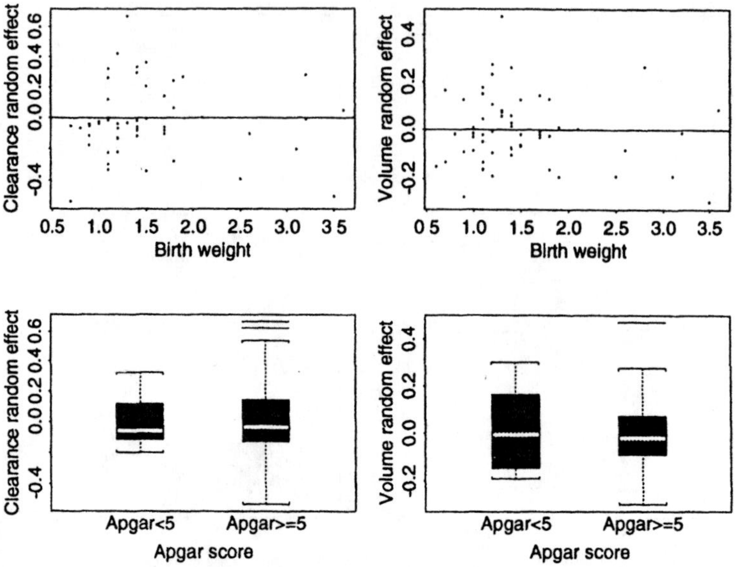

Figure 6.3. *Estimated random effects for log Cl and log V plotted against birth weight and Apgar score for the Lindstrom and Bates fit of Model 4*, phenobarbital data. The plots for birth weight have an ordinary least squares line superimposed; those for Apgar score are boxplots with the median represented by the white bar.*

Let $e_{ij} = \sigma g\{f(\boldsymbol{x}_{ij}, \boldsymbol{\beta}_{ij}), \boldsymbol{z}_{ij}, \boldsymbol{\theta}\}\epsilon_{ij}$, where the ϵ_{ij}, $j = 1, \ldots, n_i$, are independent of $\boldsymbol{b}_i$ with mean zero, variance 1, and $(n_i \times n_i)$ correlation matrix $\boldsymbol{\Gamma}_i(\boldsymbol{\alpha})$ for each i. Rewrite (6.32) in terms of ϵ_{ij}, replacing $\boldsymbol{\beta}_{ij}$ by $\boldsymbol{d}(\boldsymbol{a}_{ij}, \boldsymbol{\beta}, \boldsymbol{b}_i)$, and take a Taylor series about $\boldsymbol{b}_i = \boldsymbol{b}_i^*$. Retaining two terms in the expansion of f and one in that of $\sigma g\{f(\boldsymbol{x}_{ij}, \boldsymbol{\beta}_{ij}), \boldsymbol{z}_{ij}, \boldsymbol{\theta}\}\epsilon_{ij}$, yields

$$\begin{aligned} y_{ij} &\approx f\{\boldsymbol{x}_{ij}, \boldsymbol{d}(\boldsymbol{a}_{ij}, \boldsymbol{\beta}, \boldsymbol{b}_i^*)\} \\ &+ f'_{\beta}\{\boldsymbol{x}_{ij}, \boldsymbol{d}(\boldsymbol{a}_{ij}, \boldsymbol{\beta}, \boldsymbol{b}_i^*)\}\boldsymbol{\Delta}_{bij}(\boldsymbol{\beta}, \boldsymbol{b}_i^*)(\boldsymbol{b}_i - \boldsymbol{b}_i^*) \\ &+ \sigma g[f\{\boldsymbol{x}_{ij}, \boldsymbol{d}(\boldsymbol{a}_{ij}, \boldsymbol{\beta}, \boldsymbol{b}_i^*)\}, \boldsymbol{z}_{ij}, \boldsymbol{\theta}]\epsilon_{ij}, \end{aligned} \tag{6.34}$$

where $\boldsymbol{\Delta}_{bij}(\boldsymbol{\beta}, \boldsymbol{b}_i^*)$ is the $(p \times k)$ matrix of derivatives of $\boldsymbol{d}(\boldsymbol{a}_{ij}, \boldsymbol{\beta}, \boldsymbol{b}_i)$ with respect to $\boldsymbol{b}_i$ evaluated at $\boldsymbol{b}_i = \boldsymbol{b}_i^*$. Examination of (6.34) shows that the model for $\boldsymbol{y}_i$ may then be written as in (6.17) with the following redefinitions: $\boldsymbol{Z}_i(\boldsymbol{\beta}, \boldsymbol{b}_i^*)$ is the $(n_i \times k)$ matrix with jth row $f'_{\beta}\{x_{ij}, d(a_{ij}, \beta, b_i^*)\}\Delta_{bij}(\beta, b_i^*)$, and $e_i^* = R_i(\beta, b_i^*)\epsilon_i$, where $R_i(\beta, b_i^*) = \sigma^2 G_i^{1/2}(\beta, \theta, b_i^*)\Gamma_i(\alpha)G_i^{1/2}(\beta, \theta, b_i^*)$ and $G_i(\beta, \theta, b_i^*)$ is the n_i-dimensional diagonal matrix with jth diagonal element $g^2[f\{x_{ij}, d(a_{ij}, \beta, b_i^*)\}, z_{ij}, \theta]$. It follows that both the first-order ($b_i^* = 0$) and conditional first-order linearization methods may be adapted to the situation of time-dependent covariates in (6.32), (6.33); for example, the NONMEM package and the NLME suite of Splus functions discussed in section 6.4 readily accommodate this feature.

Individual empirical Bayes estimates for first-order methods

We did not discuss estimation of the individual random effects b_i in connection with the ML and GLS methods based on the first-order approximation, considered in section 6.2. The objective function (6.26) based on the posterior density may be minimized in b_i for any fixed values of β and ω. Thus, when estimation of these parameters is based on the first-order linearization, empirical Bayes estimates of the random effects commonly are found *ex post* by inserting the final estimates for β and ω into (6.26) and minimizing with respect to $b_i, i = 1, \ldots, m$. Estimates for the β_i may then be obtained.

Other approximations

Attempts to improve upon the first-order approximation by including additional terms in the Taylor series expansion of the hierarchical nonlinear model about $b_i = 0$ are discussed by Beal (1984a), Lindstrom and Birkes (1984), and Solomon and Cox (1992). In pharmacokinetic analysis, anecdotal evidence suggests there may be little advantage to offset the additional complexity introduced into the form of the approximate marginal mean and covariance model (6.3).

Inter-individual covariance models

Although our development has been in the context of a completely unstructured inter-individual covariance matrix D, all of the methods discussed may be adapted to the case where this matrix is taken to have a specified form. For example, it may be assumed that the k components of the random effects vectors b_i are uncorrelated, in which case D would be a diagonal matrix. If the number of random effects is large, allowing $\boldsymbol{D}$ to be completely unstructured may lead to computational difficulties.

Two-stage estimation

As discussed in section 5.3.4, for a general nonlinear model for $\boldsymbol{\beta}_i$ of the form (5.17),

$$\boldsymbol{\beta}_i = \boldsymbol{d}(\boldsymbol{a}_i, \boldsymbol{\beta}) + \boldsymbol{b}_i,$$

as long as the asymptotic theory for the individual estimates $\boldsymbol{\beta}_i^* | \boldsymbol{\beta}_i$ holds, one may specify an approximate hierarchical nonlinear model for the $\boldsymbol{\beta}_i^*$. In the notation of Chapter 5, letting $\boldsymbol{e}_i^*$ denote errors with mean $\mathbf{0}$ and covariance matrix $\boldsymbol{C}_i$, this may be written as

$$\boldsymbol{\beta}_i^* = \boldsymbol{d}(\boldsymbol{a}_i, \boldsymbol{\beta}) + \boldsymbol{b}_i + \boldsymbol{e}_i^*. \tag{6.35}$$

(6.35) is reminiscent of the first-order linearization model (6.3), with 'responses' $\boldsymbol{\beta}_i^*$. Thus, the procedures appropriate for estimation under (6.3) may be used to implement two-stage estimation as well. The software must allow the user to specify known intra-individual covariance matrices.

6.8 Bibliographic notes

An extensive literature exists on linearization of the hierarchical nonlinear model applied to the analysis of pharmacokinetic data; see Sheiner, Rosenberg, and Marathe (1977), Beal (1984a,b), Beal and Sheiner (1982, 1985, 1992), Sheiner and Beal (1980, 1981, 1983) and Sheiner and Ludden (1992). Recent work on linearization methods in the statistical literature includes Lindstrom and Bates (1990), Vonesh (1992), Vonesh and Carter (1992), and Davidian and Giltinan (1993a). Gumpertz and Pantula (1992) use a first-order linearization and GLS estimation for a nonlinear regression model with a specific variance component structure. Other related work includes Rotnitzky and Jewell (1990), Thall and Vail (1990), Waclawiw and Liang (1993), Wolfinger and O'Connell (1993), and Breslow and Clayton (1993); these last authors consider the special case of the hierarchy where $\boldsymbol{y}_i$ is univariate and the nonlinear model has the particular form of a generalized linear model, referred to as a generalized linear mixed model. They offer an extensive bibilography.

CHAPTER 7

Nonparametric and semiparametric inference

7.1 Introduction

Inferential methods based on individual estimates and linearization of the model, discussed in Chapters 5 and 6, respectively, are appropriate under the fully parametric model specification given in section 4.3.1. The assumption that the random effects $\boldsymbol{b}_i$ in a parametric model for the $\boldsymbol{\beta}_i$ arise from a specific parametric family of distributions, the normal, is fundamental to these techniques. Estimates of the $\boldsymbol{\beta}_i$ obtained using empirical Bayes methods may be quite nonrobust to nonnormality. If the distribution of the random parameters has heavier tails than the normal, or is skewed or multimodal, the 'shrinkage' toward the mean is likely to be misleading. The utility of graphical model-building strategies based on empirical Bayes estimates may thus be compromised. Moreover, as discussed in section 4.3.3, identification of misspecified systematic components of the model may be difficult under a restrictive parametric distributional assumption.

The *nonparametric* and *semiparametric* model specifications in sections 4.3.2 and 4.3.3 address these issues by providing a more flexible framework for inference. In this chapter, we discuss several approaches to inference under these model specifications. In section 7.2, we consider the nonparametric case. One approach to inference for the semiparametric model is described in section 7.3; other proposals are briefly addressed in section 7.4. Section 7.5 gives a review of available software. Semiparametric inference is illustrated using the phenobarbital data considered in Chapter 6 in section 7.6. The chapter concludes with a discussion of the practical limitations of these methods.

The mathematics underlying the theory of nonparametric and semiparametric inferential techniques are more difficult than those used elsewhere in this book. Consequently, the exposition in sections 7.2–7.4 requires a higher level of mathematical sophistication on the part of the reader. This is somewhat unavoidable because of the nature of the material. For the reader who wishes to skip the technical aspects but get a sense for the use of these methods, we suggest reading section 7.3 without dwelling on the mathematical details and proceeding to the example in section 7.6.

7.2 Nonparametric maximum likelihood (NPML)

In this section, we consider inference for the nonparametric model specification in section 4.3.2. For $i = 1, \dots, m$,

$$\boldsymbol{y}_i = \boldsymbol{f}_i(\boldsymbol{\beta}_i) + \boldsymbol{e}_i, \quad \boldsymbol{e}_i|\boldsymbol{\beta}_i \sim \Big(\boldsymbol{0}, \boldsymbol{R}_i(\boldsymbol{\beta}_i, \boldsymbol{\xi})\Big),$$
$$\boldsymbol{\beta}_i \sim H, \tag{7.1}$$

where H, the distribution of the random parameters, remains completely unrestricted. Our initial development is in the context of a model with no covariate information; incorporation of covariates is considered in section 7.2.3. This material originally appeared in Mallet (1986); our presentation closely follows that of Schumitzky (1993), who gives a very clear exposition.

7.2.1 *Basic ideas*

The conditional density of $\boldsymbol{y}_i$ given $\boldsymbol{\beta}_i$ for (7.1) will depend on the Stage 1 (intra-individual) model through the $\boldsymbol{\beta}_i$, the design vectors $\boldsymbol{x}_{i1}, \dots, \boldsymbol{x}_{in_i}$, and the within-individual covariance parameter $\boldsymbol{\xi}$. Write this as

$$p_{y|\beta}(\boldsymbol{y}_i|\boldsymbol{x}_{i1}, \dots, \boldsymbol{x}_{in_i}, \boldsymbol{\beta}_i, \boldsymbol{\xi}).$$

The marginal likelihood for the full data may then be written as the product of the individual marginals:

$$\prod_{i=1}^{m} \int p_{y|\beta}(\boldsymbol{y}_i|\boldsymbol{x}_{i1}, \dots, \boldsymbol{x}_{in_i}, \boldsymbol{\zeta}, \boldsymbol{\xi})\, dH(\boldsymbol{\zeta}). \tag{7.2}$$

Here, and throughout section 7.2, to avoid confusion, we use $\boldsymbol{\zeta}$ as function argument and dummy variable of integration for functions involving $\boldsymbol{\beta}_i$. Assume $\boldsymbol{\xi}$ is known; the implications of this are discussed in section 7.2.2. Under this condition, the only unknown quantity in the likelihood (7.2) is the distribution H. Examination of (7.2) shows that estimation of the distribution H is equivalent to the problem of estimating a mixing distribution. Several authors have considered maximum likelihood estimation of the mixing distribution (Laird, 1978; Lindsay, 1983). Application in the context of population pharmacokinetic modeling was first discussed by Mallet (1986). Here, we summarize the relevant results. Details may be found in Mallet (1986) and Schumitzky (1991a, 1993).

For simplicity, write

$$p_i(\zeta) = p_{y|\beta}(y_i | x_{i1}, \ldots, x_{in_i}, \zeta, \xi).$$

Then, highlighting the dependence on H, the likelihood in (7.2) may be written as

$$L(H) = \prod_{i=1}^{m} \int_B p_i(\zeta)\, dH(\zeta), \qquad (7.3)$$

where the support of the distribution H is assumed to be in a compact subset B of k-dimensional Euclidean space $\Re^k$. Let $\mathcal{H}$ be the class of all probability distributions on B. Then, any distribution $H^* \in \mathcal{H}$ for which $L(H^*) \geq L(H)$ for all $H \in \mathcal{H}$ is a maximum likelihood estimate of the (mixing) distribution. Let $\mathcal{H}^D$ denote the class of all distributions in $\mathcal{H}$ whose measures are discrete with support on at most m points. Any distribution H in $\mathcal{H}^D$ may thus be associated with a set of m points $\boldsymbol{Z} = \{\zeta_1, \ldots, \zeta_m\}$ in B and a set of m weights $\boldsymbol{w} = \{w_1, \ldots, w_m\}$, where $w_\ell \geq 0$ and $\Sigma w_\ell = 1$. That is, any H in $\mathcal{H}^D$ satisfies

$$H = \sum_{\ell=1}^{m} w_\ell \delta(\zeta_\ell) \qquad (7.4)$$

for some pair $(\boldsymbol{w}, \boldsymbol{Z})$, where $\delta(\zeta_\ell)$ is the Dirac delta function placing point mass at the single point ζ_ℓ. Writing $\phi = (\boldsymbol{w}, \boldsymbol{Z})$, any probability measure on $\mathcal{H}^D$ may be indexed by ϕ. We shall write H^ϕ for the distribution in $\mathcal{H}^D$ satisfying (7.4).

The main theoretical result pertaining to maximum likelihood estimation of H is as follows.

Result 1. The maximum likelihood estimate, H_{ML}, belongs to $\mathcal{H}^D$; that is, it is a discrete distribution with support on at most m points in B.

Despite the discrete nature of the solution implied by Result 1, optimization by direct methods would be challenging, because a discrete measure supported on m points in B still involves $(mk +$

$m-1$) free parameters. For example, in a data set with 100 subjects and two parameters per subject, the dimension of the problem would be 299.

Discreteness of the maximum likelihood estimate follows from a duality between the problem of estimating the mixing distribution H and a problem in optimal design theory. Here, we give a heuristic presentation of this duality. The connection is important for two reasons: it not only allows characterization of the maximum likelihood estimate, but by appealing to well-known algorithms for construction of D-optimal designs, it leads to a computational strategy for derivation of the nonparametric maximum likelihood estimate of H.

Following the derivation in Mallet (1986, section 3), recall the form of the likelihood given in (7.3):

$$L(H) = \prod_{i=1}^{m} \int_B p_i(\zeta)\, dH(\zeta).$$

Write $\boldsymbol{P}(\zeta)$ for the $(m \times m)$ matrix diag$[\sqrt{p_1(\zeta)}, \ldots, \sqrt{p_m(\zeta)}]$, and consider the linear regression model

$$\mathrm{E}(\boldsymbol{z}|\zeta) = \boldsymbol{P}(\zeta)\psi, \quad \mathrm{Cov}(\boldsymbol{z}|\zeta) = \boldsymbol{I}_m. \tag{7.5}$$

In the model specified by (7.5), $\boldsymbol{z}$ is a random $(m \times 1)$ vector of responses, and $\zeta \in B$ is the $(k \times 1)$ vector of regressors. We shall be concerned with the design aspects of this constructed regression model.

Specifically, suppose we wish to choose placement of the regressors ζ to give the D-optimal estimate of ψ; that is, to minimize the determinant of the covariance matrix of the estimate of ψ, or equivalently, maximize the determinant of the Fisher information matrix. The normalized design on B may be thought of as a probability measure H defined on B. For a design characterized by H, the Fisher information matrix is

$$\mathcal{I}(H) = \int_B \boldsymbol{P}'(\zeta)\boldsymbol{P}(\zeta)\, dH(\zeta).$$

By definition of $\boldsymbol{P}(\zeta)$, $\mathcal{I}(H)$ is diagonal with elements $\int p_i(\zeta)\, dH(\zeta)$, and thus the determinant is given by

$$|\mathcal{I}(H)| = \prod_{i=1}^{m} \int p_i(\zeta)\, dH(\zeta) = L(H),$$

the second equality coming from (7.3). Thus choosing the design measure, H, that is D-optimal for the regression problem in (7.5) is equivalent to choosing the probability measure that maximizes the likelihood in (7.3). This equivalence in turn allows us to invoke the Kiefer–Wolfowitz theorem on the equivalence of D-optimality and G-optimality (defined below) (Kiefer and Wolfowitz, 1960; Fedorov, 1972; Silvey, 1980). We state without proof the following results from optimal design theory:

Result 2. For a diagonal regression matrix $\boldsymbol{P}(\zeta)$, the information matrix $\mathcal{I}(H^*)$ for a D-optimal design H^* may be represented in the form

$$\mathcal{I}(H^*) = \sum_{\ell=1}^{m} w_\ell \boldsymbol{P}'(\zeta_\ell)\boldsymbol{P}(\zeta_\ell), \tag{7.6}$$

where $\zeta_\ell \in B$, $0 < w_\ell \leq 1$, and $\sum_{\ell=1}^{m} w_\ell = 1$ for some pair $\boldsymbol{w} = \{w_1, \ldots, w_m\}$, $\boldsymbol{Z} = \{\zeta_1, \ldots, \zeta_m\}$.

This implies Result 1 above; the upper bound on the determinant of the information matrix is attained with discrete designs involving at most m points, because the right-hand side of (7.6) is the information matrix associated with the design matrix $\sum_{\ell=1}^{m} w_\ell \delta(\zeta_\ell)$ indexed by $\boldsymbol{\phi} = (\boldsymbol{w}, \boldsymbol{Z})$. For $\zeta \in B$ and a design measure H, define

$$\Delta(\zeta, H) = \text{trace}\{\boldsymbol{P}(\zeta)\mathcal{I}^{-1}(H)\boldsymbol{P}'(\zeta)\} = \sum_{i=1}^{m} p_i(\zeta)/L_i(H),$$

where $L_i(H) = \int_B p_i(\zeta)\, dH(\zeta)$. Then, the Kiefer–Wolfowitz theorem may be stated as follows.

Result 3. The following statements are equivalent:

(i) The design measure H^* maximizes $L(H)$ (D-optimality).

(ii) The design measure H^* minimizes

$$\max_{\zeta \in B} \sum_{i=1}^{m} p_i(\zeta)/L_i(H) \quad \text{(G-optimality)}.$$

(iii) H^* satisfies

$$\max_{\zeta \in B} \sum_{i=1}^{m} p_i(\zeta)/L_i(H^*) = m.$$

The importance of this result is that it not only characterizes the D-optimal design measure (and hence the maximum likelihood estimate of H), it also provides a starting point for algorithms to construct the optimal design.

7.2.2 Algorithms for finding the maximum likelihood estimate

Here, we sketch the main features of two methods for finding the nonparametric maximum likelihood estimate. The reader who wishes to implement these techniques should consult the original references for a more detailed presentation.

Design-based algorithms

One approach to the optimization problem, which exploits the correspondence with optimal design, is the following sequential algorithm of Fedorov (1972), often referred to as the Basic Algorithm. At the cth iteration:

1. Let $H^{(c)} = \sum_{\ell=1}^{J(c)} w_\ell^{(c)} \delta(\zeta_\ell^{(c)})$, where $J(c)$, the number of support points at the current iteration, may be different from m.
2. If $\max_\zeta \Delta(\zeta, H^{(c)}) = m$, then $H^{(c)} = H_{ML}$, stop.
3. Let $\zeta^{\dagger(c)} = \arg\max\{\Delta(\zeta, H^{(c)}) : \zeta \in B\}$, where arg max is the value of the argument that maximizes the expression in braces, and define $H_s^{(c)} = (1-s)H^{(c)} + s\delta(\zeta^{\dagger(c)})$.
4. Let $s^{(c)} = \arg\max\{L(H_s^{(c)}) : s \in [0,1]\}$, and define $H^{(c+1)} = H_{s^{(c)}}$.

It may be shown (Fedorov, 1972) that iterating steps 1–4 generates a sequence of measures that increases the likelihood at each iteration and that converges to the maximum likelihood estimate H_{ML} as $c \to \infty$. The number of support points with nonzero weight w_ℓ may exceed m at a particular iteration; in the limit, however, the number of nonzero weights will be less than or equal to m. Silvey (1980) suggests starting this algorithm with a design measure $H^{(0)}$ placing equal probability at each of a finite number (less than m) of linearly independent vectors in B.

Refinement of the Basic Algorithm is suggested by Mallet (1986); specifically, he advocates restricting the number of support points to be less than or equal to m at each iteration, while still increasing the likelihood.

Nonparametric EM algorithms

An alternative approach to finding the maximum likelihood estimate, based on the EM algorithm, is described by Schumitzky

(1991a). Again, let $H^{\phi} \in \mathcal{H}^D$ be the measure $\sum_{\ell=1}^{m} w_\ell \delta(\zeta_\ell)$ indexed by $\phi = (w, Z)$, $Z = \{\zeta_1, \ldots, \zeta_m\}$ in B, and a set of m weights $w = \{w_1, \ldots, w_m\}$. Schumitzky (1993) suggests the following iterative scheme. At iteration c:

1. Let $H^{(c)} = \sum_{\ell=1}^{J(c)} w_\ell^{(c)} \delta(\zeta_\ell^{(c)})$, where $J(c)$, the number of support points at the current iteration, may be different from m.
2. Define $H^{(c+1)}$ as the distribution indexed by $(w^{(c+1)}, Z^{(c+1)})$, where, for $\ell = 1, \ldots, J(c)$,

$$w_\ell^{(c+1)} = \frac{1}{m} \sum_{i=1}^{m} w_\ell^{(c)} \frac{p_i(\zeta_\ell^{(c)})}{L_i(H^{(c)})},$$

$$\zeta_\ell^{(c+1)} = \arg\max \left\{ \sum_{i=1}^{m} \log p_i(\zeta) \frac{p_i(\zeta_\ell^{(c)}) w_\ell^{(c)}}{L_i(H^{(c)})} : \zeta \in B \right\},$$

and $Z^{(c+1)}) = \{\zeta_1^{(c)}, \ldots, \zeta_m^{(c)}\}$.

It may be shown (Schumitzky, 1993) that the transformation defined in step 2 above mapping $H^{(c)}$ to $H^{(c+1)}$ improves the likelihood at each iteration, which suggests iterating steps 1 and 2 to convergence to obtain the maximum likelihood estimate H_{ML}.

Both design-based and EM-type algorithms converge to a discrete solution by construction. This is advantageous in some contexts; for instance, a discrete distribution is in the form required of the prior distribution for modeling stochastic control of pharmacokinetic systems (Schumitzky, 1991b). In general, however, the actual distribution of pharmacokinetic parameters is likely to be smooth. Thus, it is fairly standard to 'smooth' the discrete estimate, for instance, by convolution with a normal kernel function (Mallet, 1986; Schumitzky, 1993). Mallet (1992) suggests adaption of the basic estimation procedure to impose a smoothness requirement, which restricts the distribution H to belong to the class $H * K$, where H is an arbitrary distribution, $*$ denotes convolution, and K is a symmetric kernel. An alternative method of smooth nonparametric density estimation is discussed in section 7.3.

The preceding development assumes that the intra-individual covariance parameters, $\boldsymbol{\xi}$, are known. This allows one to focus on H as the only unknown in the estimation problem. In practice, the assumption may be restrictive; to implement the algorithms as described above, some estimate of $\boldsymbol{\xi}$ is needed. Mallet *et al.* (1988b) suggest that such an estimate may be obtained in a separate, preliminary analysis. For example, the intra-individual covariance structure might be estimated based on data from those individuals for whom relatively complete profile information is available. A similar approach is advocated by Jelliffe, Gomis and Schumitzky (1990), who suggest using historical assay control data to derive the intra-subject error pattern in the context of pharmacokinetic analysis. If an estimate of $\boldsymbol{\xi}$ is not available, the approach described above must be extended. Mallet *et al.* (1988b) suggest modification of the likelihood; a consequence is that a separate maximization is required to estimate unknown fixed parameters, resulting in an increase in computation time.

A limitation of the NPML method, noted by Mallet *et al.* (1988b), is that no estimate of the precision of the distribution estimate is available. They suggest that care be taken in interpretation of results, particularly with small samples.

7.2.3 Incorporation of covariates

Extension of the estimation scheme described in sections 7.2.1 and 7.2.2 to accommodate individual-specific covariates does not involve a parametric specification for the $\boldsymbol{\beta}_i$. Here, we outline a nonparametric approach described by Mallet *et al.* (1988a), Mallet (1992), and Mentré and Mallet (1994). In implementing this technique it is necessary to distinguish three cases: (i) continuous covariates measured with error (e.g. creatinine clearance or α_1-acid glycoprotein concentration), (ii) continuous covariates measured without error (e.g. age), and (iii) categorical covariates (e.g. race or sex). The ultimate goal is to characterize the conditional distribution of the regression parameters given observed covariate values.

For continuous covariates measured with error, case (i), let $\boldsymbol{a}_i$ denote the observed covariate vector for individual i. Mallet *et al.* (1988a) assume that $\boldsymbol{a}_i$ is equal to some true covariate value $\boldsymbol{a}_i^*$, contaminated by measurement error, where the $\boldsymbol{a}_i^*$ arise from a distribution of such values in the population of individuals, jointly with the $\boldsymbol{\beta}_i$. Suppressing dependence on $\boldsymbol{x}_{i1}, \ldots, \boldsymbol{x}_{in_i}$ and $\boldsymbol{\xi}$ and writing $p_{y,a|\beta,a^*}(\boldsymbol{y}_i, \boldsymbol{a}_i|\boldsymbol{\beta}_i, \boldsymbol{a}_i^*)$ to denote the joint density of $(\boldsymbol{y}_i, \boldsymbol{a}_i)$ given the random regression parameters and true covariate values, the marginal likelihood of the observed m data pairs $(\boldsymbol{y}_i, \boldsymbol{a}_i)$ is given by

$$\prod_{i=1}^{m} \int p_{y,a|\beta,a^*}(\boldsymbol{y}_i, \boldsymbol{a}_i|\boldsymbol{\zeta}, \boldsymbol{a}^*)\, dM(\boldsymbol{\zeta}, \boldsymbol{a}^*),$$

where M is the joint distribution of $(\boldsymbol{\beta}_i, \boldsymbol{a}_i^*)$. In obvious notation, $p_{y,a|\beta,a^*}(\boldsymbol{y}_i, \boldsymbol{a}_i|\boldsymbol{\beta}_i, \boldsymbol{a}_i^*) = p_{y|\beta,a^*,a}(\boldsymbol{y}_i|\boldsymbol{\beta}_i, \boldsymbol{a}_i^*, \boldsymbol{a}_i) \times p_{a|\beta,a^*}(\boldsymbol{a}_i|\boldsymbol{\beta}_i, \boldsymbol{a}_i^*)$. Assuming that knowledge of $\boldsymbol{a}_i$ and $\boldsymbol{a}_i^*$ contributes no additional information on $\boldsymbol{y}_i$ beyond that in $\boldsymbol{\beta}_i$, and similarly that the measurement error does not depend on $\boldsymbol{\beta}_i$, the likelihood may be rewritten as

$$\prod_{i=1}^{m} \int p_{y|\beta}(\boldsymbol{y}_i|\boldsymbol{x}_{i1}, \ldots, \boldsymbol{x}_{in_i}, \boldsymbol{\zeta}, \boldsymbol{\xi}) q_i(\boldsymbol{a}_i|\boldsymbol{a}^*)\, dM(\boldsymbol{\zeta}, \boldsymbol{a}^*), \qquad (7.7)$$

where $q_i(\boldsymbol{a}_i|\boldsymbol{a}^*)$ is the density of $\boldsymbol{a}_i$ given $\boldsymbol{a}_i^*$. A multivariate normal density with diagonal covariance matrix has been used for q_i (Mallet, 1992), where the diagonal elements corresponding to the measurement error variances of each component of $\boldsymbol{a}_i$ are known, as is $\boldsymbol{\xi}$. The likelihood (7.7) is of the form considered previously; thus, Mallet *et al.* (1988a) consider nonparametric maximum likelihood estimation of M. By Results 1–3 of section 7.2.1, the nonparametric maximum likelihood estimate of the joint distribution M is discrete and may be obtained by implementation of the algorithms given in section 7.2.2; see Mallet *et al.* (1988a).

Once the estimate of M has been obtained, predictive inference given a new observed covariate value, $\boldsymbol{a}$, can be based on the derived conditional distribution of $\boldsymbol{\beta}_i$ given $\boldsymbol{a}$. Practical implementation of the method presupposes knowledge of the measurement error variance for each covariate, which may need to be derived from external information (Mallet, 1992).

The inclusion of covariates measured without error, case (ii), may be accomplished by setting the relevant measurement error variances to zero. Incorporation of discrete covariate information, case (iii), is not entirely straightforward; see Mallet (1992) for one possible approach.

A potential drawback of incorporation of covariates as above is the need to estimate a potentially high-dimensional distribution, the joint distribution of the random parameters and covariates, with the limited information likely to be available in a typical data set. An alternative method for inclusion of time-dependent covariate information has been used by Jelliffe *et al.* (1990). In their approach, $\boldsymbol{a}_i$ is treated as fixed, a parametric model for $\boldsymbol{\beta}_i$ is substituted directly into the regression function $\boldsymbol{f}(\boldsymbol{\beta}_i)$ from the first stage, and the parameters in the resulting model assume the role of the random parameters. This is best illustrated by example. Consider a simple linear model for scalar β_i as a function of single time-varying covariate a_{ij}, $\beta_i = \beta^*_{1i} + \beta^*_{2i}a_{ij}$, where the intercept and slope β^*_{1i} and β^*_{2i} vary across individuals. The distribution of the parameter $\boldsymbol{\beta}^*_i = [\beta^*_{1i}, \beta^*_{2i}]'$ would then be estimated nonparametrically.

7.3 Smooth nonparametric maximum likelihood (SNP)

In this section, we focus on a particular inferential method for the semiparametric model specification in section 4.3.3, the seminonparametric (SNP) procedure proposed by Davidian and Gallant (1993), referred to as *smooth nonparametric maximum likelihood* in the pharmacokinetics literature (Davidian and Gallant, 1992). Under this specification, the model is

$$\boldsymbol{y}_i = \boldsymbol{f}_i(\boldsymbol{\beta}_i) + \boldsymbol{e}_i, \quad \boldsymbol{e}_i|\boldsymbol{b}_i \sim \Big(\boldsymbol{0}, \boldsymbol{R}_i(\boldsymbol{\beta}_i, \boldsymbol{\xi})\Big),$$

$$\boldsymbol{\beta}_i = \boldsymbol{d}(\boldsymbol{a}_i, \boldsymbol{\beta}, \boldsymbol{b}_i); \quad \boldsymbol{b}_i \sim h, \;\; h \in \mathcal{H}, \tag{7.8}$$

where h is a density belonging to a class $\mathcal{H}$ of 'smooth' densities. This model was first applied in the context of pharmacokinetic analysis; because pharmacokinetic parameters are likely to vary smoothly and continuously in the population, it makes sense to restrict attention to distributional models for the random component with regularly-behaved densities. In contrast to the fully nonparametric specification of the previous section, this model allows explicit parametric modeling of the $\boldsymbol{\beta}_i$ through the function $\boldsymbol{d}$, so that the effect of potential covariates is accommodated as in the fully parametric case. There is thus no need to address incorporation of covariates separately. The random component of variation in $\boldsymbol{\beta}_i$ is accounted for by $\boldsymbol{b}_i$, and it is the distribution of $\boldsymbol{b}_i$ that is estimated directly rather than the implied distribution of $\boldsymbol{\beta}_i$.

7.3.1 Estimation

From the arguments in section 4.4, under the semiparametric model specification (7.8), writing the density of the conditional distribution of $\boldsymbol{y}_i$ given $\boldsymbol{b}_i$ as $p_{y|b}(\boldsymbol{y}_i|\boldsymbol{x}_{i1}, \ldots, \boldsymbol{x}_{in_i}, \boldsymbol{a}_i, \boldsymbol{\beta}, \boldsymbol{\xi}, \boldsymbol{b}_i)$, the likelihood for the full data may be written as the product of the marginal densities:

$$\prod_{i=1}^{m} \int p_{y|b}(\boldsymbol{y}_i|\boldsymbol{x}_{i1}, \ldots, \boldsymbol{x}_{in_i}, \boldsymbol{a}_i, \boldsymbol{\beta}, \boldsymbol{\xi}, \boldsymbol{b})\, h(\boldsymbol{b})\, d\boldsymbol{b}.$$

In the SNP approach, the likelihood is maximized simultaneously in the fixed parameters $\boldsymbol{\beta}$ and $\boldsymbol{\xi}$ and the k-variate random effects density h; equivalently, the estimates minimize in $\boldsymbol{\beta}$, $\boldsymbol{\xi}$, and h

$$L_{SNP}(\boldsymbol{\beta},\boldsymbol{\xi},h) = -\sum_{i=1}^{m} \log \int p_{y|b}(\boldsymbol{y}_i|\boldsymbol{x}_{i1},\ldots,\boldsymbol{x}_{in_i},\boldsymbol{a}_i,\boldsymbol{\beta},\boldsymbol{\xi},\boldsymbol{b})\, h(\boldsymbol{b})\, d\boldsymbol{b}, \quad (7.9)$$

where the estimate of h minimizes $L_{SNP}(\boldsymbol{\beta},\boldsymbol{\xi},h)$ over a particular class of densities $\mathcal{H}$.

The SNP approach differs from both procedures based on individual estimates and linearization in two key respects: the distribution of the random effects is not restricted to belong to a parametric family such as the normal, but rather is estimated in its entirety, and no analytic approximation is made to avoid computation of the integral in (7.9).

Estimation of h

The class $\mathcal{H}$ underlying the SNP procedure was proposed originally in the econometrics literature by Gallant and Nychka (1987), who give a full mathematical description of $\mathcal{H}$. Here, we confine our attention to practical aspects. The most important requirement is that a density in $\mathcal{H}$ must satisfy the smoothness restriction that it be at least $k/2$ times differentiable. A consequence of this restriction is that densities in $\mathcal{H}$ may not exhibit unusual behavior such as kinks, jumps, or oscillation; however, densities in $\mathcal{H}$ may be skewed, multi-modal, and fat-tailed or thin-tailed relative to the k-variate normal density, and the class contains the normal. Thus, the assumption that the true density belongs to $\mathcal{H}$ allows for the possibility of a wide range of behavior, but rules out densities with unusual features that are unlikely to be associated with biological phenomena such as pharmacokinetic parameters.

Under the smoothness restriction and other regularity conditions, Gallant and Nychka (1987) show that a density in the class $\mathcal{H}$ may be represented for practical applications as a series expansion. Specifically, a density in $\mathcal{H}$ may be expressed as the square of a polynomial of infinite degree times a k-variate normal density, so that $h \in \mathcal{H}$ may be written as

$$h(\boldsymbol{b}) = \{P_\infty(\boldsymbol{L}^{-1}\boldsymbol{b})\}^2 n_k(\boldsymbol{b}|\boldsymbol{0},\boldsymbol{L}\boldsymbol{L}'), \quad (7.10)$$

where $n_k(\boldsymbol{b}|\boldsymbol{\mu},\boldsymbol{\Sigma})$ is the k-variate normal density with mean $\boldsymbol{\mu}$ and covariance $\boldsymbol{\Sigma}$, $\boldsymbol{L}$ is a $(k \times k)$ upper triangular matrix, and $P_\infty(\boldsymbol{z})$

is the infinite sum of terms consisting of unknown constants times all nonnegative powers and cross-products of the k components of $\boldsymbol{z}$. For example, if $k = 2$, so that $\boldsymbol{z} = [z_1, z_2]'$,

$$\begin{aligned} P_\infty(\boldsymbol{z}) &= a_{00} + a_{10}z_1 + a_{01}z_2 + a_{20}z_1^2 + a_{11}z_1z_2 + a_{02}z_2^2 \\ &\quad + a_{30}z_1^3 + a_{21}z_1^2z_2 + a_{12}z_1z_2^2 + a_{03}z_2^3 \\ &\quad + a_{40}z_1^4 + a_{31}z_1^3z_2 + a_{22}z_1^2z_2^2 + a_{13}z_1z_2^3 + a_{04}z_2^4 + \cdots \end{aligned}$$

where $(a_{00}, a_{10}, a_{01}, \ldots)$ are the coefficients of the infinite polynomial.

Because it is an infinite series, the representation (7.10) is impractical for applications, but it suggests a convenient approximation to a density in $\mathcal{H}$. This approximation is simply an appropriately scaled truncation of the infinite expansion obtained by retaining only a finite number of leading terms of $P_\infty(\boldsymbol{z})$. For some nonnegative integer K, these would be the nonnegative integer powers and cross-products of the components of $\boldsymbol{z}$ up to and including order K. Let $P_K(\boldsymbol{z})$ be a polynomial of degree K in the components of $\boldsymbol{z}$; that is, the sum of all powers and cross-products of the components of $\boldsymbol{z}$ up to and including degree K. Writing the resulting approximation to a density $h \in \mathcal{H}$ as $h_K(\boldsymbol{b})$, we have

$$h_K(\boldsymbol{b}) = \frac{\{P_K(\boldsymbol{L}^{-1}\boldsymbol{b})\}^2 n_k(\boldsymbol{b}|\boldsymbol{0}, \boldsymbol{L}\boldsymbol{L}')}{\int P_K(\boldsymbol{z}) n_k(\boldsymbol{z}|\boldsymbol{0}, \boldsymbol{I}_k)\, d\boldsymbol{z}}. \tag{7.11}$$

For example, if $k = 2$ and $K = 2$, then

$$P_2(\boldsymbol{z}) = a_{00} + a_{10}z_1 + a_{01}z_2 + a_{20}z_1^2 + a_{11}z_1z_2 + a_{02}z_2^2.$$

Division by the scaling factor in the denominator ensures that the truncated approximation is a density; that is, $h_K(\boldsymbol{b})$ integrates to one. Technically, the coefficients of $P_K(\boldsymbol{z})$ can only be determined to within a scalar multiple; thus, to achieve a unique representation, the constant term a_{00} is set to one in computations ($a_{00} \equiv 1$). Note that for $K = 0$, $P_0(\boldsymbol{z})$ is equal to one for all $\boldsymbol{z}$ and thus $h_0(\boldsymbol{b})$ is a normal density.

For given K, let $\boldsymbol{\eta}_{(1)}$ be the vector whose elements are the coefficients $(a_{00}, a_{10}, a_{01}, \ldots)$ of P_K, let $\boldsymbol{\eta}_{(2)}$ be the vector whose elements are the distinct elements of $\boldsymbol{L}$, and let $\boldsymbol{\eta} = [\boldsymbol{\eta}_{(1)}', \boldsymbol{\eta}_{(2)}']'$ be the Q-dimensional vector of all parameters in h_K. Q is determined solely by K and k; in the above example with $K = 2$ and $k = 2$, $Q = 9$. Note that the vector $\boldsymbol{\eta}$ thus describes completely the truncated approximation h_K; we emphasize this by writing $h_K(\boldsymbol{b}|\boldsymbol{\eta})$ where it is relevant to highlight this dependence. Moreover, for a particular value for $\boldsymbol{\eta}$, the denominator of h_K is simply a weighted sum of products of moments of the standard normal distribution; for example, for $K = 2$, the denominator is a weighted sum of moments to the fourth order of the standard normal distribution. Thus, for a particular K and $\boldsymbol{\eta}$, $h_K(\boldsymbol{b}|\boldsymbol{\eta})$ may be calculated easily at any value of the argument $\boldsymbol{b}$.

Densities having the form of the infinite expansion (7.10) exhibit a wide range of regular behavior, as described above. Even for choices of K as small as one or two, the form of the truncated approximation (7.11) is sufficiently rich to capture this behavior, reflecting features such as skewness, fat- or thin-tails, and multiple modes.

This suggests a natural approach to estimation of h: approximate h by h_K, estimate $\boldsymbol{\eta}$ from available data by $\hat{\boldsymbol{\eta}}$, say, and consequently estimate $h(\boldsymbol{b})$ by $\hat{h}_K(\boldsymbol{b}) = h_K(\boldsymbol{b}|\hat{\boldsymbol{\eta}})$. The density estimator $\hat{h}_K(\boldsymbol{b})$ has been studied extensively in the econometrics literature (e.g. Gallant and Tauchen, 1989). Because $\hat{h}_K(\boldsymbol{b})$ is determined completely by a finite number of estimated parameters, it may not be apparent that this estimator for h should be classified as nonparametric. This designation is appropriate, however, because the form of the estimator does not correspond to any particular distributional family, such as the normal, lognormal, or t, and is capable of tracking features of the true underlying density in the same way as other nonparametric estimators such as kernel estimators (e.g. Silverman, 1986). K acts as a tuning parameter to control the degree of smoothness in a manner similar to the bandwidth used with kernel density estimators; choice of K is discussed shortly. Consequently, the estimator has been referred to as 'SemiNonParametric' to highlight both its construction using parametric estimation and its nonparametric properties. We prefer that the acronym SNP be interpreted as 'Smooth NonParametric,' given the basis on a class of smooth densities.

Estimation for the hierarchical model

Davidian and Gallant (1992, 1993) proposed use of this approach to density estimation in the context of the semiparametric hierarchical nonlinear model specification. In the objective function $L_{SNP}(\boldsymbol{\beta}, \boldsymbol{\xi}, h)$ in (7.9), approximate h by the truncated expansion $h_K(\boldsymbol{b}|\boldsymbol{\eta})$ given in (7.11). Thus, consider the objective function

$$L_{SNP,K}(\boldsymbol{\beta}, \boldsymbol{\xi}, \boldsymbol{\eta}) = \tag{7.12}$$
$$-\sum_{i=1}^{m} \log \int p_{y|b}(\boldsymbol{y}_i|\boldsymbol{x}_{i1}, \ldots, \boldsymbol{x}_{in_i}, \boldsymbol{a}_i, \boldsymbol{\beta}, \boldsymbol{\xi}, \boldsymbol{b})\, h_K(\boldsymbol{b}|\boldsymbol{\eta})\, d\boldsymbol{b}.$$

Note that the arguments of $L_{SNP,K}$ in (7.12) are the parameters $\boldsymbol{\beta}$, $\boldsymbol{\xi}$, and $\boldsymbol{\eta}$, because minimization in h_K corresponds to minimization with respect to $\boldsymbol{\eta}$. Thus, minimization of the objective function $L_{SNP,K}$ is a standard finite-dimensional optimization problem.

In the SNP approach, then, one minimizes $L_{SNP,K}(\boldsymbol{\beta}, \boldsymbol{\xi}, \boldsymbol{\eta})$ for a given choice of K to obtain estimates $\hat{\boldsymbol{\beta}}_{SNP,K}$, $\hat{\boldsymbol{\xi}}_{SNP,K}$, and $\hat{\boldsymbol{\eta}}_{SNP,K}$. The estimate of $h(\boldsymbol{b})$ is then $\hat{h}_{SNP,K}(\boldsymbol{b}) = h_K(\boldsymbol{b}|\hat{\boldsymbol{\eta}}_{SNP,K})$.

The construction of the SNP estimator for h does not require the assumption that $\mathrm{E}(\boldsymbol{b}_i) = \boldsymbol{0}$. This restriction may be imposed in computations if desired; in this case, the estimators with $K = 0$ and $K = 1$ are the same. See Davidian and Gallant (1992, 1993) for further details.

Once the estimate of h has been obtained, moments of h, in particular its covariance matrix $\boldsymbol{D}$, are easily estimated. For illustration, consider estimation of $\boldsymbol{D}$ under the restriction that $\mathrm{E}(\boldsymbol{b}_i) = \boldsymbol{0}$. The estimate of $\boldsymbol{D}$ is then given by the integral

$$\hat{\boldsymbol{D}}_{SNP,K} = \int (\boldsymbol{b}\boldsymbol{b}')\, \hat{h}_{SNP,K}(\boldsymbol{b})\, d\boldsymbol{b}. \tag{7.13}$$

From the form of h_K, this expression may be rewritten as the ratio of two integrals, each of which is an integral of a polynomial times a normal density. Both integrals are then easily obtained by using standard formulæ, as weighted sums of products of the moments of the standard normal distribution.

Other features of the population that may not be expressed in terms of integrals of polynomials may also be calculated using Monte Carlo integration, by drawing a large sample of size M from the estimated density $\hat{h}_{SNP,K}(\boldsymbol{b})$; designate this sample as $\boldsymbol{b}_m^*, m = 1, \ldots, M$. An estimate of any function $\psi(\boldsymbol{b})$ of $\boldsymbol{b}$ may then be obtained as the average of the M values $\psi(\boldsymbol{b}_m^*)$. Similarly, a percentile is computed as the corresponding percentile of the M $\boldsymbol{b}_m^*$. By taking the number of Monte Carlo drawings M large enough, these estimates can be made as accurate as desired; suitable values for M are likely to be at least 1000.

Using the estimates $\hat{\boldsymbol{\beta}}_{SNP,K}$, $\hat{\boldsymbol{\xi}}_{SNP,K}$, and $\hat{h}_{SNP,K}(\boldsymbol{b})$, empirical Bayes estimates of the random effects are obtained by maximizing the posterior mode of $\boldsymbol{b}_i$; thus, the empirical Bayes estimates $\hat{\boldsymbol{b}}_{i,SNP}$ maximize in $\boldsymbol{b}$

$$p_{y|b}(\boldsymbol{y}_i|\boldsymbol{x}_{i1},\ldots,\boldsymbol{x}_{in_i},\boldsymbol{a}_i,\hat{\boldsymbol{\beta}}_{SNP},\hat{\boldsymbol{\xi}}_{SNP},\boldsymbol{b})\hat{h}_{SNP,K}(\boldsymbol{b}). \quad (7.14)$$

From these, empirical Bayes estimates of the individual random parameters $\boldsymbol{\beta}$ may then be estimated by

$$\hat{\boldsymbol{\beta}}_{i,SNP} = \boldsymbol{d}(\boldsymbol{a}_i,\hat{\boldsymbol{\beta}}_{SNP},\hat{\boldsymbol{b}}_{i,SNP}). \quad (7.15)$$

One may estimate the population density of the $\boldsymbol{\beta}_i$ by appropriate transformation of the estimate of h.

Computation of the integral

Minimization of the objective function $L_{SNP,K}$ requires a way of computing integrals of the form

$$\int \psi(\boldsymbol{b})h_K(\boldsymbol{b})\,d\boldsymbol{b}. \quad (7.16)$$

One approach to this problem is by Monte Carlo integration, as described above. If minimization is accomplished through, for example, the Newton–Raphson procedure, this would entail drawing and averaging a Monte Carlo sample of size M at each internal iteration of the algorithm, which may lead to prohibitively excessive computation times.

Alternatively, one may appeal to standard approaches to numerical integration. The integral (7.16) is a ratio, with numerator given by $\int \psi(\boldsymbol{b})\{P_K(\boldsymbol{L}^{-1}\boldsymbol{b})\}^2 n_k(\boldsymbol{b}|\boldsymbol{0},\boldsymbol{LL}')\,d\boldsymbol{b}$ and denominator $\int P_K(\boldsymbol{z})n_k(\boldsymbol{z}|\boldsymbol{0},\boldsymbol{I}_k)\,d\boldsymbol{z}$. The denominator is easily computed as noted above. The numerator, upon the change of variables $\boldsymbol{t} = \boldsymbol{L}^{-1}\boldsymbol{b}\sqrt{2}$, may be written as

$$\int\cdots\int \psi(\sqrt{2}\boldsymbol{L}\boldsymbol{b})P_K(\sqrt{2}\boldsymbol{b})\pi^{-k/2}\exp(-t_1^2)\cdots\exp(-t_k^2)\,dt_1\cdots dt_k.$$

Gauss–Hermite quadrature, a numerical technique of evaluating integrals of the form $\int \psi^*(t)\exp(-t^2)\,dt$, may then be applied to perform the integration. Use of this approach may reduce computation times considerably relative to Monte Carlo integration without loss of accuracy. Further details may be found in Davidian and Gallant (1993).

7.3.2 Confidence intervals and hypothesis testing

Standard inferential procedures

It is possible to use standard maximum likelihood theory, allowing the number of individuals m to become large, to construct standard errors and confidence intervals for the elements of $\boldsymbol{\beta}$ and $\boldsymbol{\xi}$; uncertainty due to estimation of the unknown density h is automatically taken into account (e.g. Eastwood, 1991). The covariance matrix for all estimated parameters may be estimated by the inverse of the final value of the Hessian from the fit, and standard errors and confidence intervals may be obtained in the usual fashion.

To carry out inference on features of the random effects density h, the same approach may be employed, because $h_K(\boldsymbol{b}|\boldsymbol{\eta})$ is a function of $\boldsymbol{\eta}$. For example, standard errors for $\hat{\boldsymbol{\omega}}_{SNP,K}$, the vector of distinct elements of $\hat{\boldsymbol{D}}_{SNP,K}$, may be obtained from the estimated covariance matrix for all parameters in the usual way by noting that (7.13) is simply a function of $\hat{\boldsymbol{\eta}}$.

As discussed in connection with the two-stage and linearization techniques, for a d-dimensional vector-valued function $\boldsymbol{a}(\boldsymbol{\beta})$ of $\boldsymbol{\beta}$, letting $\boldsymbol{A}$ be the $(d \times p)$ matrix of derivatives of $\boldsymbol{a}(\boldsymbol{\beta})$, a test of the hypotheses

$$H_0 : \boldsymbol{a}(\boldsymbol{\beta}) = \mathbf{0} \text{ vs. } H_1 : \boldsymbol{a}(\boldsymbol{\beta}) \neq \mathbf{0}, \tag{7.17}$$

may be based on the test statistic

$$T^2 = \boldsymbol{a}'(\hat{\boldsymbol{\beta}}_{SNP})(\hat{\boldsymbol{A}}\hat{\boldsymbol{\Sigma}}_{SNP}\hat{\boldsymbol{A}}')^{-1}\boldsymbol{a}(\hat{\boldsymbol{\beta}}_{SNP}), \tag{7.18}$$

where $\hat{\boldsymbol{A}}$ is the matrix $\boldsymbol{A}$ with $\boldsymbol{\beta}$ replaced by $\hat{\boldsymbol{\beta}}_{ANP}$ and $\hat{\boldsymbol{\Sigma}}_{SNP}$ is the estimated asymptotic covariance matrix for $\hat{\boldsymbol{\beta}}_{SNP}$. The test is conducted by comparing T^2 to the appropriate χ^2_d critical value.

For model selection, one may invoke standard likelihood ratio procedures for selection of an appropriate model $\boldsymbol{d}$ within a series of nested models; see section 6.2.2. This approach tends to lead to overparameterized models in the context of SNP estimation.

Alternatively, to compare two nested or non-nested models, several standard statistical model selection criteria are favored in connection with the SNP method. For given K, and two models A and B with parameter dimensions Q_A and Q_B, respectively, these are based on a a penalized objective function; the rationale for model selection criteria was given in section 6.2.2. For model A, say, the Schwarz or Bayesian Information Criterion (BIC) is, in obvious notation,

$$BIC_A = \hat{L}_{SNP,K,A} + (1/2)\log NP_{net,A},$$

and similarly for B, where $P_{net,A} = (Q_A - 1)$ if the constraint that h have mean zero is not imposed and $P_{net,A} = (Q_A - k - 1)$ if it is; subtraction of 1 accounts for the restriction $a_{00} \equiv 1$. The Hannan–Quinn criterion is $HQ_A = \hat{L}_{SNP,K,A} + \log\log NP_{net,A}$, and the Akaike Information Criterion discussed in section 6.2.2 takes the form $AIC_A = \hat{L}_{SNP,K,A} + P_{net,A}$, and similarly for B. To select a model based on a particular criterion, one would compare, for example, BIC_A to BIC_B, and prefer the model with the smaller value. Note that the penalty decreases as one goes from BIC to HQ to AIC. Thus, the three criteria represent a series of successively more conservative model selection strategies, in the sense that the AIC will generally favor inclusion of more terms in the model.

Davidian and Gallant (1992, 1993) recommend an *ad hoc* approach to model selection based on critical examination of all three with this ordering in mind, but recommend the HQ criterion if an automatic selection rule is desired. If the three model selection criteria agree, there is reasonably persuasive statistical evidence for the favored model.

Choice of K

The choice of the appropriate tuning parameter K may also be addressed using the model selection criteria described above; again, this approach is preferred to the conservative likelihood ratio procedure. To choose between two fits of a model using different values of K, say K_A and K_B, one would compute, for example, BIC under each model and compare BIC_A to BIC_B; the same procedure would be applied using the other criteria. Because of the decreasing penalty, the favored value of K may, but need not increase, as one goes from BIC to HQ to AIC. Davidian and Gallant (1992, 1993) recommend inspection of plots of the estimated density for all choices of K indicated by the three criteria, from which a visual selection may be made. As with kernel density estimation, examination of the fit over this range of values of K will provide insight into the shape of h. Again, the intermediate HQ criterion is preferred as an automatic rule.

A convenient feature of the SNP approach deserves comment. If $K = 0$ then h_K is the normal density. This is an advantage in applications when the normal distribution is a reasonable first approximation and one expects only modest departures from the normal such as an extra mode. One may thus use the model selection strategy above to test the hypothesis that the true random effects density is normal.

7.4 Other approaches

We note briefly some related approaches, each of which offers a different strategy for relaxing the way in which inter-individual variation is characterized. Further details may be found in the relevant references.

Beal and Sheiner (1992) assume a mixture model for the intra-individual model for individual i. Specifically, they assume that each individual arises from one of r_i subpopulations, indexed by $s = 1, \ldots, r_i$, with probability p_{is}, where r_i is known. The probabilities p_{is} are unknown but are taken to depend on covariate information and fixed effects to be estimated. This approach thus allows more flexibility in the characterization of the population of individuals. The method is implemented using a linearization of the resulting model, and is supported by the software package NONMEM (see section 6.4).

Magder (1994) has suggested an extension of the approach of section 7.2 that may be viewed as a smoothed version of nonparametric maximum likelihood. In this framework, which parallels that giving rise to the discrete estimator, the estimate for the distribution of the random parameters, H, is a finite mixture of at most m normal distributions whose covariance matrices have determinants greater than or equal to some fixed value ν. The constant ν, specified by the user, plays the role of a tuning parameter similar to K in the SNP procedure of section 7.3. Discussion of the choice of ν, limitations of the method for $k > 1$, and examples of application of the approach to the hierarchical linear model, are provided by Magder (1994). To our knowledge, this approach has not been applied to the hierarchical nonlinear model, but extension would, in principle, be straightforward.

Fattinger Bachmann *et al.* (1993) propose an inferential strategy for the semiparametric model specification of section 4.3.3. In their approach, the random effects are assumed to have density $h \in \mathcal{H}$, where $\mathcal{H}$ is a class of transformations of the normal distribution. Each component of the random effect $\boldsymbol{b}_i$ is represented as an unknown monotone transformation of the corresponding component of a random vector $\boldsymbol{b}_i^{\dagger}$, which has a k-variate normal distribution. The transformation for each component is taken to be a natural cubic spline constrained to be monotone non-decreasing. The unknown parameters characterizing the spline for each component are then estimated along with the other model parameters.

7.5 Software implementation

Nonparametric maximum likelihood estimation using a continuous version of the EM algorithm described in section 7.2.2 is implemented in a program, NPEM, which is available as part of the USC*PACK suite of PC programs (Jelliffe, Schumitzky and Van Guilder, 1994). The NPEM program has been developed for the particular purpose of population pharmacokinetic analysis and accommodates compartmental modeling up to a three-compartment model with either oral or intravenous dosing. The ability to incorporate covariate information is limited to a single, scalar covariate at this time, most usually weight or creatinine clearance, and is performed as described at the end of section 7.2.3. Nonparametric maximum likelihood estimation using the design-based algorithm in section 7.2.2, in the context of population pharmacokinetic analysis, is implemented in a program, NPML, developed by Mallet and co-workers.

A Fortran program called NLMIX (Davidian and Gallant, 1994) implements the SNP method described in section 7.3. The program and documentation are available from its authors or from the Statlib collection maintained by Carnegie-Mellon University. NLMIX computes parameter estimates, empirical Bayes estimates of the random effects, data for plotting, and simulations from the estimated density. The program requires a fair amount of effort by the user. Specifically, the user must construct a Fortran subroutine that performs data input and output and computes the conditional density $p_{y|b}$ and its derivatives. A template routine is provided as part of the code, the form of which may be exploited to reduce user effort; however, proficiency in Fortran is a strong requirement for use of the program. Convergence may be difficult to achieve; it is recommended that a 'wave' of runs from different starting values be employed to give more assurance that the maximum has been found. These computational issues arise with use of software for the nonparametric method as well.

7.6 Example

Recall the pharmacokinetic data for phenobarbital in neonates introduced in section 6.6. We use these data to illustrate the SNP method discussed in section 7.3; the analyses presented were constructed using the NLMIX program. As before, the one-compartment open model with intravenous administration and first-order elimination given in (6.31) is used to describe drug disposition:

$$C_i(t) = \sum_{d:t_{id}<t} \frac{D_{id}}{V_i} \exp\left\{-\frac{Cl_i}{V_i}(t - t_{id})\right\},$$

where $C_i(t)$ is mean phenobarbital concentration at time t for subject i following a single dose D_{id} administered at time t_{id}, $t > t_{id}$, and Cl_i and V_i are clearance and volume of distribution, respectively, for subject i. As before, $\boldsymbol{b}_i$ is taken to have mean $\mathbf{0}$ and covariance $\boldsymbol{D}$. The covariance matrix $\boldsymbol{D}$ was not constrained to be diagonal in the fits reported below. The components of the intra-individual error term $\boldsymbol{e}_i$ are taken to be uncorrelated, so that $\boldsymbol{R}_i(\boldsymbol{\beta}_i, \boldsymbol{\xi})$ is a diagonal matrix, with diagonal elements $\text{Var}(e_{ij}|\boldsymbol{\beta}_i) = \sigma^2 f^2(\boldsymbol{x}_{ij}, \boldsymbol{\beta}_i)$, where $f^2(\boldsymbol{x}_{ij}, \boldsymbol{\beta}_i)$ is determined by the function $C_i(t)$ above. Let w_i denote the birth weight of the ith subject.

Davidian and Gallant (1992) point out that SNP inference is particularly well-suited to accommodate graphical model-building, as outlined in section 6.5. Omission of important covariates from the inter-individual regression model $\boldsymbol{d}(\boldsymbol{a}_i, \boldsymbol{\beta}, \boldsymbol{b}_i)$ for $\boldsymbol{\beta}_i$ tends to induce dispersion or multi-modality of nonparametric estimates of h. Empirical Bayes estimates of the $\boldsymbol{b}_i$ will thus be separated, the separation being related to the omitted factors. Accordingly, the following strategy is sensible. Fit models with no covariates, increasing K until empirical Bayes estimates of the random effects separate. Next, plot these estimates against each potential covariate. The graphical evidence may be used to guide covariate selection and suggest refinements of the model. Formal criteria, discussed in section 7.3.2, may then be used to validate final model selection.

We now illustrate with the phenobarbital data. Initial fits were carried out using the following model for values of $K = 0, 2$ (because $\text{E}(\boldsymbol{b}_i) = \mathbf{0}$, $K = 0$ and $K = 1$ are equivalent).

Model 1. (No covariates, multiplicative error structures)

$$\boldsymbol{\beta}_i = [Cl_i, V_i]', \quad \boldsymbol{\beta} = [\beta_1, \beta_2]'$$

$$\beta_{1i} = \beta_1 \exp(b_{1i}), \quad \beta_{2i} = \beta_2 \exp(b_{2i})$$

Table 7.1. *Model selection criteria of section 7.3.2 for the fits of models 1–3 to the phenobarbital data.*

Model	K	P_{net}	*BIC*	*HQ*	*AIC*
1	0	6	507.118	501.696	497.988
	2	9	502.224	494.092	489.493
2	0	6	450.334	444.913	441.204
	2	9	453.939	445.806	440.243
3	0	7	451.621	445.295	440.969
	2	10	454.746	445.710	439.529

Pronounced separation of the estimated random effects was seen for the fit with $K = 2$. Table 7.1 summarizes values of the model selection criteria for $K = 0$ and $K = 2$. In the absence of any explanatory covariates, all three selection criteria favor the richer $K = 2$ fit. Figure 7.1 shows empirical Bayes estimates from this fit plotted against birth weight and Apgar category for each subject. Inspection of the plots suggests the same covariate relationships seen in the analyses of section 6.6. Accordingly, Model 1 was extended to incorporate dependence of clearance and volume on birth weight. Inclusion of weight as an explanatory variable was found to obviate the need for an intercept term in both clearance and volume, in agreement with the results in section 6.6. Omission of these intercept terms yields the following:

Model 2. (Birth weight as covariate for clearance and volume, multiplicative error structures)

$$\boldsymbol{\beta}_i = [Cl_i, V_i]', \quad \boldsymbol{\beta} = [\beta_1, \beta_2]'$$

$$\beta_{1i} = \beta_1 w_i \exp(b_{1i}), \quad \beta_{2i} = \beta_2 w_i \exp(b_{2i})$$

Figure 7.2 shows estimated random effects from the fit with $K = 2$ plotted against the covariates. As expected, no dependence on weight is observed; however, a more pronounced dependence on Apgar category now becomes evident, especially for volume. Here we report on the addition of dichotomized Apgar score to Model 2 as a covariate for volume, which corresponds to the final analysis given by Grasela and Donn (1985).

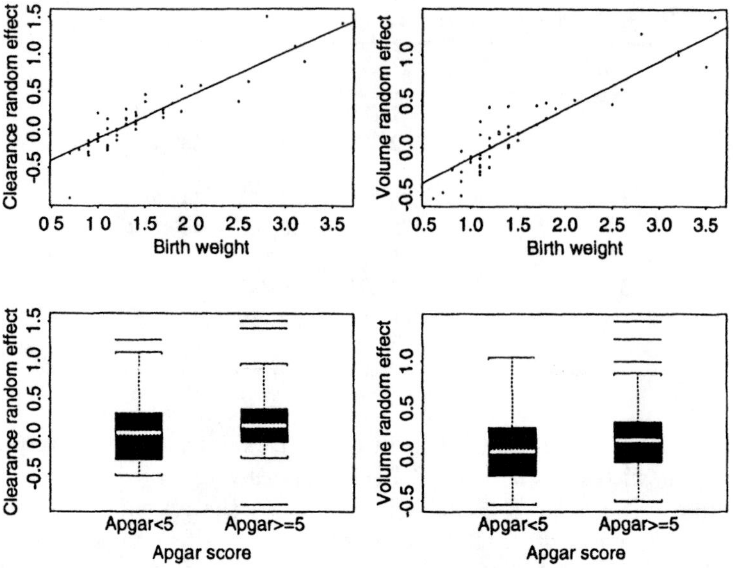

Figure 7.1. *Estimated random effects for Cl and V plotted against birth weight and Apgar score for the SNP (K = 2) fit of Model 1, phenobarbital data. The plots for birth weight have an ordinary least squares line superimposed; those for Apgar score are boxplots with the median represented by the white bar.*

Model 3. (Birth weight as covariate for clearance, birth weight and dichotomized Apgar score as covariates for volume, multiplicative error structures)

$$\boldsymbol{\beta}_i = [Cl_i, V_i]', \quad \boldsymbol{\beta} = [\beta_1, \beta_2, \beta_3]'$$

$$\beta_{1i} = \beta_1 w_i \exp(b_{1i}),$$

$$\beta_{2i} = \beta_2 w_i (1 + \beta_3 I_{[\text{Apgar}<5]}) \exp(b_{2i}),$$

where $I_{[A]}$ is the indicator function for the event A.

Table 7.1 indicates that, for these data, the AIC criterion is more conservative in two ways. In contrast to the BIC and HQ criteria, it favors Model 3 over Model 2 for both choices of K. Furthermore, for all three models, the AIC selects the $K = 2$ fit. This behavior reflects the generally conservative nature of this procedure. Overall,

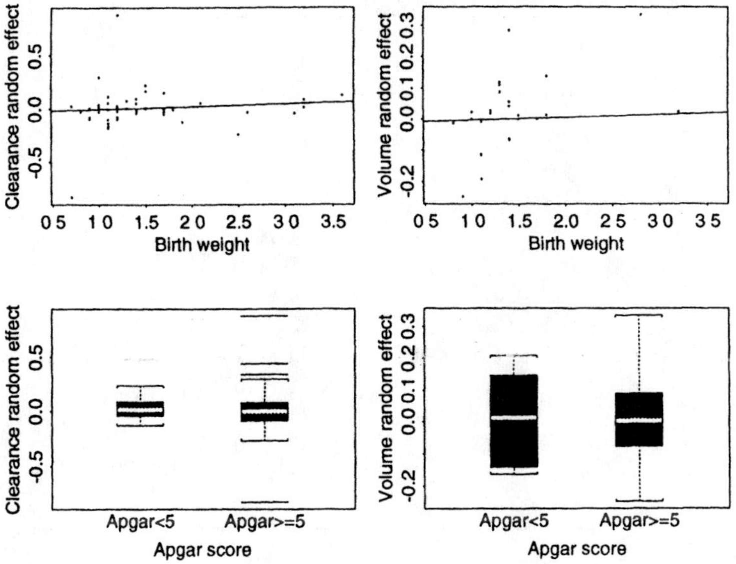

Figure 7 2. *Estimated random effects for Cl and V plotted against birth weight and Apgar score for the SNP ($K = 2$) fit of Model 2, phenobarbital data. The plots for birth weight have an ordinary least squares line superimposed; those for Apgar score are boxplots with the median represented by the white bar.*

these data suggest a weak association between volume and Apgar category, which bears out the conclusions of the analyses in section 6.6. Table 7.2 summarizes the parameter estimates for both fits of Model 3.

The advantage of adopting a less restrictive characterization of the random effects density is illustrated in Figure 7.3, which displays estimated densities and associated empirical Bayes estimates of the random effects for the $K = 0$ and $K = 2$ fits of Model 3. The most interesting feature is the bimodality of the estimate for $K = 2$, which divides the sample into two groups. Evidently, the seven infants represented by the diamonds in panels (c) and (d) had low measured concentrations after the loading dose. Because the initial concentration measurement is highly influential for estimation of volume in this model, the observed pattern is not unexpected. The low initial concentrations did not seem to be associated

Table 7.2. *Parameter estimates for the SNP fits of Model 3, $K = 0$ and $K = 2$.*

	Model with $K = 0$	$K = 2$
β_1	4.76×10^{-3}	4.69×10^{-3}
(SE)	(2.12×10^{-4})	(3.10×10^{-4})
β_2	0.977	0.981
(SE)	(0.028)	(0.036)
β_3	0.147	0.105
(SE)	(0.060)	(0.060)
D_{11}	3.51×10^{-2}	4.45×10^{-2}
(SE)	(1.74×10^{-2})	(3.59×10^{-2})
D_{12}	1.96×10^{-2}	1.44×10^{-2}
(SE)	(1.14×10^{-2})	(1.76×10^{-2})
D_{22}	2.09×10^{-2}	2.71×10^{-2}
(SE)	(7.94×10^{-3})	(1.43×10^{-2})
σ	0.117	0.110
(SE)	(0.012)	(0.013)

with any measured attribute; a relevant, unmeasured attribute may be an explanation. Comparison of panels (c) and (d) confirms the expected behavior of SNP estimation. For $K = 0$, implying the unimodal normal distribution, effects for these seven infants are pulled in towards those for the remaining subjects. Choice of $K = 2$ affords sufficient flexibility that this potentially informative feature of the data is not obscured in the estimation process.

7.7 Discussion

The lack of restrictive assumptions underlying the inferential methods discussed in this chapter is very appealing. However, several questions arise in the implementation and interpretation of these techniques. Issues of starting value selection, model parameterization, and convergence arise in the context of non- and semiparametric inference just as in the parametric case. It is intuitively clear that successful estimation of a distribution is possible only if sample sizes are sufficiently large. In the context of repeated measurement data, this corresponds to a requirement that the number of individuals be large. A common feature of all the methods in this chapter is that the dimension of the optimization problem can

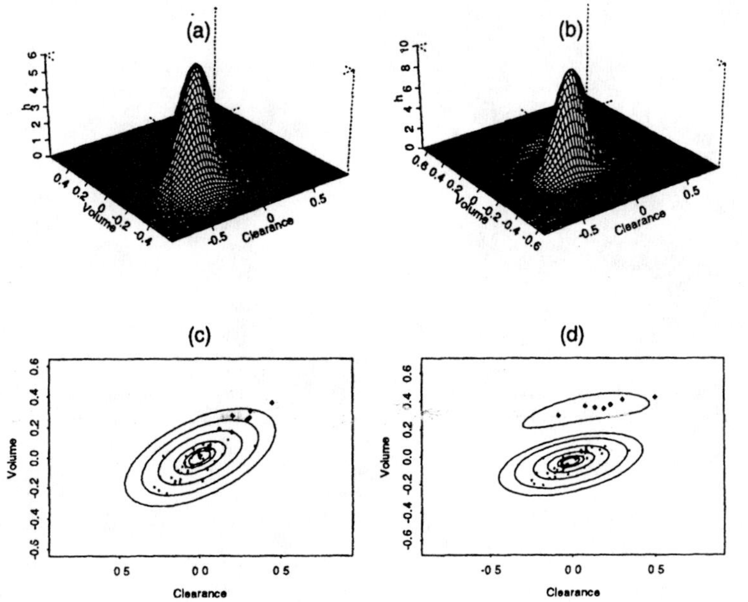

Figure 7.3. *Estimated random' effects densities and empirical Bayes estimates for the phenobarbital data. (a) perspective plot of the estimated density for $K = 0$; (b) perspective plot of the estimated density for $K = 2$; (c) contour plot of the estimated density for $K = 0$ at quantiles 10%, 25%, 50%, 75%, 90%, and 95%, with empirical Bayes estimates superimposed; (d) same as (c), $K = 2$. In (c) and (d), the diamonds represent empirical Bayes estimates for the same seven individuals.*

become very large quite rapidly. This issue is familiar to statisticians in many contexts and is often referred to as the 'curse of dimensionality.' It may translate to formidable computational challenges.

The techniques of this chapter are sufficiently recent that experience in their use is still somewhat limited. They may be useful in an adjunctive role, particularly for exploratory analysis. As experience with the methods accumulates, a better understanding of their advantages and limitations should facilitate their judicious use.

CHAPTER 8

Bayesian inference

8.1 Introduction

In this chapter, we give a brief overview of the application of Bayesian methods to the hierarchical nonlinear model described in section 4.3.4. Computational advances in the last decade have led to considerable interest in the application of Bayesian approaches to complex modeling problems; see Smith and Roberts (1993) for a review. It would be impossible to treat all of this material in a single chapter. Our aim here is modest; we hope to illustrate the potential utility of Bayesian methods within the hierarchical modeling framework and to convey the spirit of the associated computational techniques.

We begin in section 8.2 by reviewing the three-stage Bayesian hierarchy presented in Chapter 4. Recall from section 3.3.3 that for the linear version of this hierarchy, analytic derivation of posterior quantities of interest is possible under appropriate prior specifications. The possibility of analytic solution in this case arises not only from the assumption of normality, it also depends strongly on the linearity of the problem. Just as nonlinearity of the regression function complicates inference in likelihood-based approaches, it also makes implementation of Bayesian methods more difficult. Recent advances in computational techniques may be exploited to circumvent these difficulties. Specifically, Markov chain Monte Carlo methods, especially the Gibbs sampler, have been shown to be useful in Bayesian analysis of hierarchical models (Gelfand and Smith, 1990; Wakefield, 1995). The effect of nonlinearity is shown in section 8.3, and a brief review of the salient features of the Gibbs sampler is presented. Greater detail on Gibbs sampling is provided in section 8.4; this section may be omitted without loss of continuity on a first reading. The availability of these powerful computational methods allows realistic consideration of more complicated hierarchical models; section 8.5 illustrates this flexibility.

Available software is considered in section 8.6, and the chapter concludes with a discussion in section 8.7.

8.2 Model framework

In this section, we briefly review the general three-stage hierarchy for the Bayesian model specification, first described in section 4.3.4, and we consider two specific simple cases. More complicated examples are discussed in sections 8.5.

8.2.1 General three-stage hierarchical model

We consider the Bayesian model framework specified in section 4.3.4. Throughout this chapter, we use the notation $p_{c|d}$ and $p_{c,d}$ to denote the implied conditional and joint distributions, respectively, for collections of random variables c and d, and we suppress dependence on the covariates $\boldsymbol{x}_{i1}, \ldots, \boldsymbol{x}_{in_i}$ for $i = 1, \ldots, m$.

Stage 1 (intra-individual variation)

$$\boldsymbol{y}_i = \boldsymbol{f}_i(\boldsymbol{\beta}_i) + \boldsymbol{e}_i, \quad \boldsymbol{e}_i|\boldsymbol{\beta}_i \sim \Big(\boldsymbol{0}, \boldsymbol{R}_i(\boldsymbol{\beta}_i, \boldsymbol{\xi})\Big)$$

so that

$$(\boldsymbol{y}_i|\boldsymbol{\beta}_i, \boldsymbol{\xi}) \sim p_{y|\beta_i,\xi}(\boldsymbol{y}_i|\boldsymbol{\beta}_i, \boldsymbol{\xi}). \tag{8.1}$$

Stage 2 (inter-individual variation)

$$\boldsymbol{\beta}_i = \boldsymbol{d}(\boldsymbol{a}_i, \boldsymbol{\beta}, \boldsymbol{b}_i), \quad \boldsymbol{b}_i \sim p_{b|D}(\boldsymbol{b}_i|\boldsymbol{D}) \tag{8.2}$$

Stage 3 (hyperprior distribution)

$$(\boldsymbol{\beta}, \boldsymbol{\xi}, \boldsymbol{D}) \sim p_{\beta,\xi,D}(\boldsymbol{\beta}, \boldsymbol{\xi}, \boldsymbol{D}). \tag{8.3}$$

The distribution $p_{\beta,\xi,D}$ in (8.3) is generally taken to reflect weak knowledge at this stage of the hierarchy. If historical information is available from prior studies this may be incorporated in the specification of (8.3).

Stage 2 of the hierarchy gives an explicit relationship between the random parameters $\boldsymbol{\beta}_i$ and the random effects $\boldsymbol{b}_i$. This relationship allows flexibility in specifying the relevant conditional distributions; it may be done in terms of $\boldsymbol{\beta}_i$ or $\boldsymbol{b}_i$, whichever proves more convenient. We shall take advantage of this flexibility in the development below.

8.2.2 Simple examples

Normal linear hierarchy

A simple particular case is given by the normal linear random coefficient model:

Stage 1

$$(\boldsymbol{y}_i|\boldsymbol{\beta}_i, \sigma) \sim \mathcal{N}\{\boldsymbol{X}_i\boldsymbol{\beta}_i, \sigma^2 \boldsymbol{I}_{n_i}\}. \tag{8.4}$$

Stage 2

$$\boldsymbol{\beta}_i = \boldsymbol{\beta} + \boldsymbol{b}_i, \quad \boldsymbol{b}_i \sim \mathcal{N}(\boldsymbol{0}, \boldsymbol{D}). \tag{8.5}$$

Stage 3

$$\boldsymbol{\beta} \sim \mathcal{N}(\boldsymbol{\beta}^*, \boldsymbol{H}), \quad \boldsymbol{D}^{-1} \sim \text{Wi}\{(\rho \boldsymbol{D}^*)^{-1}, \rho\},$$
$$\sigma^{-2} \sim \text{Ga}(\nu_0/2, \nu_0\tau_0/2). \tag{8.6}$$

Here, Wi and Ga denote the Wishart and gamma distributions, respectively. The parameters $\boldsymbol{\beta}^*$, $\boldsymbol{H}$, ρ, $\boldsymbol{D}^*$, ν_0, and τ_0 that characterize the hyperprior distribution are known.

Note that this is a particularly simple version of the general linear hierarchy described in section 3.2.3 in that attention is limited to uncorrelated within-individual errors with constant variance. Thus, in Stage 1, the intra-individual variance parameter $\boldsymbol{\xi} = \sigma$.

Normal nonlinear hierarchy

The nonlinear analog of the normal linear model described above is as follows:

Stage 1

$$(\boldsymbol{y}_i|\boldsymbol{\beta}_i, \xi) \sim \mathcal{N}\{\boldsymbol{f}_i(\boldsymbol{\beta}_i), \sigma^2 \boldsymbol{I}_{n_i}\}. \tag{8.7}$$

Stage 2

$$\boldsymbol{\beta}_i = \boldsymbol{\beta} + \boldsymbol{b}_i, \quad \boldsymbol{b}_i \sim \mathcal{N}(\boldsymbol{0}, \boldsymbol{D}). \tag{8.8}$$

Stage 3

$$\boldsymbol{\beta} \sim \mathcal{N}(\boldsymbol{\beta}^*, \boldsymbol{H}), \quad \boldsymbol{D}^{-1} \sim \text{Wi}\{(\rho \boldsymbol{D}^*)^{-1}, \rho\},$$
$$\sigma^{-2} \sim \text{Ga}(\nu_0/2, \nu_0\tau_0/2). \tag{8.9}$$

This model differs from the normal linear hierarchy only in the introduction of nonlinearity in the specification of the conditional mean of y_i in Stage 1. This nonlinearity is sufficient to complicate inference, as we shall see shortly.

8.3 Inference

The Bayesian paradigm bases inference on the posterior distribution of the parameters, given the data. Section 8.3.1 begins by identifying the posterior quantities of interest for the general hierarchical model structure (8.1)–(8.3). Calculation of the relevant posterior distributions is discussed in section 8.3.2 for the simple normal linear and nonlinear examples. The computational difficulties involved in the necessary high-dimensional integration motivate discussion of the Gibbs sampler, introduced in section 8.3.3. We conclude this section by showing how the Gibbs sampler allows one to perform Bayesian inference for the linear and nonlinear examples.

8.3.1 Inference for the general hierarchical model

Recall the notation used in the description of the combined hierarchical linear model for all m individuals given in section 3.3.2: let $\boldsymbol{y} = [\boldsymbol{y}_1', \ldots, \boldsymbol{y}_m']'$ $(N \times 1)$, and let $\boldsymbol{b} = [\boldsymbol{b}_1', \ldots, \boldsymbol{b}_m']'$ $(km \times 1)$. Under the model assumptions in (8.1)–(8.3), the joint posterior distribution of all parameters is

$$\begin{aligned} &p_{\beta,\xi,b,D|y}(\boldsymbol{\beta}, \boldsymbol{\xi}, \boldsymbol{b}, \boldsymbol{D}|\boldsymbol{y}) \\ &\quad = p_{y|\beta,\xi,b}(\boldsymbol{y}|\boldsymbol{\beta}, \boldsymbol{\xi}, \boldsymbol{b}) p_{b|D}(\boldsymbol{b}|\boldsymbol{D}) p_{\beta,\xi,D}(\boldsymbol{\beta}, \boldsymbol{\xi}, \boldsymbol{D}) / p_y(\boldsymbol{y}). \end{aligned} \tag{8.10}$$

Calculation of marginal posterior distributions and associated moments and modes of interest will generally require evaluation of multi-dimensional integrals, using (8.10) as a starting point. For instance, inferences about the 'fixed effects' parameters, $\boldsymbol{\beta}$, might be based on the posterior marginal $p_{\beta|y}(\boldsymbol{\beta}|\boldsymbol{y})$, which, using (8.10), may be written

$$\frac{\int\int\int p_{y|\beta,\xi,b}(\boldsymbol{y}|\boldsymbol{\beta}, \boldsymbol{\xi}, \boldsymbol{b}) p_{b|D}(\boldsymbol{b}|\boldsymbol{D}) p_{\beta,\xi,D}(\boldsymbol{\beta}, \boldsymbol{\xi}, \boldsymbol{D})\, d\boldsymbol{\xi}\, d\boldsymbol{b}\, d\boldsymbol{D}}{\int\int\int\int p_{y|\beta,\xi,b}(\boldsymbol{y}|\boldsymbol{\beta}, \boldsymbol{\xi}, \boldsymbol{b}) p_{b|D}(\boldsymbol{b}|\boldsymbol{D}) p_{\beta,\xi,D}(\boldsymbol{\beta}, \boldsymbol{\xi}, \boldsymbol{D}) d\boldsymbol{\beta}\, d\boldsymbol{\xi}\, d\boldsymbol{b}\, d\boldsymbol{D}}.$$

Similarly, to make inferences about the ith random parameter $\boldsymbol{\beta}_i$, one would need to calculate the marginal posterior $p_{\beta_i|y}(\boldsymbol{\beta}_i|\boldsymbol{y})$. It is clear that the multi-dimensional integration required for explicit calculation of the requisite marginals may be prohibitive, even if one adopts an empirical Bayes strategy, replacing the covariance components $\boldsymbol{\xi}$ and $\boldsymbol{D}$ by point estimates.

It is often the case, however, that explicit expressions are available for certain conditional distributions. We illustrate this in the next section for the simple examples described in section 8.2.2.

8.3.2 Conditional distributions

Normal linear hierarchy

For the normal linear hierarchy specified in (8.4)–(8.6), one approach is to proceed as described in section 3.3.3; that is, to derive the marginal posterior distributions of $\boldsymbol{\beta}_i$ and $\boldsymbol{\beta}$ explicitly, assuming $\boldsymbol{D}$ and $\boldsymbol{\sigma}$ are known. Practical implementation proceeds according to an empirical Bayes strategy, replacing $\boldsymbol{D}$ and $\boldsymbol{\sigma}$ by estimates. Alternatively, one might wish to pursue a fully Bayesian analysis. Calculation of the required posterior distributions is prohibitive; however, writing $\bar{\boldsymbol{\beta}} = m^{-1}\sum_{i=1}^{m}\boldsymbol{\beta}_i$, $\boldsymbol{H}_i = \sigma^{-2}\boldsymbol{X}_i'\boldsymbol{X}_i + \boldsymbol{D}^{-1}$, and $\boldsymbol{U}^{-1} = m\boldsymbol{D}^{-1} + \boldsymbol{H}^{-1}$, it is relatively straightforward to derive the following conditional distributions (Wakefield *et al.*, 1994):

$$(\boldsymbol{\beta}_i|\boldsymbol{y},\boldsymbol{\beta},\sigma,\boldsymbol{D},\boldsymbol{\beta}_\ell,\ell\neq i) \sim \mathcal{N}\{\boldsymbol{H}_i(\sigma^{-2}\boldsymbol{X}_i'\boldsymbol{y}_i + \boldsymbol{D}^{-1}\boldsymbol{\beta}),\boldsymbol{H}_i\},$$

$$(\boldsymbol{\beta}|\boldsymbol{y},\sigma,\boldsymbol{D},\boldsymbol{\beta}_\ell,\ell=1,\ldots,m) \sim \mathcal{N}\{\boldsymbol{U}(m\boldsymbol{D}^{-1}\bar{\boldsymbol{\beta}} + \boldsymbol{H}^{-1}\boldsymbol{\beta}^*),\boldsymbol{U}\},$$

$$(\boldsymbol{D}^{-1}|\boldsymbol{y},\boldsymbol{\beta},\sigma,\boldsymbol{\beta}_\ell,\ell=1\ldots m) \sim \mathrm{Wi}\Big([\sum_{i=1}^{m}(\boldsymbol{\beta}_i-\boldsymbol{\beta})(\boldsymbol{\beta}_i-\boldsymbol{\beta})' + \rho\boldsymbol{D}^*]^{-1}, m+\rho\Big),$$

$$(\sigma^{-2}|\boldsymbol{y},\boldsymbol{\beta},\boldsymbol{D},\boldsymbol{\beta}_\ell,\ell=1\ldots m) \sim \mathrm{Ga}\Big(\frac{1}{2}(\nu_0+N),\frac{1}{2}\{\sum_{i=1}^{m}(\boldsymbol{y}_i-\boldsymbol{X}_i\boldsymbol{\beta}_i)'(\boldsymbol{y}_i-\boldsymbol{X}_i\boldsymbol{\beta}_i)' + \nu_0\tau_0\}\Big).$$

Thus, for this model, one can write the *full* conditional distribution of any of the parameters, given *all* the remaining parameters and the data, explicitly. This ability stems in large part from the hierarchical linear structure and the conjugacy of the priors. It is not immediately evident why this ability is useful, but we remark for the moment that, if there were a simple method of obtaining marginal (posterior) distributions from conditionals of this form, Bayesian inference would be straightforward.

Normal nonlinear hierarchy

For the nonlinear hierarchy given in (8.7)–(8.9), the full conditionals for $\boldsymbol{\beta}$, $\boldsymbol{D}^{-1}$, and σ^{-2} may be written explicitly as in the linear case:

$$(\boldsymbol{\beta}|\boldsymbol{y},\sigma,\boldsymbol{D},\boldsymbol{\beta}_\ell,\ell=1,\ldots,m) \sim \mathcal{N}\{\boldsymbol{U}(m\boldsymbol{D}^{-1}\bar{\boldsymbol{\beta}}+\boldsymbol{H}^{-1}\boldsymbol{\beta}^*),\boldsymbol{U}\},$$

$$(\boldsymbol{D}^{-1}|\boldsymbol{y},\boldsymbol{\beta},\sigma,\boldsymbol{\beta}_\ell,\ell=1\ldots m) \\ \sim \mathrm{Wi}\Big([\sum_{i=1}^{m}(\boldsymbol{\beta}_i-\boldsymbol{\beta})(\boldsymbol{\beta}_i-\boldsymbol{\beta})'+\rho\boldsymbol{D}^*]^{-1},m+\rho\Big),$$

$$(\sigma^{-2}|\boldsymbol{y},\boldsymbol{\beta},\boldsymbol{D},\boldsymbol{\beta}_\ell,\ell=1\ldots m) \\ \sim \mathrm{Ga}\Big(\frac{1}{2}(\nu_0+N),\frac{1}{2}[\sum_{i=1}^{m}\{\boldsymbol{y}_i-\boldsymbol{f}_i(\boldsymbol{\beta}_i)\}'\{\boldsymbol{y}_i-\boldsymbol{f}_i(\boldsymbol{\beta}_i)\}+\nu_0\tau_0]\Big).$$

Here, however, due to the nonlinearity in the regression function, the full conditional density of each $\boldsymbol{\beta}_i$, given the remaining parameters and the data, cannot be calculated explicitly. The distribution of $(\boldsymbol{\beta}_i|\boldsymbol{y},\boldsymbol{\beta},\sigma,\boldsymbol{D},\boldsymbol{\beta}_\ell,\ell\neq i)$ may, however, be written, up to a proportionality constant, as

$$\exp\Big(-\frac{\sigma^{-2}}{2}\{\boldsymbol{y}_i-\boldsymbol{f}_i(\boldsymbol{\beta}_i)\}'\{\boldsymbol{y}_i-\boldsymbol{f}_i(\boldsymbol{\beta}_i)\}-\frac{1}{2}(\boldsymbol{\beta}_i-\boldsymbol{\beta})'\boldsymbol{D}^{-1}(\boldsymbol{\beta}_i-\boldsymbol{\beta})\Big).$$

These conditionals are relatively easy to specify, due mainly to the hierarchical nature of the model. Again, we note that if there were a simple way of determining marginal posteriors, given the availability of these conditionals, Bayesian inference would be greatly simplified. It is this observation that motivates use of the Gibbs sampler.

8.3.3 From conditionals to marginals: the Gibbs sampler

We now give a heuristic exposition of the Monte Carlo method known as the Gibbs sampler, emphasizing its potential utility within the hierarchical modeling framework. A more detailed description is given in section 8.4. The literature on Gibbs sampling is extensive; our exposition is based on that in Gelfand *et al.* (1990) and Casella and George (1992).

Assume that we are dealing with a collection $\{U_1,\ldots,U_J\}$ of J random variables for which the joint density $p_{U_1,\ldots,U_J}(u_1,\ldots,u_J)$

is determined uniquely by the full conditional densities

$$p(u_s|u_1,\ldots,u_{s-1},u_{s+1},\ldots,u_J)$$
$$= p_{U_s|U_1,\ldots,U_{s-1},U_{s+1},\ldots,U_J}(u_s|u_1,\ldots,u_{s-1},u_{s+1},\ldots,u_J),$$

for $s = 1,\ldots,J$. Interest focuses on the marginal distributions $p_{U_s}(u_s)$, $s = 1,\ldots,J$. The Gibbs sampler is a method for obtaining the marginals of interest from a set of full conditionals. We assume that sampling from $p(u_s|u_1,\ldots,u_{s-1},u_{s+1},\ldots,u_J)$ may be carried out easily and efficiently. Methods for achieving this in cases where the conditional distribution cannot be specified in closed form are discussed in section 8.4.3.

The Gibbs sampler is an iterative algorithm that proceeds as follows. Start with an initial set of J values $\{u_1^{(0)},\ldots,u_J^{(0)}\}$. To complete the first iteration of the Gibbs sampling scheme, generate the following J random variates:

$$U_1^{(1)} \sim p(u_1|u_2^{(0)},u_3^{(0)},\ldots,u_J^{(0)})$$
$$U_2^{(1)} \sim p(u_2|u_1^{(1)},u_3^{(0)},\ldots,u_J^{(0)})$$
$$\vdots$$
$$U_J^{(1)} \sim p(u_J|u_1^{(1)},u_2^{(1)},\ldots,u_{J-1}^{(1)}).$$

At completion of the first iteration, we have a new set of J values $\{u_1^{(1)},\ldots,u_J^{(1)}\}$. After T such iterations, we have a realization of the random vector $\boldsymbol{U}^{(T)} = \{U_1^{(T)},\ldots,U_J^{(T)}\}$. It may be shown that the sequence thus generated is a Markov chain with equilibrium distribution $p_{U_1,\ldots,U_J}(u_1,\ldots,u_J)$, the joint distribution of $\{U_1,\ldots,U_J\}$. It follows (Geman and Geman, 1984) that, under mild conditions, as $T \to \infty$, $\boldsymbol{U}^{(T)}$ tends in distribution to a drawing from the joint distribution $p_{U_1,\ldots,U_J}(u_1,\ldots,u_J)$. Furthermore, the average of the T realizations, $T^{-1}\sum_{\ell=1}^{T}\psi(\boldsymbol{u}^{(\ell)})$, is a consistent estimator of the expected value (with respect to the joint distribution of $\{U_1,\ldots,U_J\}$) of any integrable function $\psi(\boldsymbol{u})$.

The first of these large-T results suggests that one can obtain a sample of size t, say, that may be viewed as a simulated sample from the joint density of $\{U_1,\ldots,U_J\}$ either by performing one long run of length T of the chain and collecting t suitably spaced realizations after an initial 'burn-in' period (to eliminate correlations between successive $\boldsymbol{U}^{(\ell)}$) or by performing t parallel independent runs, each of length T, and collecting the final realization from each. Issues pertaining to monitoring the convergence are discussed further in section 8.4. Given a sample of t such replicates from the Gibbs sampler, $\boldsymbol{u}^{(1)}, \ldots, \boldsymbol{u}^{(t)}$, the marginal distribution $p_{U_s}(u_s)$ for any component of $\{U_1, \ldots, U_J\}$ may be estimated by

$$t^{-1} \sum_{\ell=1}^{t} p(u_s | u_1^{(\ell)}, \ldots, u_{s-1}^{(\ell)}, u_{s+1}^{(\ell)}, \ldots, u_J^{(\ell)}).$$

This procedure is more efficient than the estimate obtained by smoothing the t replicate realizations of U_s, for reasons discussed in section 8.4.

A key aspect of hierarchical models is that the parameters separate naturally into a small number of distinct groups, i.e. the fixed effects $\boldsymbol{\beta}$, the covariance components $\boldsymbol{\xi}$ and $\boldsymbol{D}$, and the random parameters $\boldsymbol{\beta}_i$. In applying the results above, each element of the collection of random variables $\{U_1, \ldots, U_J\}$ corresponds to one of these distinct parameter groups. The marginal distribution, $p_{U_s}, s = 1, \ldots, J$, of interest corresponds to the posterior marginal, given the data.

8.3.4 Sampling from conditional distributions

Inference in hierarchical models employing conjugate priors is particularly amenable to the Gibbs sampling approach. We have seen in section 8.2 that the hierarchical structure means that the necessary full conditional densities may be completely specified in the linear case. In contrast, the nonlinear case in (8.7)–(8.9) is less tractable, and the required full conditionals may be specified only up to a normalizing constant. As noted by Gelfand *et al.* (1990), this is sufficient to allow implementation of the Gibbs sampler using appropriate techniques for random variate generation. In section 8.4.3, we briefly discuss methods of random variate generation that have been used in the hierarchical modeling context. In the overall development, these may be viewed as tools that allow implementation of the Gibbs sampler. The important issue is that it is possible to construct efficient methods for sampling from conditional distributions. The details of how this may be accomplished, while of general interest and critical to successful implementation, are technical and may be omitted by the reader concerned mainly with basic ideas.

8.4 Description of the Gibbs sampler

In this section, we discuss the Gibbs sampler in a little more detail. Our aim is to review the most important features and to provide a heuristic understanding of the main properties of the technique. The reader who does not wish to explore the technical aspects of the material at this level of detail may proceed to section 8.5 without loss of continuity.

There has been considerable interest in applications of Markov chain Monte Carlo techniques in general, and of the Gibbs sampler in particular, in recent years (see, for example, Smith and Roberts, 1993). A comprehensive review is beyond the scope of this book; we focus on the main ideas in order to help the reader gain sufficient understanding to appreciate the utility of these methods in hierarchical modeling. Casella and George (1992) offer a lucid introduction to the Gibbs sampler, which we recommend. Our presentation relies heavily on their exposition.

For simplicity of exposition, we confine attention in this section to the case of two or three variables, which we denote by U, V, and W.

8.4.1 Bivariate case

Implementation of the Gibbs sampler for the two-variable case (U, V) involves generation of a 'Gibbs sequence' of random variables $U_0, V_0, U_1, V_1, \ldots, U_T, V_T$, where successive iterates are generated from the conditionals

$$\begin{aligned} V_\ell &\sim p_{V|U}(v|u_\ell), \\ U_{\ell+1} &\sim p_{U|V}(u|v_\ell). \end{aligned}$$

Under general conditions, it is possible to show that the distribution of U_T converges to $p_U(u)$, the true marginal of U, as $T \to \infty$. Thus, generating t independent Gibbs sequences, each of length T, and using the final realization u_T from each yields an approximate independent and identically distributed sample from $p_U(u)$. Similarly, the t final values for V in each Gibbs sequence yield a sample of size t from the marginal distribution $p_V(v)$.

Note that the expected value of the conditional density $p_{U|V}(u|V)$ is

$$\mathrm{E}\{p_{U|V}(u|V)\} = \int p_{U|V}(u|v)p_V(v)\,dv = p_U(u),$$

suggesting estimation of the marginal $p_U(u)$ by

$$\hat{p}_U(u) = t^{-1} \sum_{\ell=1}^{t} p_{U|V}(u|v_\ell). \tag{8.11}$$

Estimation of $p_U(u)$ based only on the values $\{u_\ell\}, \ell = 1, \ldots, t$, would involve a kernel density estimate. Because such a kernel estimate makes no use of knowledge of the form $p_{U|V}(u|V)$ used in (8.11), the mixture estimate in (8.11) is more efficient. This gain in efficiency is a consequence of the Rao–Blackwell theorem; see Gelfand and Smith (1990, section 2.6) for details. This superiority extends to evaluating characteristics of $p_U(u)$ beyond the density itself. For instance, one possible estimate of the mean of U is $t^{-1}\sum_{\ell=1}^{t} u_\ell$, but a better estimate is provided by $t^{-1}\sum_{\ell=1}^{t} \mathrm{E}(U|v_\ell)$, as long as these conditional expectations are readily available. This reflects the fact that the quantities $p_{U|V}(u|v_1), \ldots, p_{U|V}(u|v_t)$ carry more information about $p_U(u)$ than the realizations $u_1, \ldots, u_t$ alone.

The Gibbs sampler is a particular example of what are usually referred to as Markov chain Monte Carlo (MCMC) techniques. The origin of the designation 'Monte Carlo' should be evident from our discussion so far. We now comment on the Markovian nature of the process in the simple two-variable case; this provides insight into the convergence properties of the scheme. Recall that a system is said to satisfy the Markovian property if, given the present state, the past states have no influence on the future; systems having this property are called Markov chains.

First consider the simple case where the joint distribution of U and V has finite support on a grid of $(n_U \times n_V)$ points. Assuming that the marginal of V is of primary interest, focus on the behavior of the V values in the Gibbs sequence. To go from V_0 to V_1, the iteration sequence must pass through U_1. Thus, the probability of transition from V_0 to V_1 is given by

$$P(V_1 = v_1|V_0 = v_0) = \sum_u P(V_1 = v_1|U_1 = u)P(U_1 = u|V_0 = v_0), \tag{8.12}$$

where the sum is taken over the n_U support points for U_1. In obvious notation, write $\boldsymbol{A}_{u|v}$ to denote the 'transition matrix' whose (h, l) entry denotes the probability of going to $U = u_h$ from $V = v_l$, and define $\boldsymbol{A}_{v|u}$ analogously. The iteration sequence $V_0, V_1, \ldots, V_T$ forms a Markov chain with transition probability matrix $\boldsymbol{A}_{v|v}$ satisfying $\boldsymbol{A}_{v|v} = \boldsymbol{A}_{v|u}\boldsymbol{A}_{u|v}$ from (8.12). Thus, the transition matrix that gives $P(V_\ell = v_\ell|V_0 = v_0)$ is given by $(\boldsymbol{A}_{v|v})^\ell$. If we define

$\boldsymbol{p}_\ell$ to be the n_V-vector of marginal probabilities of V_ℓ, then it is straightforward to show that, for any ℓ,

$$\boldsymbol{p}_\ell = \boldsymbol{p}_0(\boldsymbol{A}_{v|v})^\ell = \boldsymbol{p}_0(\boldsymbol{A}_{v|v})^{\ell-1}\boldsymbol{A}_{v|v} = \boldsymbol{p}_{\ell-1}\boldsymbol{A}_{v|v}. \tag{8.13}$$

From standard results on finite-dimensional Markov chains, as long as all entries of $\boldsymbol{A}_{v|v}$ are positive, for any initial set of probabilities $\boldsymbol{p}_0$, (8.13) implies that, as $\ell \to \infty$, $\boldsymbol{p}_\ell$ converges to the unique distribution $\boldsymbol{p}$ that is a stationary point of (8.13); that is, $\boldsymbol{p}$ satisfies

$$\boldsymbol{p}\boldsymbol{A}_{v|v} = \boldsymbol{p}. \tag{8.14}$$

Thus, for large enough T, the Gibbs sequence will satisfy (8.14), converging to a distribution $\boldsymbol{p}$, which must be the marginal distribution of V.

If either, or both, of U and V are continuous, then these finite-dimensional arguments do not apply. However, it is nonetheless possible to show that the Gibbs sampler still converges to the marginal density of V. The continuous version of (8.13) is

$$p_{V_\ell|V_0}(v|v_0) = \int p_{V_\ell|V_{\ell-1}}(v|t)p_{V_{\ell-1}|V_0}(t|v_0)\,dt, \tag{8.15}$$

and the stationary point of this equation is the marginal density of V, the density to which $p_{V_\ell|V_{\ell-1}}(v_\ell|v_{\ell-1})$ converges.

Convergence may be deduced from the following arguments. From the conditional densities $p_{U|V}(u|v)$ and $p_{V|U}(v|u)$, one may determine the marginal density $p_V(v)$, and hence the joint density $p_{U,V}(u,v)$, as follows. By definition, $p_V(v) = \int p_{U,V}(u,v)\,du$. Writing $p_{U,V}(u,v) = p_{V|U}(v|u)p_U(u)$, we have

$$\begin{aligned} p_V(v) &= \int p_{V|U}(v|u)p_U(u)\,du \\ &= \int p_{V|U}(v|u)\int p_{U|V}(u|t)p_V(t)\,dt\,du \\ &= \int\int \Big(p_{V|U}(v|u)p_{U|V}(u|t)\,du\Big)p_V(t)\,dt \\ &= \int h(v,t)p_V(t)\,dt, \end{aligned} \tag{8.16}$$

where $h(v,t) = \int p_{V|U}(v|u)p_{U|V}(u|t)\,du$. Equation (8.16) is a fixed point integral equation, for which $p_V(v)$ is a solution. It is also the limiting form of (8.15); as $\ell \to \infty$ in that equation, $p_{V_\ell|V_0}(v|v_0) \to p_V(v)$ and $p_{V_\ell|V_{\ell-1}}(v|t) \to h(v,t)$. Thus, the stationary point of (8.15) is the marginal density. The arguments above depend on the assumption that the conditional densities $p_{U|V}(u|v)$ and $p_{V|U}(v|u)$

determine the marginals. Casella and George (1992) give an example to show that a set of proper bivariate conditionals does not always determine a proper marginal distribution.

8.4.2 More than two variables

For three or more variables, the relationships among conditionals, marginals, and joint distributions are more complex. For example, the relationship (conditional × marginal) = joint does not hold for all conditionals and marginals. A consequence is that it is possible to use different sets of conditionals to calculate the marginal of interest. Following Casella and George (1992), we illustrate for the case of three variables U, V, and W.

Suppose we would like to calculate the marginal distribution $p_V(v)$. One possibility is to consider the pair (U, W) as a single random variable and proceed in a manner like that of the previous section. The key integral equation, analogous to (8.16), would be

$$p_V(v) = \int\int \Big(p_{V|U,W}(v|u,w)p_{U,W|V}(u,w|t)\,du\,dw\Big)p_V(t)\,dt. \tag{8.17}$$

This suggests cycling between the pair of conditional distributions $p_{V|U,W}(v|u,w)$ and $p_{U,W|V}(u,w|v)$ to obtain a sequence of random variables converging in distribution to $p_V(v)$.

In contrast, the Gibbs sampler would be based on the full set of conditionals $p_{U|V,W}(u|v,w)$, $p_{V|U,W}(v|u,w)$ and $p_{W|U,V}(w|u,v)$. At the ℓth iteration, variates are generated from the conditionals

$$\begin{aligned} W_\ell &\sim p_{W|U,V}(w|u_\ell, v_\ell), \\ U_{\ell+1} &\sim p_{U|V,W}(u|v_\ell, w_\ell), \\ V_{\ell+1} &\sim p_{V|U,W}(v|u_{\ell+1}, w_\ell); \end{aligned}$$

for large T, the Tth value v_T may be considered a realization from $p_V(v)$. The defining characteristic of the Gibbs sampler is that it always uses full conditionals to determine the marginals of interest.

Casella and George (1992) note that it is possible to show that this Gibbs sequence will also solve the fixed-point equation (8.17). Sampling based on the pair of conditionals $p_{V|U,W}(v|u,w)$ and $p_{U,W|V}(u,w|v)$ is an example of the Data Augmentation Algorithm (see Tanner, 1993, chapter 5). A discussion of the interrelationships between this method, Gibbs sampling, and other approaches is beyond the scope of this book; the reader is referred to Gelfand and

Smith (1990), Smith and Roberts (1993), and Tanner (1993) for further details.

8.4.3 Sampling from conditional distributions

Implementation of the Gibbs sampler relies on the ability to sample from the relevant conditional distributions. This is straightforward when the necessary full conditionals are given explicitly, allowing standard methods for random variate generation to be employed. For example, for the simple linear normal hierarchy given in (8.4)–(8.6), one may appeal to random variate generation routines intrinsic to many high-level computing languages to sample from distributions such as the normal or gamma. Algorithms for generation from less common distributions are also available; for instance, sampling from a Wishart distribution may be accomplished using the method of Odell and Fieveson (1966).

Under more complex models, such as the normal nonlinear hierarchy in (8.7)–(8.9), sampling from some conditionals is complicated by the fact that they may be specified only up to a normalizing constant. A number of techniques are available for random variate generation under these conditions; here, we describe one such method, that of rejection sampling, which has been used extensively in the context of the Gibbs sampler, and remark briefly on some other approaches.

Rejection sampling (see, for example, Smith and Gelfand, 1992; Tanner, 1993, chapter 3) is a general method for generating random variates from a density $p(y)$, say, where $p(y)$ may be specified only up to an integration constant. That is, we have available $p^*(y)$, where $p(y) = cp^*(y)$ for unknown $c > 0$; the case $c^{-1} = \int p^*(y)dy > 0$ is relevant in our context. Suppose we have available a function $p_e(y)$ such that $p^*(y)/p_e(y) \leq M$, where $M > 0$ is known, for all y in the support of $p(y)$. In its simplest form, rejection sampling then consists of the following scheme:

(i) Sample a value y^* from $p_e(y)$, and sample independently a value u from a uniform distribution on $(0, 1)$.

(ii) If $u \leq p^*(y^*)/\{Mp_e(y^*)\}$, accept y^* as a random variate generated from $p(y)$; otherwise, reject y^* and return to (i).

The accepted values from successive iterates form a random sample from the density $p(y)$; see Smith and Gelfand (1992). The function $p_e(y)$ is sometimes referred to as the 'envelope' function.

Despite the apparent simplicity of this scheme, sensible use of rejection sampling requires some sophistication on the part of the user. An important practical consideration is choice of the envelope function. Sampling from $p_e(y)$ must be straightforward in order for the method to be feasible; however, depending on the problem, determining an appropriate function $p_e(y)$ may be difficult. Moreover, a poor choice of $p_e(y)$ may result in a low acceptance rate, which may compromise the integrity of the Gibbs sequence. This situation is ameliorated to some extent by taking advantage of commonalities that exist among densities. As noted by Smith and Roberts (1993), the property of log-concavity may be shown to hold for a large class of common distributions, which may be exploited to achieve an extremely efficient extension to the rejection sampling scheme described above (Gilks and Wild, 1992).

Other methods of random variate generation that have been used with the Gibbs sampler include the related idea of importance sampling (see Smith and Gelfand, 1992; Tanner, 1993, chapter 3) and the generalized ratio-of-uniforms method (Wakefield *et al.*, 1994). As with rejection sampling, implementation of these methods requires some depth of understanding from the data analyst. An alternative technique is to embed a random variate simulation scheme based on the Metropolis–Hastings algorithm (see Smith and Roberts, 1993; Tanner, 1993, chapter 6; Tierney, 1995) within the Gibbs sequence; use of this procedure appears to require a high degree of sophistication on the part of the user, as it involves fine tuning to address the specific situation; an example is given in Wakefield (1995). Discussion of the details of these methods is beyond the scope of this book; however, we remark that a reasonable understanding of the issues involved in random variate generation is critical for successful implementation of the Gibbs sampler for complex nonlinear problems and that there is no 'all-purpose' approach.

8.4.4 Convergence and implementation

Theoretical results on the convergence of the Gibbs sequence and derived quantities, such as estimates of moments or predictive densities, do not answer practical questions of implementation. For instance, how long should a chain be run? Is it preferable to run several shorter chains or to perform a very long run of a single chain?

No consensus exists on the answers to these questions; extensive discussion of the issues involved may be found in the articles (and subsequent discussions) of Gelman and Rubin (1992) and Smith and Roberts (1993). Three points emerge clearly from these discussions:

(i) There is a danger in using runs that are too short, as results may be overly dependent on the starting point, particularly if one has started in a region from which it is hard for the chain to 'escape.' Use of a single very long run would reduce this dependence on the starting point.

(ii) Use of several chains has some advantage in addressing the issue of robustness to starting point.

(iii) Choice of parameterization can be critical. Wakefield (discussions of Smith and Roberts, 1993, and Gilks *et al.*, 1993) gives an example that illustrates the importance of parameterizing in such a way as to minimize posterior correlation, thereby improving convergence behavior of the Gibbs sequence.

As experience with the method accumulates, it is likely that guidelines for use in specific contexts will be established. For now, these issues suggest that care be taken when using Gibbs sampling techniques.

8.5 More complex models

The relative simplicity of implementation of the Gibbs sampler makes consideration of more sophisticated modeling approaches a realistic option, at least for exploratory purposes. In the words of Clifford (discussions of Smith and Roberts, 1993, and Gilks *et al.*, 1993), 'From now on we can compare our data with the model that we actually want to use rather than with a model which has some mathematically convenient form.' In this section, we sketch some recent applications in the context of nonlinear hierarchical models that illustrate the flexibility afforded by use of the Gibbs sampler. We encourage the reader to refer to the original source for a more detailed description in each case. Further uses of Gibbs sampling in biomedical applications are reviewed in Gilks *et al.* (1993).

Normal nonlinear hierarchy with model for heterogeneous intra-individual variation

This extends the model specified by (8.7)–(8.9) to accommodate heteroscedastic within-individual variation. Wakefield (1995) considers the power variance model, so that (8.7) is replaced by

$$(\boldsymbol{y}_i|\boldsymbol{\beta}_i,\theta) \sim \mathcal{N}\{\boldsymbol{f}_i(\boldsymbol{\beta}_i), \sigma^2 \boldsymbol{S}_i(\boldsymbol{\beta}_i,\theta)\},$$

$$\boldsymbol{S}_i(\boldsymbol{\beta}_i,\theta) = \text{diag}[f^{2\theta}(\boldsymbol{x}_{i1},\boldsymbol{\beta}_i)\ldots, f^{2\theta}(\boldsymbol{x}_{in_i},\boldsymbol{\beta}_i)].$$

The second stage of the hierarchy remains unchanged from (8.8), and a prior for the power parameter θ is added at Stage 3; for instance, θ may be taken to be uniformly distributed on $(0,\theta_0)$ for a suitable choice of θ_0, assumed known.

For this extension of the model in (8.7)–(8.9), conditionals for $\boldsymbol{\beta}$ and $\boldsymbol{D}^{-1}$ are as before. Conditionals for σ^{-2}, $\boldsymbol{\beta}_i$ and θ may be shown to satisfy the following relationships:

$$(\sigma^{-2}|\boldsymbol{y},\boldsymbol{\beta},\boldsymbol{D},\boldsymbol{\beta}_\ell,\ell=1,\ldots,m) \sim \text{Ga}\{(\nu_0+N)/2, A_0\}, \quad (8.18)$$

where

$$A_0 = \frac{1}{2}\Big(\sum_{i=1}^{m}\{\boldsymbol{y}_i - \boldsymbol{f}_i(\boldsymbol{\beta}_i)\}'\boldsymbol{S}_i^{-1}(\boldsymbol{\beta}_i,\theta)\{\boldsymbol{y}_i - \boldsymbol{f}_i(\boldsymbol{\beta}_i)\} + \nu_0\tau_0\Big),$$

The distribution of $(\boldsymbol{\beta}_i|\boldsymbol{y},\boldsymbol{\beta},\sigma,\boldsymbol{D},\boldsymbol{\beta}_\ell,\ell\neq i)$ is proportional to

$$\exp\Big(-\frac{\sigma^{-2}}{2}\{\boldsymbol{y}_i - \boldsymbol{f}_i(\boldsymbol{\beta}_i)\}'\boldsymbol{S}_i^{-1}(\boldsymbol{\beta}_i,\theta)\{\boldsymbol{y}_i - \boldsymbol{f}_i(\boldsymbol{\beta}_i)\}\Big)$$
$$\times \exp\Big(-\frac{1}{2}(\boldsymbol{\beta}_i - \boldsymbol{\beta})'\boldsymbol{D}^{-1}(\boldsymbol{\beta}_i - \boldsymbol{\beta})\Big)\sigma|\boldsymbol{S}_i(\boldsymbol{\beta}_i,\theta)|^{-1/2}, \quad (8.19)$$

and the distribution of $(\theta|\boldsymbol{y},\boldsymbol{\beta},\sigma,\boldsymbol{D},\boldsymbol{\beta}_\ell,\ell = 1,\ldots,m)$ is proportional to

$$\prod_{i=1}^{m}\exp\Big(-\frac{\sigma^{-2}}{2}\{\boldsymbol{y}_i - \boldsymbol{f}_i(\boldsymbol{\beta}_i)\}'\boldsymbol{S}_i^{-1}(\boldsymbol{\beta}_i,\theta)\{\boldsymbol{y}_i - \boldsymbol{f}_i(\boldsymbol{\beta}_i)\}\Big)$$
$$\times\, \sigma|\boldsymbol{S}_i(\boldsymbol{\beta}_i,\theta)|^{-1/2}. \quad (8.20)$$

Simulating from the conditional given in (8.18) is straightforward. Wakefield *et al.* (1994) and Wakefield (1995) discuss methods for efficient random variate generation from the conditionals (8.19) and (8.20).

Nonlinear hierarchy with multivariate t distribution at the second stage

Extension of the specification in (8.7)–(8.9) at the second stage to that of a multivariate t distribution is discussed by Wakefield *et al.* (1994) and Wakefield (1995). There are two potential advantages. Inferences for population moments should be more robust to outlying individuals; furthermore, the technique that is used to represent the t distribution yields an outlier detection mechanism. Although the t distribution affords greater flexibility in modeling inter-individual variation than the normal, in some cases it may still impose too much structure on the random parameter distribution; for instance, in the case where omission of an important covariate induces multimodality (see section 4.3.3).

A convenient representation of a t distribution with ν degrees of freedom is as a scale mixture of normals:

$$\prod_{i=1}^{m} \mathcal{N}(\boldsymbol{\beta}_i | \boldsymbol{\beta}, \lambda_i^{-1} \boldsymbol{D}), \tag{8.21}$$

where λ_i, $i = 1, \ldots, m$, are each distributed as a χ^2_ν random variable divided by its degrees of freedom ν, with ν known. Details on this representation may be found in Johnson and Kotz (1972, chapter 27).

Wakefield *et al.* (1994) comment that the ith scale parameter, λ_i, may be interpreted as a global indicator of outliers. The prior expectation of λ_i is 1, so that a value of λ_i substantially below 1 indicates that the ith individual regression parameter $\boldsymbol{\beta}_i$ is likely to be far away from the population mean $\boldsymbol{\beta}$. Because Mahalanobis distance is effectively used to measure distance from $\boldsymbol{\beta}$, λ_i provides only a global diagnostic; to investigate which specific elements of $\boldsymbol{\beta}_i$ may be outlying, examination of graphical summaries may be helpful.

For this extension of the model in (8.7)–(8.9), with (8.8) replaced by (8.21), the set of full conditionals required to implement the Gibbs sampler satisfies:

$$(\boldsymbol{\beta} | \boldsymbol{y}, \sigma, \boldsymbol{D}, \lambda_\ell, \boldsymbol{\beta}_\ell, \ell = 1, \ldots, m)$$
$$\sim \mathcal{N}\{\boldsymbol{U}(m\boldsymbol{D}^{-1} \sum_{i=1}^{m} \lambda_i \boldsymbol{\beta}_i + \boldsymbol{H}^{-1}\boldsymbol{\beta}^*), \boldsymbol{U}\},$$

where now $\boldsymbol{U}^{-1} = \boldsymbol{D}^{-1} \sum_{i=1}^{m} \lambda_i + \boldsymbol{H}^{-1}$,

$$(\boldsymbol{D}^{-1}|\boldsymbol{y}, \boldsymbol{\beta}, \sigma, \lambda_\ell, \boldsymbol{\beta}_\ell, \ell = 1, \ldots, m)$$
$$\sim \text{Wi}\Big([\sum_{i=1}^{m} \lambda_i(\boldsymbol{\beta}_i - \boldsymbol{\beta})(\boldsymbol{\beta}_i - \boldsymbol{\beta})' + \rho \boldsymbol{D}^*]^{-1}, m + \rho\Big),$$

the distribution of $(\boldsymbol{\beta}_i|\boldsymbol{y}, \boldsymbol{\beta}, \boldsymbol{\sigma}, \boldsymbol{D}, \lambda_\ell, \ell = 1, \ldots, m, \boldsymbol{\beta}_\ell, \ell \neq i)$ is proportional to

$$\exp\Big(-\frac{\sigma^{-2}}{2}\{\boldsymbol{y}_i - \boldsymbol{f}_i(\boldsymbol{\beta}_i)\}'\{\boldsymbol{y}_i - \boldsymbol{f}_i(\boldsymbol{\beta}_i)\} - \frac{1}{2}(\boldsymbol{\beta}_i - \boldsymbol{\beta})'\lambda_i \boldsymbol{D}^{-1}(\boldsymbol{\beta}_i - \boldsymbol{\beta})\Big),$$

and

$$(\lambda_i|\boldsymbol{y}, \boldsymbol{\beta}, \sigma, \boldsymbol{D}, \lambda_\ell, \ell \neq i, \boldsymbol{\beta}_\ell, \ell = 1, \ldots, m) \sim \chi^2_{\nu+p}/\nu.$$

The conditional distribution for σ^{-2} is as before.

Incorporation of covariate information

Within the nonlinear Bayesian hierarchy, incorporating covariates in the model for inter-individual variation in the $\boldsymbol{\beta}_i$ is straightforward. For covariates, $\boldsymbol{a}_i$, that are constant within an individual, and with a linear additive model

$$\boldsymbol{\beta}_i = \boldsymbol{A}_i\boldsymbol{\beta} + \boldsymbol{b}_i, \tag{8.22}$$

where $\boldsymbol{A}_i$ is a design matrix based on $\boldsymbol{a}_i$, the extension is trivial. Even if the dependence of $\boldsymbol{a}_i$ is nonlinear in $\boldsymbol{\beta}$, extension is straightforward as long as $\boldsymbol{b}_i$ enters in an additive, linear fashion, as in (8.22). Generalization to dependence nonlinear in the $\boldsymbol{b}_i$ and to time-dependent covariates is conceptually straightforward, although the technical details are difficult; see Wakefield (1995).

Sensitivity analysis

Smith and Roberts (1993, section 4.3) point out that, because Bayesian inference imposes more modeling structure than other approaches, it should offer a greater facility for diagnostics and model validation. This is easily accomplished within a Monte Carlo sampling framework. It is of interest to investigate how inference may change under the following perturbations:

(i) omission of a subset of observations, assuming the same distributional form for the data

(ii) changes to the distributional assumptions at Stages 1 and 2

(iii) changes to the hyperprior distributional assumptions at Stage 3.

Investigation of the effects of such perturbations is straightforward because, as pointed out by Smith and Roberts (1993), it is not necessary to restart the whole computation from scratch for each change to the specification. Instead, reanalysis proceeds on the basis of reweighting or resampling output from the original analysis. Details are beyond the scope of this discussion; see Smith and Roberts (1993).

8.6 Software implementation

A program called BUGS (Spiegelhalter, Thomas and Gilks, 1993) that carries out Bayesian inference using the Gibbs sampler is being developed at the Medical Research Council Biostatistics Unit (Cambridge, UK). The program is intended to provide a simple language to facilitate specification of full conditional distributions for fairly general Bayesian models, including hierarchical modeling. The program is still under development; further information may be obtained from the authors.

A software package to implement Bayesian inference for population pharmacokinetic data, POPKAN, has been developed jointly by CIBA-GEIGY and Imperial College. The program allows fitting of a variety of pharmacokinetic models and offers flexibility in the choice of prior distributions.

8.7 Discussion

The use of the Gibbs sampler to fit hierarchical nonlinear models under the Bayesian model specification is an appealing approach that will undoubtedly see more widespread application. One area for which the method is particularly well-suited is in the analysis of population pharmacokinetic data. Wakefield (1995) has used the Gibbs sampler to implement Bayesian inference for the quinidine data of section 1.1.2; see Wakefield *et al.* (1994) for another example. Zeger and Karim (1991) have used a Gibbs sampling approach within the context of a Bayesian hierarchal generalized linear model; their work offers insight into practical implementation of a rejection algorithm for random variate generation.

In our development, we have confined attention to specification of a parametric family of distributions for the random component of inter-individual variation; that is, the distribution of the random parameters $\boldsymbol{\beta}_i$ (or the random effects $\boldsymbol{b}_i$). It is possible to relax this restriction in the spirit of the nonparametric and semiparametric approaches of Chapter 7. In the linear case, Escobar and West (1992) have proposed a Bayesian nonparametric approach where the distribution of the β_i is taken to arise from a rich, flexible class of distributions provided by the Dirichlet process. Recently, Wakefield and Walker (1994) have extended their work to the nonlinear hierarchy. This work offers potential for detection of multimodality and skewness of the population distribution at the possible expense of a more complex implementation of the Gibbs sampler.

Because of their flexibility, the Gibbs sampling methods described in this chapter show great promise for analysis of complicated hierarchical nonlinear models. However, caution is needed with respect to issues of parameterization, model checking, sampling from conditional distributions, and monitoring for convergence; as remarked by Spiegelhalter *et al.* (1993), 'Beware – Gibbs sampling can be dangerous!' It seems fair to conclude that, as with any new technique, guidance on its most effective use can come only with experience. The recent growth of interest in the Gibbs sampler has already led to several applications in the area of hierarchical nonlinear modeling. As work in this area continues, we look forward to further refinement of these methods as well as a better understanding of their appropriate use.

CHAPTER 9

Pharmacokinetic and pharmacodynamic analysis

9.1 Introduction

Hierarchical nonlinear models provide a natural setting for the analysis of data from pharmacokinetic and pharmacodynamic studies. Indeed, several of the inferential techniques discussed in earlier chapters were originally proposed in this context. For instance, the pioneering work of Sheiner, Rosenberg and Melmon (1972) provided impetus for development of many of the linearization methods described in Chapter 6.

In this chapter, we focus on pharmacokinetic and pharmacodynamic applications of inferential techniques for the hierarchical nonlinear model by considering three case studies. For convenience, we shall use the abbreviations 'PK' and 'PD' for 'pharmacokinetic' and 'pharmacodynamic,' respectively, throughout this chapter. Section 9.2 provides a brief overview of the design and analysis goals of typical PK/PD studies. In section 9.3, we describe population analysis of the quinidine data first introduced in section 1.1.2. The following section considers data from a pharmacokinetic study of insulin-like growth factor I, or IGF-I, in severe head trauma patients. Population PK and PD analyses of data from a dose escalation study of argatroban, an anti-coagulant, are summarized in section 9.5. This study exemplifies the type of PK/PD modeling that is of interest in early clinical investigations of a new therapeutic agent.

Analyses presented in this chapter are meant to be illustrative rather than exhaustive. Space limitations preclude presentation of results from applying all possible methods to each of the three data sets. Instead, we have attempted to illustrate one or two approaches for each; our aim is to provide further insight into the practical implementation of the various methods by showing how they perform in specific cases. For convenience, the description of each case study presented here is self-contained.

9.2 Background

In this section, we summarize the main issues that typically arise in the evaluation of population PK and PD studies. Specific examples will be covered in sections 9.3–9.5. An excellent overview of these issues may be found in Rowland, Sheiner and Steimer (1985).

9.2.1 Population pharmacokinetics

In previous chapters, we have made the distinction between 'experimental' and 'routine clinical' PK studies. The former typically involve collection of full PK profiles on a relatively small number of subjects, often healthy volunteers. This information may be used to characterize the appropriate PK model and the pattern of intra-subject variation; for instance, using the methods described in Chapter 5. The small number and relative homogeneity of the subjects limit the scope of possible inference, however. Data from clinical studies, in contrast, *are* collected in the patient population of interest. Characteristic features of PK data obtained in a clinical setting include

- relatively sparse data, collected from each of a large number of subjects
- subjects who are representative of the actual patient population of interest
- availability of demographic or physiological patient attributes such as age, race, smoking status, and renal function. These covariate values may be constant for a given subject over the course of the study (e.g. race, baseline serum chemistry values) or may vary over time (e.g. serum chemistry values obtained during the dosing interval).

The term *population pharmacokinetics* has been used to refer both to studies with these features and the analysis of the resulting data.

Generally, the correct form of the PK model has been established from previous studies. If compartmental model methods are invoked, the relevant model will typically be nonlinear in the PK parameters of interest. A review of compartmental modeling techniques is beyond the scope of this book; in what follows, we shall assume the form of the PK model without derivation. Further details on compartmental modeling may be found in Gibaldi and Perrier (1982).

The primary analytic goal in population PK studies is characterization of inter-subject variation in PK behavior. Within the hierarchical model framework considered in this book, this amounts to understanding population variation in the subject-specific PK parameters $\boldsymbol{\beta}_i$. This involves two primary aspects: (i) characterization of the systematic factors affecting drug disposition by obtaining a model relating systematic variation in $\boldsymbol{\beta}_i$ to available covariate information, and (ii) characterization of the random component of inter-subject variation in $\boldsymbol{\beta}_i$. The utility of the techniques described in Chapters 5–8 in addressing these analytic goals is obvious.

Correct understanding of inter-subject variation in drug disposition may have important clinical advantages. The dosage regimen for a particular patient may be individualized based on relevant physiological information identified by population PK analysis. Such individualization of dosing is particularly important for drugs with a low therapeutic index, where the window of desirable serum concentrations is relatively narrow. Using population PK models to derive optimal dosing strategies has been carried out successfully for antibiotics such as gentamicin (Jelliffe *et al.*, 1991) and cardiovascular agents such as lidocaine (Rodman *et al.*, 1984).

9.2.2 Pharmacodynamic modeling

Common practice in early clinical studies of a drug is to randomize patients to one of several dose groups and use the results to elucidate dose-response. This information is clearly useful; in some instances, however, the dose-response profile may be blurred by substantial inter-patient variation in drug kinetics. The underlying rationale for PD modeling is that a better understanding of the physiologic response to the drug should be obtained by relating drug effect (the PD response) to the concentration of the drug at its site of action. Thus, ideally one would like to model PD response in terms of an 'effect site' concentration of the drug. In most cases, there is an obvious practical problem; drug effect may take place in organs or tissues, so that sampling drug concentration at the effect site is not feasible. Of necessity, therefore, PD modeling usually focuses on the relationship between drug effect and concentrations in plasma or serum, where levels *can* be measured.

Directly relating PD effect to plasma concentrations should allow better resolution in studying drug effect by removing the impact of variable inter-subject kinetics. Peck (1990) and others have suggested taking this argument one step further, advocating the conduct of concentration-controlled efficacy trials. The idea behind such a trial is that response variability within treatment groups should be lowered by randomizing subjects in a particular group to a therapy that targets maintenance of drug levels within a specified concentration range, rather than to a particular fixed dose. Results from a concentration-controlled trial should provide a good understanding of the concentration-effect relationship. Logistics of implementation and extrapolation of results to standard clinical and prescribing practices may be difficult, however.

Data from a PD study typically take the form of concurrent measurements of PD response and plasma drug concentrations for each of a number of subjects. The response-concentration relationship is often nonlinear in parameters of interest, and thus the hierarchical nonlinear model is useful. Concentrations may not vary widely for measurements within a given subject; thus, successful modeling of the response usually requires a reasonable range of concentrations across subjects. Given data of this type, a number of challenges arise. These include:

(i) Choice of pharmacologic endpoint: just as specifying a suitable endpoint can be difficult in a clinical trial, choice of an appropriate response for PD modeling may not be easy. Factors to be considered in making this choice include clinical relevance, ease of measurement, variability in the response, and overall sensitivity and temporal relationship to changes in drug concentration.

(ii) Possible lag between changes in plasma concentration and PD effect: this reflects the difficulty that may arise because PK measurements are not taken directly at the effect site. Denote plasma drug concentration by C_p, concentration at the effect site by C_e, and plasma concentration at steady state by C_{pss}. At steady state, equilibrium conditions dictate that C_{pss} and C_e differ by a constant proportion; in general, the difference between C_p and C_e may be due to a delay in the drug reaching the effect site. Sheiner (1985) suggests a method for diagnosing such a lag. One plots effect versus C_p measurements in time order; typically, this so-called 'hysteresis curve' forms a counter-clockwise loop with two limbs, one ascending and one descending. The degree of hysteresis, or separation of the two limbs, provides an indication of the extent of the lag between changes in plasma concentration and that at the effect site. There are two common approaches to dealing with hysteresis in PD modeling. One is to use only plasma concentrations at steady state, in which case C_p and C_e are in constant ratio. The other is to extend the compartment model for PK to include a hypothetical 'effect' compartment. This device allows one to include all PK measurements in modeling. A full discussion is beyond the scope of this book; further details may be found in Sheiner *et al.* (1979) and Sheiner (1985).

(iii) Derivation of a suitable PD model: unlike PK modeling, this is usually done in a somewhat empirical fashion. By far the most commonly used model is the so-called 'E_{max}' model, or some variation thereof (Holford and Sheiner, 1981). This has the general form

$$y = E_0 + \frac{E_{max} - E_0}{1 + EC_{50}/C_e}. \tag{9.1}$$

Parameters in the model have the following interpretation: E_0 is the response at zero concentration (baseline), E_{max} is the maximal response, and EC_{50} is that concentration eliciting a response halfway between E_0 and E_{max}.

(iv) Error in concentration measurements: in general, concentration values are measured with nonnegligible error. Ignoring this measurement error may bias estimates of the PD parameters. Two approaches to this problem are common. In the first of these, when fitting the PD model, actual concentrations are replaced by predicted concentrations from a preliminary pharmacokinetic analysis. The resulting smoothing of concentration values presumably mitigates the potential for attenuation bias in the estimates of PD model parameters. The second approach involves *joint* modeling of PK and PD responses. This method is illustrated in section 9.4 below.

Other issues that may arise in PD modeling include accommodation of diurnal or circadian variation in response, the possibility of feedback loops causing up- or down-regulation of response, or the formation of metabolites that may exert inhibitory effects on response. This is not an exhaustive list – other issues may come up, depending on the particular situation in question.

9.3 Population pharmacokinetics of quinidine

Data description

Data for $m = 136$ hospitalized participants (135 men, 1 woman) in a clinical study of the pharmacokinetics of the anti-arrhythmic agent quinidine (Verme *et al.*, 1992) were introduced in section 1.1.2. Subjects in the study received quinidine by oral administration for treatment of atrial fibrillation or ventricular arrhythmias.

Quinidine concentrations (mg/L) were obtained for routine clinical purposes from these patients by enzyme immunoassay. A total of $N = 361$ measurements, ranging from $n_i = 1$ to 11 observations per patient, were taken within a range of 0.08 hours to 70.5 hours after dose. These data are sparse; most patients had only one to four measurements. For all 136 participants, concentrations ranged from 0.4 mg/L to 9.4 mg/L, with mean and standard deviation 2.45 mg/L and 1.22 mg/L, respectively, and patients were observed over a period ranging from 0.13 to 8095.0 hours.

As is often the case in a routine clinical study, a large number of demographic and physiological variables were recorded for each subject. This information is summarized in Table 9.1. For characteristics measured repeatedly over the course of study on all patients, α_1-acid glycoprotein concentration, weight, age, and creatinine clearance, the table summarizes the initial values for the first three; only dichotomized creatinine clearance was available, which changed over the observation period for 11 subjects. Although quinidine was given in different forms, doses were adjusted for difference in salt content between the two forms by conversion to milligrams of quinidine base; converted doses ranged from 83 to 603 mg, with mean steady state dosing intervals of 6.25 hr for the sulfate form and 7.70 hr for the gluconate form. Consequently, dosage form was not taken into account in the analyses reported below, nor was albumin concentration, which was not measured on some patients.

Partial data records for two of the patients in the study were displayed in Table 1.1.

Objective

The goal of the analyses given here is to characterize the pharmacokinetics of quinidine in this patient population and to identify associations between kinetics and demographic and physiological patient attributes.

Table 9.1. *Demographic and physiological characteristics for 136 patients, population pharmacokinetic study of quinidine.*

	Mean	*(SD)*	*Range*
Initial body weight (kg)	79.58	(15.64)	41–119
Initial age (yr)	66.88	(8.92)	42–92
Height (in)	69.63	(2.37)	60–79
Initial α_1-acid glycoprotein conc. (mg/dL)	118.54	(46.23)	39–316
Creatinine	> 50		84 pts
Clearance (ml/min)	< 50		41 pts
	> and < 50		11 pts
Dosage form	gluconate		57 pts
	sulfate		53 pts
	both		26 pts
Race	Caucasian		91 pts
	Black		10 pts
	Latin		35 pts
Smoking status	no		91 pts
	yes		45 pts
Ethanol abuse	none		90 pts
	current		16 pts
	former		30 pts
Congestive heart failure	none or mild		56 pts
	moderate		40 pts
	severe		40 pts
Albumin conc. (g/dL)	Incomplete for some patients[1]		

[1] not considered in analyses

Methods

In the literature, a one-compartment model with first-order absorption and a two-compartment model with zero-order absorption are the two most common characterizations of quinidine disposition (Ochs, Greenblatt and Woo, 1980; Fattinger *et al.*, 1991; Verme *et al.*, 1992). For the analyses presented here, we followed the latter authors and used the one-compartment open model with first-order absorption, given in (1.2)–(1.4): Let $C(t)$ be the concentration of quinidine and let $C_a(t)$ be the apparent concentration of quinidine in the absorption depot at time t. Written in recursive form, the model is:

For the non-steady state at a dosage time $t = t_\ell$

$$\begin{aligned} C_a(t_\ell) &= C_a(t_{\ell-1})\exp\{-k_a(t_\ell - t_{\ell-1})\} + FD_\ell/V \\ C(t_\ell) &= C(t_{\ell-1})\exp\{-k_e(t_\ell - t_{\ell-1})\} + C_a(t_{\ell-1})\frac{k_a}{k_a - k_e} \\ &\quad \times \Big[\exp\{-k_e(t_\ell - t_{\ell-1})\} - \exp\{-k_a(t_\ell - t_{\ell-1})\}\Big]. \end{aligned} \tag{9.2}$$

For the steady state at a dosage time, $t = t_\ell$

$$\begin{aligned} C_a(t_\ell) &= (FD_\ell/V)(1 - \exp\{-k_a\tau_{ss}\})^{-1} \\ C(t_\ell) &= (FD_\ell/V)\frac{k_a}{k_a - k_e} \\ &\quad \times \Big[(1 - \exp\{-k_e\tau_{ss}\})^{-1} - (1 - \exp\{-k_a\tau_{ss}\})^{-1}\Big]. \end{aligned} \tag{9.3}$$

Between dosage times, $t_\ell < t < t_{\ell+1}$

$$\begin{aligned} C(t) = C(t_\ell)\exp\{-k_e(t - t_\ell)\} + C_a(t_\ell)\frac{k_a}{k_a - k_e} \\ \times\Big[\exp\{-k_e(t - t_\ell)\} - \exp\{-k_a(t - t_\ell)\}\Big]. \end{aligned} \tag{9.4}$$

Here, t_ℓ, $\ell = 0, 1, \ldots$, are the times at which doses D_ℓ are administered, $C_a(t_0) = FD_0/V$, $C(t_0) = 0$, F is the fraction of dose available, k_a is the absorption rate constant, $k_e = Cl/V$ is the elimination rate constant, Cl is the clearance, V is the apparent volume of distribution, $Cl, V, k_a > 0$, and τ_{ss} is the steady-state dosing interval.

From (9.2)–(9.4), the nonlinear model $f(\boldsymbol{x}, \boldsymbol{\beta})$ of interest has $\boldsymbol{x} = [D, t, \tau_{ss}]'$ and parameter vector $\boldsymbol{\beta}$ consisting of the PK parameters k_a, Cl, and V. These data do not permit estimation of the fraction of dose available, F; thus, we take $F = 1$, so that, in reality, the pharmacokinetic parameters being estimated are k_a, Cl/F and V/F. The lowest value for F reported for quinidine is 0.7 (Ochs *et al.*, 1980); thus, the possible upward bias in estimates for Cl and V reported here may be as high as 40%.

The number of samples collected during the absorption phase for these data were not sufficient to allow precise estimation of k_a for all subjects; this is a common limitation, because it is often not feasible to obtain concentration measurements soon enough following a dose for drugs that are absorbed rapidly. In this situation, it is common practice to treat k_a as fixed across the population, and either assign to it a prespecified value based on previous reports in the literature (Verme *et al.*, 1992) or treat it as a fixed parameter to be estimated. Both approaches represent a model misspecification, forcing a parameter known to vary in the population to be fixed across subjects; however, intuition suggests that the estimate obtained by treating it as fixed may be similar to the mean value in the population. Accordingly, for the analyses below, we adopted the second strategy.

To account for intra-individual heterogeneity of variance, we took the within-patient errors to be uncorrelated with variance proportional to the square of the mean response, the so-called proportional error model. Writing subject-specific values for the PK parameters as Cl_i, V_i, and k_{ai}, letting $\boldsymbol{\beta}_i = [Cl_i, V_i, k_{ai}]'$, and letting $\boldsymbol{x}_{ij}$ denote the values for dose, time, and steady-state dosing interval, if applicable, for the ith patient at the jth observation time, the basic two-stage model, in the notation of Chapter 4, is as follows.

Stage 1 (within-subject)

$$y_{ij} = f(\boldsymbol{x}_{ij}, \boldsymbol{\beta}_i) + e_{ij},$$
$$(\boldsymbol{e}_i|\boldsymbol{\beta}_i) \sim \{\boldsymbol{0}, \boldsymbol{R}_i(\boldsymbol{\beta}_i, \boldsymbol{\xi})\}, \tag{9.5}$$

where y_{ij} represents the jth concentration measurement taken at time t_{ij} for the ith subject, $f(\boldsymbol{x}, \boldsymbol{\beta})$ is the model given in (9.2)–(9.4), and within-subject covariance matrix $\boldsymbol{R}_i(\boldsymbol{\beta}_i, \boldsymbol{\xi}) = \sigma^2 \boldsymbol{G}_i(\boldsymbol{\beta}_i, \theta)$, where $\boldsymbol{G}_i(\boldsymbol{\beta}_i, \theta) = \text{diag}[f^2(\boldsymbol{x}_{i1}, \boldsymbol{\beta}_i), \ldots, f^2(\boldsymbol{x}_{in_i}, \boldsymbol{\beta}_i)]$.

Stage 2 (subject-to-subject)

$$\boldsymbol{\beta}_i = [\beta_{1i}, \beta_{2i}, \beta_{3i}]' = [Cl_i, V_i, k_{ai}]',$$
$$\boldsymbol{\beta}_i = \boldsymbol{d}(\boldsymbol{a}_i, \boldsymbol{\beta}, \boldsymbol{b}_i), \quad \boldsymbol{b}_i \sim (\boldsymbol{0}, \boldsymbol{D}), \tag{9.6}$$

where $\boldsymbol{\beta} = [\beta_1, \beta_2, \beta_3]'$, and $\boldsymbol{b}_i = [b_{1i}, b_{2i}]'$, and the components of $\boldsymbol{d}$ are given by

$$Cl_i = \exp(\beta_1 + b_{1i}), \quad V_i = \exp(\beta_2 + b_{2i}), \quad k_{ai} = \exp(\beta_3).$$

The specification in (9.6), which corresponds to parameterization of the model (9.2)–(9.4) in terms of the logarithms of the PK parameters, ensures necessary positivity of these quantities. Moreover, the distributions of PK parameters are often skewed, with approximately constant coefficient of variation, which is likely to be fulfilled if the random effects vector $\boldsymbol{b}_i$ is normally, or at least symmetrically, distributed.

We shall refer to the basic model given in (9.5) and (9.6) as Model 1. This model makes no allowance for the dependence of individual kinetics on subject-specific demographic and physiological characteristics at Stage 2; in the analyses below, we investigated this possibility by extending the model to incorporate individual covariate information in the inter-subject regression model $\boldsymbol{d}(\boldsymbol{a}_i, \boldsymbol{\beta}, \boldsymbol{b}_i)$. Because some covariate values changed across the study period, these model extensions allow the possibility that Cl and V may vary within a subject over time; k_a was not altered, as it is taken to be fixed for all patients. Following section 4.2.5, the general form of these model extensions is given below:

Stage 1 (within-subject)

$$y_{ij} = f(\boldsymbol{x}_{ij}, \boldsymbol{\beta}_{ij}) + e_{ij},$$

$$(\boldsymbol{e}_i | \boldsymbol{\beta}_{ij}) \sim \{\boldsymbol{0}, \boldsymbol{R}_i(\boldsymbol{\beta}_{ij}, \boldsymbol{\xi})\}, \qquad (9.7)$$

where y_{ij}, $f(\boldsymbol{x}, \boldsymbol{\beta})$, and $\boldsymbol{R}_i(\boldsymbol{\beta}_{ij}, \boldsymbol{\xi})$ are as above.

Stage 2 (subject-to-subject)

$$\boldsymbol{\beta}_{ij} = [\beta_{1ij}, \beta_{2ij}, \beta_{3ij}]' = [Cl_{ij}, V_{ij}, k_{ai}]',$$

$$\boldsymbol{\beta}_{ij} = \boldsymbol{d}(\boldsymbol{a}_{ij}, \boldsymbol{\beta}, \boldsymbol{b}_i), \quad \boldsymbol{b}_i \sim (\boldsymbol{0}, \boldsymbol{D}), \qquad (9.8)$$

where $\boldsymbol{\beta}$ is an $(r \times 1)$ vector of fixed effects, $\boldsymbol{b}_i = [b_{i1}, b_{2i}]'$ as before, and $\boldsymbol{a}_{ij}$ are values of possibly time-dependent covariates.

The specification in (9.8) requires that covariate values are available at the same times as observations, indexed by j within subjects, or at least are measured in, for example, the dosing interval immediately preceding a PK concentration. This is generally not the case; a standard convention is to adopt some type of interpolation method to obtain an appropriate value (Boeckmann *et al.*, 1992). For these data, a 'last value carried forward' strategy was used. This is discussed further in the context of the IGF-I data in section 9.4. In contrast to the IGF-I example, covariates here did not change markedly and were measured relatively infrequently within most subjects, so the time-varying aspect is not a major feature of the analyses here.

To fit Model 1 and extensions of the form (9.7), (9.8), we used both the SNP method of section 7.3 and full maximum likelihood assuming normality of within- and among-subject errors and the first-order linearization described in section 6.2.2. The NLMIX program and NONMEM software, respectively, were used to carry out the fitting. The former required writing of a Fortran subroutine to calculate the one-compartment model (9.2)–(9.4) and its derivatives; the latter used the existing intrinsic subroutine for this model.

Results

For the SNP fit, we adopted the graphical model-building strategy discussed in section 6.5. Following the approach taken with the phenobarbital data in section 7.6, we began by fitting Model 1 with a progression of values for the tuning parameter $K = 0, 2, \ldots$; as discussed in section 7.3.1, because we assume $\mathrm{E}(\boldsymbol{b}_i) = \mathbf{0}$, $K = 0$ and $K = 1$ both represent the assumption of normality.

Inspection of the estimated joint density of the components of $\boldsymbol{b}_i$ for $K = 2$ revealed multi-modality and corresponding separation of the empirical Bayes estimates of the random effects based on this fit. Consequently, we plotted the empirical Bayes estimates for Cl and V against potential covariates, shown in Figures 9.1 and 9.2, using initial values for attributes changing over time within subjects. Examination of individual panels in Figure 9.1 reveals that creatinine clearance appears to be the most important categorical variable, followed by race; the race relationship, however, is due entirely to the 10 Black subjects. The visual evidence for relationships with the other categorical covariates is not strong. Among the continuous covariates, α_1-acid glycoprotein concentration exhibits the strongest association with estimated Cl random effects, followed by weaker relationships with age, height and weight. In Figure 9.2, there do not seem to be any strong relationships between estimated volume effects and covariates.

We also plotted the estimated random effects against measured concentrations as an *ad hoc* device for assessing the importance of all omitted covariates taken as a group, mimicking the use of residual plots in linear regression analysis for models with covariates and replicates, with empirical Bayes estimates replacing residuals. The model in (9.6) is analogous to a regression model with no covariates, for which residuals and observations are by construction perfectly correlated, so the results of this plot should be interpreted with caution; however, the technique seems informative here. This plot for the Cl random effects in Figure 9.1 shows a

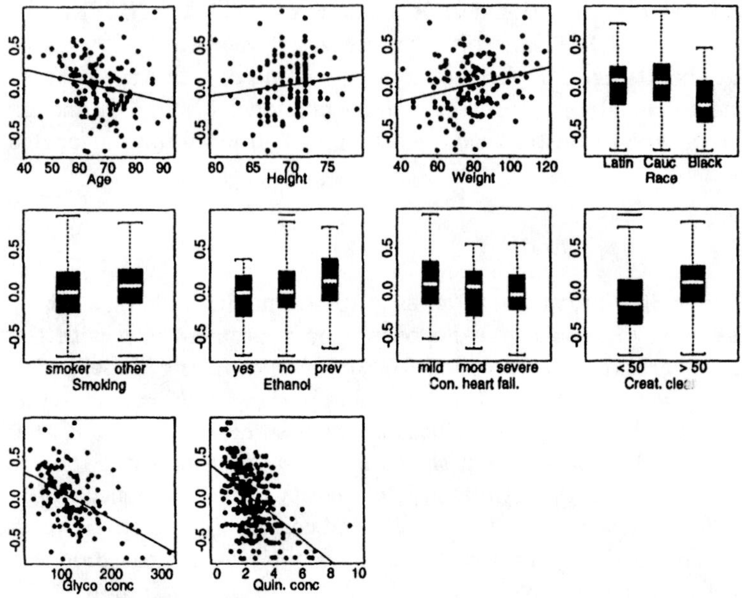

Figure 9.1. *Empirical Bayes estimates of random effects for Cl plotted against potential covariates for the SNP fit of Model 1 to the quinidine data. The plots for continuous covariates have an ordinary least squares line superimposed. In the case of time-varying covariates, plots are based on initial values.*

strong association, raising the possibility of omission of important covariates from the second stage of Model 1. The same plot for V in Figure 9.2 is almost flat, suggesting that the specification for β_{2i} in (9.5) may be adequate.

Based on these observations, we chose to extend Stage 2 to the following model of the form (9.8), where w_{ij} and g_{ij} represent the values for weight and α_1-acid glycoprotein concentration, respectively, associated with the jth observation on subject i, and $c_{ij} = 0$ (1) if creatinine clearance is $<$ ($>$) 50 ml/min:

$$Cl_{ij} = \exp(\beta_1 + \beta_4 w_i + \beta_5 g_{ij} + \beta_6 c_{ij} + b_{1i})$$

$$V_{ij} = \exp(\beta_2 + b_{2i}), \quad k_{ai} = \exp(\beta_3),$$

$$\boldsymbol{\beta} = [\beta_1, \beta_2, \beta_3, \beta_4, \beta_5, \beta_6]';$$

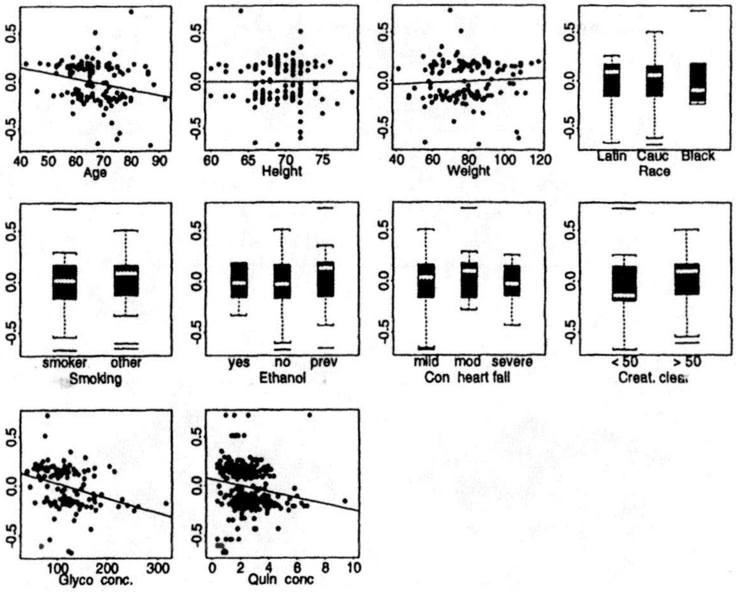

Figure 9.2. *Empirical Bayes estimates of random effects for V plotted against potential covariates for the SNP fit of Model 1 to the quinidine data. See Figure 9.1 for a description.*

we refer to this as Model 2. In Table 9.2, we report the values of the BIC, HQ, and AIC model selection criteria, described in section 7.3.2, for the fit of this model for a progression of K values (smaller values of the criteria are preferred). The inclusion of weight, α_1-acid glycoprotein concentration, and dichotomized creatinine clearance together is strongly supported for each value of K. The conservative BIC criterion selects $K = 0$, normality of the random effects, while HQ and AIC favor the richer $K = 2$ specification. The preference for $K = 2$ over $K = 0$ in Model 2 may be due to reduction in the amount of inter-individual variation being attributed to random error, thus allowing a more precise estimate of the random effects density.

Figure 9.3 shows the estimated density of $\boldsymbol{b}_i$ for the $K = 0$ and $K = 2$ fits of Model 2. In contrast to the fit of Model 1, the estimate for $K = 2$ is not multi-modal, suggesting that inclusion of further covariates at the second stage of the model may be unnecessary. To

Table 9.2. *Model selection criteria for NLMIX fits of Model 2 to the quinidine data; values are divided by* $N = 361$. P_{net} *is the total number of fixed parameters for each fit.*

K	P_{net}	*BIC*	*HQ*	*AIC*
		Model 2		
0	10	1.183	1.150	1.129
2	13	1.191	1.149	1.121
3	17	1.214	1.159	1.122

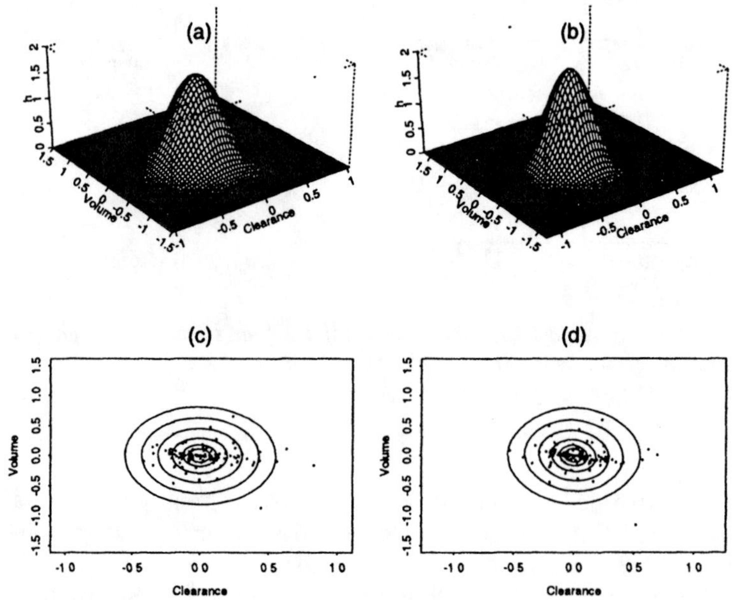

Figure 9.3. *Estimated random effects densities and empirical Bayes estimates for fits of Model 2 to the quinidine data. (a) perspective plot of the estimated density for* $K = 0$*; (b) perspective plot of the estimated density for* $K = 2$*; (c) contour plot of the estimated density for* $K = 0$ *at quantiles 10%, 25%, 50%, 75%, 90%, and 95%, with empirical Bayes estimates superimposed; (d) same as (c),* $K = 2$.

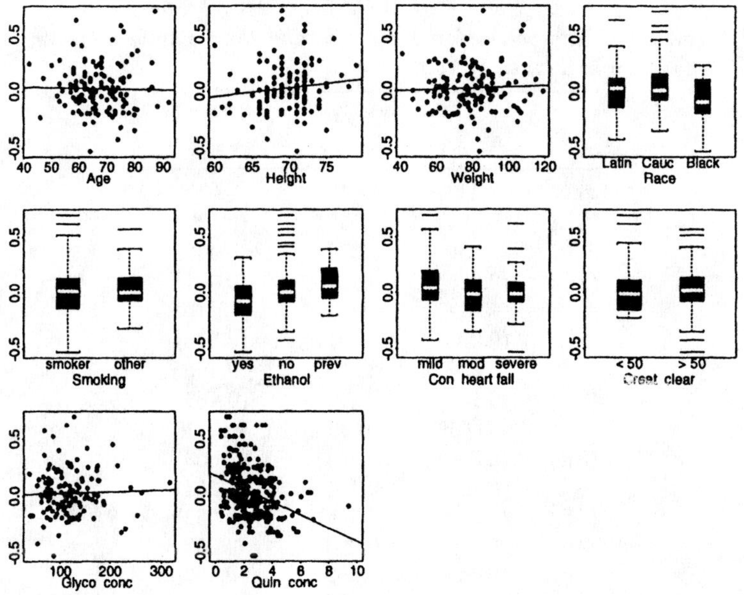

Figure 9.4. *Empirical Bayes estimates of random effects for Cl plotted against potential covariates for the SNP fit of Model 2 to the quinidine data. See Figure 9.1 for a description.*

assess further the effect of the extension of the model to incorporate the three chosen covariates, we plotted the same information as in Figures 9.1 and 9.2 for the fit of Model 2. Figure 9.4 shows this plot for the Cl random effect. The associations between the Cl random effects and the covariates included in Model 2 shows considerable attenuation from those in Figure 9.1. The relationships for the other covariates are weak, suggesting, along with Figure 9.3, that no further refinement is needed. The analogous plots for the V random effect, not shown, were consistently flat for all relationships, indicating no need to include covariates in the model for β_{2i}, once the systematic association of Cl with weight, α_1-acid glycoprotein concentration, and creatinine clearance was taken into account. Extensions to Model 2 including further covariates were investigated; in no case was visual or formal evidence for extension compelling; see Davidian and Gallant (1992).

A similar graphical analysis, supplemented by examination of

Table 9.3. *NLMIX and NONMEM fits to Model 2 for the quinidine data. The values for β_4 and β_5 and their standard errors have been multiplied by 100.*

	NLMIX, $K = 0$	*NLMIX,* $K = 2$	*NONMEM*
β_1	2.489	2.647	2.650
(SE)	(0.160)	(0.157)	(0.127)
β_2	5.360	5.356	5.440
(SE)	(0.085)	(0.084)	(0.072)
β_3	−0.890	−1.065	−0.079
(SE)	(0.293)	(0.301)	(0.334)
β_4	0.588	0.480	0.400
(SE)	(0.164)	(0.167)	(0.153)
β_5	−0.459	−0.495	−0.496
(SE)	(0.048)	(0.044)	(0.053)
β_6	0.165	0.112	0.154
(SE)	(0.042)	(0.046)	(0.062)
D_{11}	0.066	0.086	0.053
D_{22}	0.143	0.182	0.092
σ	0.242	0.235	0.250

the AIC model selection criterion, may also be undertaken using the NONMEM software. We do not report on such an analysis here; for comparison with the SNP fits, we report in Table 9.3 the results of a NONMEM fit to Model 2 along with the NLMIX results for $K = 0$ and $K = 2$. The estimates are fairly similar across methods, with the exception of k_a; this reflects the inherent difficulty in estimating absorption rate constant. Note that for the SNP fits, the estimated variances of the Cl and V random effects increase from $K = 0$ to $K = 2$. This may suggest that a difference in the two fits is in the spread of the random effects distribution.

Discussion

This example shows the utility of graphical techniques for screening a large number of potential covariates, as discussed in section 6.5. We have illustrated this approach in the context of SNP estimation; however, it may be used with other methods, such as those based on linearization in Chapter 6. Wakefield (1995) uses a similar strategy in a Bayesian analysis of these data using the techniques described

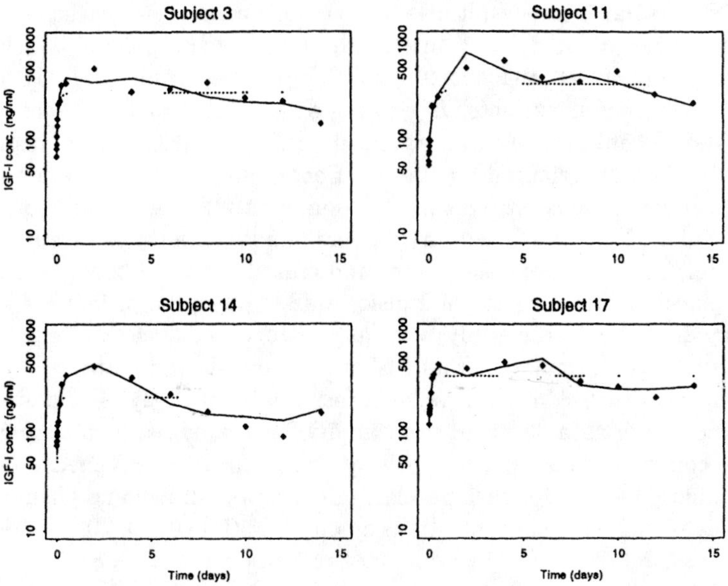

Figure 9.5. *IGF-I concentration-time profiles for four subjects. Empirical Bayes profiles from the NLME fits are superimposed; dashed line, fit of Model 1, solid line, fit of Model 2.*

in Chapter 8. The conclusions from the fits using the SNP approach are qualitatively similar both to those obtained from linearization methods as implemented in NONMEM, reported here, and those of Wakefield (1995). This supports the contention that, despite sparsity of observations on individual subjects, when the available data contain information of good quality on the population, all applicable methods are likely to lead to similar inferences.

9.4 Pharmacokinetics of IGF-I in trauma patients

Data description

Figure 9.5 shows IGF-I concentration-time profiles for four subjects taken from a clinical study of recombinant human IGF-I (rhIGF-I) in patients with severe head trauma. IGF-I, or insulin-like growth factor I, is a naturally occurring 70-amino acid protein that exhibits a high degree of homology to human insulin. It exerts a broad spectrum of effects within the body; in addition to the insulin-like effects its name implies, it also has anabolic properties. These provided the rationale for its use in the trial discussed here, which was a randomized, placebo-controlled, open label study of rhIGF-I in head trauma patients. It was hoped that the anabolic effects of IGF-I would reverse the catabolism and subsequent loss of lean body mass experienced by these patients. Subjects were randomized to continuous intravenous infusion of rhIGF-I or placebo at a rate of 0.01 mg/kg/hr within 48 hours of admission to the trauma center. Seventeen patients were randomized to IGF-I therapy, 16 to placebo. The duration of infusion was 14 days; the primary efficacy endpoint in the study was daily nitrogen balance, a measure of anabolism calculated from 24-hour urine collection data.

In our discussion here, we shall focus only on analysis of pharmacokinetic data obtained from the IGF-I treated patients during the course of the study. A relatively large number of PK measurements were taken for each patient; samples were drawn and plasma assayed for IGF-I concentrations on day 1 at time 0, 10, 20, 30, 45, 60, 120, 240, 360, and 720 minutes, and on days 2, 4, 6, 8, 10, 12, and 14 at time zero, where all times are measured relative to the beginning of infusion. Five deaths occurred in the study period, three in the placebo group and two in the IGF-I treatment group. One patient, subject #6 in the IGF-I treated group died on day 1; we have omitted that subject's data from all analyses reported below. The other death (subject #15), in that group occurred on study day 6; data for this patient are included.

A word on nomenclature is in order before proceeding. It is common to distinguish between endogenously occurring and exogenously administered protein by writing IGF-I for the former and rhIGF-I for the latter. The designation 'rh' indicates that the exogenously administered protein is the *human* form, produced by *recombinant* technology. We shall suppress this distinction, because, after a patient is dosed, for a given IGF-I measurement in plasma it is impossible to determine how much is due to exogenous administration. A further complication in studying IGF-I kinetics is that the protein may circulate in an unbound state (so-called 'free IGF-I') or bound in a secondary complex to one of six binding proteins; even, on occasion, in higher molecular weight ternary complexes. The immunoassay used in this study does not distinguish between free and bound forms of the protein; it measures total IGF-I. Thus, in reference to IGF-I concentration throughout this section, total IGF-I concentration is implied.

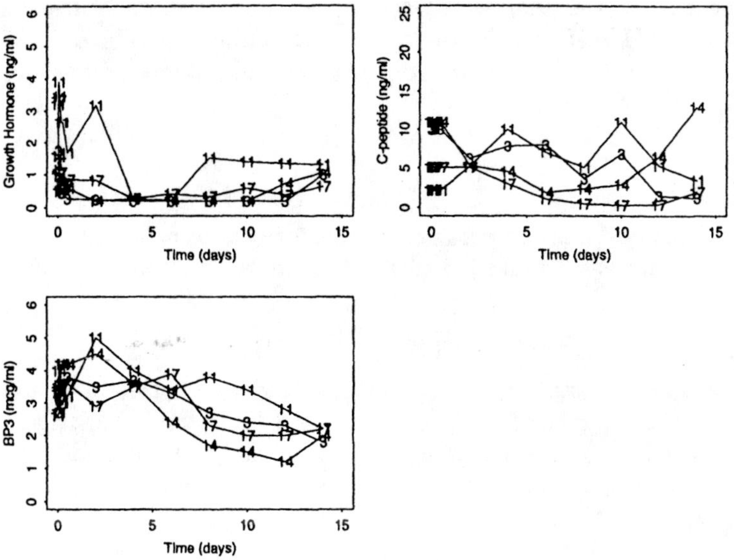

Figure 9.6. *Values of GH, C-peptide, and BP3 for four subjects, IGF-I pharmacokinetic study.*

In addition to IGF-I concentration data, information on demographic and physiological characteristics was available for each of the 16 patients included in the analysis. These covariates include time-independent quantities (sex, race, baseline serum chemistry values) as well as a number of time-dependent covariates. We shall be concerned with three variables in particular: growth hormone (GH) levels, C-peptide levels, and the levels of binding protein 3 (BP3) for IGF-I. These were measured throughout the study period, at times not necessarily coincident with the PK sampling times. C-peptide levels are often considered to be a proxy for endogenous insulin production, and BP3 is the main binding protein for IGF-I; that is, the majority of circulating IGF-I does so bound to BP3. We shall refer to GH, C-peptide, and BP3 levels as 'time-dependent covariates,' although, strictly speaking, this is an abuse of terminology, as their values are affected by the treatment. Figure 9.6 shows GH, C-peptide, and BP3 levels for four of the subjects on study.

Objective

The goal of the analyses described below is to characterize the kinetics of IGF-I over the 14-day infusion period and to investigate the relationship between kinetic behavior and potential covariates of interest.

Methods

The standard one-compartment model for concentration at time t during intravenous infusion to steady-state is

$$C(t) = \frac{D}{Cl}\left\{1 - \exp\left(-\frac{Cl}{V}t\right)\right\}, \tag{9.9}$$

where D represents dose rate, and Cl and V are clearance and volume of distribution, respectively. As a first step in describing the IGF-I concentration-time profiles, one might consider adding a term B to (9.9) to represent endogenous, or baseline, IGF-I level. Thus, the nonlinear model $f(\boldsymbol{x}, \boldsymbol{\beta})$ is given by

$$C(t) = B + \frac{D}{Cl}\left\{1 - \exp\left(-\frac{Cl}{V}t\right)\right\},$$

with $\boldsymbol{x} = [D, t]'$ and parameter vector $\boldsymbol{\beta}$ consisting of the values B, Cl, and V. Adopting subject-specific values for the PK parameters B, Cl, and V, write B_i, Cl_i, and V_i for the parameter values for the ith patient and parameterize the model in terms of $[B_i^*, Cl_i^*, V_i^*]' = [\log B_i, \log Cl_i, \log V_i]'$. Letting D_i be the (constant) infusion rate over the study period for subject i, write the following two-stage model:

Stage 1 (within-subject)

$$y_{ij} = \exp(B_i^*) + \frac{D_i}{e^{Cl_i^*}}\left\{1 - \exp\left(-\frac{e^{Cl_i^*}}{e^{V_i^*}}t_{ij}\right)\right\} + e_{ij},$$

$$(\boldsymbol{e}_i|\boldsymbol{\beta}_i) \sim \{\boldsymbol{0}, \boldsymbol{R}_i(\boldsymbol{\beta}_i, \boldsymbol{\xi})\}, \tag{9.10}$$

where y_{ij} represents the jth concentration measurement taken at time t_{ij} for the ith subject.

Stage 2 (subject-to-subject)

$$\boldsymbol{\beta}_i = [\beta_{1i}, \beta_{2i}, \beta_{3i}]' = [B_i^*, Cl_i^*, V_i^*]',$$

$$\boldsymbol{\beta}_i = \boldsymbol{\beta} + \boldsymbol{b}_i, \quad \boldsymbol{b}_i \sim (\boldsymbol{0}, \boldsymbol{D}), \tag{9.11}$$

where $\boldsymbol{\beta} = [\beta_1, \beta_2, \beta_3]'$, and $\boldsymbol{b}_i = [b_{1i}, b_{2i}, b_{3i}]'$. Thus, we have

$$\beta_{1i} = B_i^* = \beta_1 + b_{1i}, \quad \beta_{2i} = Cl_i^* = \beta_2 + b_{2i},$$

$$\beta_{3i} = V_i^* = \beta_3 + b_{3i}.$$

Because the parameters B, Cl, and V must be nonnegative to be meaningful, we have again adopted the convention of enforcing this constraint by parameterizing in terms of $\log B$, $\log Cl$, and $\log V$. This may also improve the approximate normality of the distribution of $\boldsymbol{\beta}_i$ implied in the second stage in (9.11). Furthermore, a log-linear model for the PK parameters necessarily implies the assumption of a multiplicative inter-subject error structure. As discussed in connection with the previous PK examples throughout this book, this assumption is likely to provide accurate representation of the nature of inter-individual random variation in PK parameters, whose population distributions are often skewed with constant coefficient of variation.

We shall refer to the model specified by (9.10) and (9.11) as Model 1. Inspection of the IGF-I concentration profiles in Figure 9.5 indicates that this model cannot provide adequate representation without further refinement, the model specifies that concentrations reach, and maintain, a steady-state level, which is obviously not the case for the data at hand. Nonetheless, this model provides a convenient point of departure.

One possible extension is to relate the individual PK parameters $\boldsymbol{\beta}_i = [B_i^*, Cl_i^*, V_i^*]'$ to subject-specific covariates, $\boldsymbol{a}_i$, say, at Stage 2 of the model. We have seen that this approach can be informative; for instance, in the phenobarbital and quinidine examples. For the IGF-I data, however, relating $\boldsymbol{\beta}_i$ to subject-specific *constant* covariates was not fruitful; this approach is predicated on the assumption that the components of $\boldsymbol{\beta}_i$ are themselves unchanging over time. But the failure to maintain steady state IGF-I levels in these patients suggests that this assumption is ill-founded for this data set. It seems evident, for instance, that the clearance of IGF-I does not remain constant for a given subject over the 14-day infusion period. Thus, a model which allows the PK parameters to vary over time may better suit these data.

Accordingly, we chose to extend Model 1 by allowing the possibility that clearance and volume of distribution might vary not only across subjects but across time for a given subject. Given the interpretation of B_i as endogenous IGF-I at baseline, we did not treat B_i^* as time-varying. One might, however, consider a model where B_i^* changes over time, reflecting changes in endogenous IGF-I production, though estimation in such a model might encounter identifiability problems.

Allowing clearance and volume to vary both across subjects and over time, it makes sense to try to accommodate this variation at Stage 2 of the hierarchy, possibly by introduction of time-dependent covariates. As in section 4.2.5, we considered a sequence of extensions to Model 1 of the following form:

Stage 1 (within-subject)

$$y_{ij} = \exp(B_i^*) + \frac{D_i}{e^{Cl_{ij}^*}}\left\{1 - \exp\left(\frac{e^{-Cl_{ij}^*}}{e^{-V_{ij}^*}} t_{ij}\right)\right\} + e_{ij},$$

$$(\boldsymbol{e}_i|\boldsymbol{\beta}_{ij}) \sim \{\boldsymbol{0}, \boldsymbol{R}_i(\boldsymbol{\beta}_{ij}, \boldsymbol{\xi})\}, \tag{9.12}$$

where $\boldsymbol{\beta}_{ij} = [B_i^*, Cl_{ij}^*, V_{ij}^*]'$.

Stage 2 (subject-to-subject)

$$\boldsymbol{\beta}_{ij} = \boldsymbol{A}_{ij}\boldsymbol{\beta} + \boldsymbol{b}_i, \quad \boldsymbol{b}_i \sim (\boldsymbol{0}, \boldsymbol{D}), \tag{9.13}$$

where $\boldsymbol{A}_{ij}$ is a $(p \times r)$ design matrix incorporating time-dependent covariate information with $p = 3$ and r determined by the nature of the specification, and the fixed effects vector $\boldsymbol{\beta}$ is $(r \times 1)$.

For example, one might model log clearance and log volume with a linear dependence on one or more of the time-varying covariates:

$$\begin{aligned}
\beta_{1ij} &= B_{ij}^* = \beta_1 + b_{1i} \\
\beta_{2ij} &= Cl_{ij}^* = \beta_2 + \beta_4 BP3_{ij} + \beta_5 CPEP_{ij} + b_{2i} \\
\beta_{3ij} &= V_{ij}^* = \beta_3 + \beta_6 GH_{ij} + b_{3i},
\end{aligned}$$

where $\boldsymbol{\beta} = [\beta_1, \beta_2, \beta_3, \beta_4, \beta_5, \beta_6]'$ $(r = 6)$; $BP3_{ij}$, $CPEP_{ij}$, and GH_{ij} represent the level of BP3, C-peptide, and GH, respectively, for subject i at the jth sampling time, and

$$\boldsymbol{A}_{ij} = \begin{bmatrix} 1 & 0 & 0 & 0 & 0 & 0 \\ 0 & 1 & 0 & BP3_{ij} & CPEP_{ij} & 0 \\ 0 & 0 & 1 & 0 & 0 & GH_{ij} \end{bmatrix}.$$

Note that incorporation of time-dependent covariates in this manner presupposes that covariates are measured at the same time points as the PK measurements or at least fall in a relevant interval 'close to' the PK sampling times. This was not true for this study;

covariates after day 1 of the infusion period were recorded a full day before or after PK values. Thus, some type of interpolation or imputation scheme is needed to obtain covariate values corresponding to PK measurement times t_{ij}. This can be done simply, e.g. by invoking a 'last value carried forward' or 'nearest neighbor' imputation rule. More sophisticated model-based interpolation methods could also be used and may help mitigate possible attenuation bias in fixed-effect estimates induced by measurement error in the covariates. Higgins, Davidian and Giltinan (1995) describe such an approach; for our illustrative analysis here, we used a simpler 'nearest neighbor' convention.

Despite availability of relatively full profiles for each subject, methods based on individual estimates were not used, as these do not allow incorporation of time-dependent covariates. Instead, Model 1 and its extensions were fitted using the techniques based on linearization, described in Chapter 6. We used both the NLME and NONMEM programs to carry out the fitting. In each case, a multiplicative within-subject error structure was assumed; this was accomplished differently for each approach. For estimation by the method of Lindstrom and Bates (1990) (section 6.3.2) as implemented in the NLME software, we applied a log transformation to both sides of the model for Stage 1 (see section 2.7) to accommodate the intra-individual heterogeneity of variance. Results are given below for maximum likelihood estimation of the covariance components; conclusions based on restricted maximum likelihood estimation are similar. Inference using full normal maximum likelihood was accomplished using the NONMEM package, under both the first-order and conditional first-order linearizations, referred to as the FO and FOCE methods in the software documentation; results below are based on the conditional first-order linearization. Here, a multiplicative error structure for within-subject variation was specified explicitly by taking $\boldsymbol{R}_i(\boldsymbol{\beta}_i, \boldsymbol{\xi})\}$ to be the diagonal matrix with jth diagonal element $\sigma^2 f^2(\boldsymbol{x}_{ij}, \boldsymbol{\beta}_{ij})$. The existing intrinsic subroutine within NONMEM for a one-compartment intravenous infusion model is based on (9.9), with no baseline term. It is relatively straightforward to accommodate the baseline parameter without writing a new subroutine. For NLME, the user must write an Splus function specifying the model f for mean response.

Results

Table 9.4 summarizes results for a sequence of models of the form (9.13) fitted using NLME. Inspection of the approximate AIC

Table 9.4. *Model selection criteria for NLME fits to the IGF-I data.* P_{net} *is the total number of fixed parameters for the model,* χ^2 *is the value of the 'likelihood ratio test' statistic for comparison to the indicated model: Models 2–7 are compared to Model 1, and Models 8–11 are compared to Model 2.*

Model	*Covariates included*	P_{net}	*Approx. loglike.*	*AIC*	χ^2
1	None	10	77.6	−135	-
2	BP3 in Cl^*	11	130.2	−238	105.2
3	CPEP in Cl^*	11	84.4	−146	13.6
4	GH in Cl^*	11	77.9	−133	0.6
5	BP3 in V^*	11	78.6	−135	2.0
6	CPEP in V^*	11	78.0	−134	0.8
7	CH in V^*	11	77.8	−134	0.4
8	BP3, CPEP in Cl^*	12	130.9	−237	1.4
9	BP3 in Cl^* BP3 in V^*	12	129.5	−235	-
10	BP3 in Cl^* CPEP in V^*	12	130.7	−237	1.0
11	BP3 in Cl^* GH in V^*	12	130.8	−237	1.8

model selection criteria (section 6.3.2, smaller values are favored) for each model yields the following conclusions. Considerable improvement over Model 1 may be obtained; addition of either C-peptide or BP3 to explain variation in $\log Cl$ improves the model fit significantly. BP3 is the more important of these two covariates; once it has been entered into the model, no further significant improvement is achieved (Models 8–11 versus Model 2); in particular, inclusion of C-peptide in addition to BP3 in the model for $\log Cl$ does not appear to improve the fit by very much (Model 8 versus Model 2). There was a slight irregularity in this sequence of fits – a slight *drop* in the approximate loglikelihood between Models 2 and 9, perhaps attributable to flatness of the optimization surface. Nonetheless, the overall conclusion from Table 9.4 is unambiguous; BP3 is the major factor associated with changes in IGF-I clearance over time, and the other available covariates do not add explanatory power; thus, Model 2, which specifies a linear dependence of

Table 9.5. *NLME and NONMEM fits to Model 2.* β_1, β_2, *and* β_3 *are the intercepts for* B^*_{ij}, Cl^*_{ij}, *and* V^*_{ij}, *respectively;* β_4 *is the coefficient of* $BP3_{ij}$ *in* Cl^*_{ij}. *The fit reported using NONMEM is that under the conditional first-order linearization (FOCE).*

	Fit by NLME	*Fit by NONMEM*
β_1	4.040	4.070
(SE)	(0.156)	(0.155)
β_2	−0.954	−1.620
(SE)	(0.193)	(0.260)
β_3	−4.727	−3.710
(SE)	(0.246)	(0.371)
β_4	−0.547	−0.378
(SE)	(0.050)	(0.061)
D_{11}	0.329	0.371
D_{12}	−0.034	–
D_{13}	−0.036	–
D_{22}	0.249	0.201
D_{23}	0.123	–
D_{33}	0.775	1.790
σ	0.287	0.256

$\log Cl$ on BP3, is preferred. Table 9.5 summarizes the fits for this model using both the NLME and NONMEM software packages; for the latter, the default specification of diagonal $\boldsymbol{D}$ was used. Estimates of the fixed effects are reasonably consistent across methods.

Figure 9.5 shows empirical Bayes fits to the concentration-time profiles for four subjects, based on NLME fitting of Model 1 (dashed line) and Model 2 (solid line), respectively. Inclusion of BP3 in the model for $\log Cl$ improves the fit; even so, the overall fit of Model 2 is not very impressive. It is apparent that the variation in clearance of IGF-I over time is a complicated phenomenon; the relative simplicity of incorporation of available covariate information taken here only goes part of the way toward describing it. In the absence of other possibly informative covariates, such as levels of other binding proteins and/or ternary complexes, we did not pursue further modeling of these data beyond that reported here.

Discussion

This example provides a good illustration of the utility of linearization-based methods in handling time-dependent covariates. As mentioned earlier, methods based on individual-subject parameter estimates are not equipped to deal with this case. One issue that we did not address in the preceding analysis is that of measurement error in the covariates. This feature, combined with the need to have relevant covariate values available at the times of PK sampling, suggests that modeling covariates in terms of a smooth underlying process and using appropriate smoothed values in the PK model may be an attractive approach. Results of such an analysis of the IGF-I data are reported elsewhere (Higgins *et al.*, 1995).

9.5 Dose escalation study of argatroban

Data description

Table 9.6 shows data from two participants in a PK/PD study of argatroban, an anti-coagulant. Thirty-seven subjects took part in this study; each received a four-hour intravenous infusion of one of several doses of argatroban. Infusion rates ranged from 1 to 5 μg/kg/min in increments of 0.5 μg/kg/min. Four subjects were dosed at each of the nine infusion levels. A 37th subject had a recorded infusion rate of 4.38 μg/kg/min; this irregularity probably stems from a dosing error in the highest dose group. Serial blood samples were taken from each subject at 30, 60, 90, 115, 160, 200, 240, 245, 250, 260, 275, 295, 320, and 360 minutes following initiation of the infusion and assayed for argatroban levels by HPLC. Additional blood samples were taken at five to nine time points per subject, ranging from 0 to 540 minutes following the beginning of infusion. These samples were used to measure the coagulation parameters activated partial thromboplastin time (aPTT) and prothrombin time (PT). Higher doses of argatroban should result in higher aPTT and PT values. Our analyses here focus only on aPTT measurements.

Objectives

The primary objectives of the analyses reported below were:

(i) to characterize the pharmacokinetic behavior of argatroban following intravenous infusion at the doses administered

Table 9.6. *Data for two subjects, PK/PD study of argatroban.*

	Subject 17		*Subject 33*	
time (min)	*conc.* (ng/ml)	*aPTT* (sec)	*conc.* (ng/ml)	*aPTT* (sec)
0	–	21.1	–	28.2
30	579.5	–	904.0	–
60	849.1	–	1568.0	–
90	928.8	–	1721.0	–
115	1305.2	–	1839.0	–
120	–	51.9	–	95.8
160	1272.0	–	2137.0	–
200	1301.7	–	2191.0	–
240	1288.4	58.6	2405.0	99.3
245	1064.0	–	1937.0	–
250	1041.2	–	2013.0	–
260	1197.9	–	1671.0	–
270	–	41.5	–	80.3
275	667.5	–	1362.0	–
295	534.9	–	1062.0	–
300	–	38.7	–	82.3
320	562.9	–	891.0	–
360	303.3	32.1	584.0	64.5
420	–	–	–	55.3
480	–	–	–	49.5
540	–	–	–	38.5
	Dose= 3.0 μg/kg/min		Dose= 5.0 μg/kg/min	

(ii) to elucidate the relationship between argatroban concentration and the pharmacodynamic endpoint aPTT.

Pharmacokinetic analysis – model and methods

Argatroban concentration-time profiles were fitted according to the model

$$C_p(t) = \frac{D}{Cl}\Big\{\exp\Big(-\frac{Cl}{V}t^*\Big) - \exp\Big(-\frac{Cl}{V}t\Big)\Big\}, \qquad (9.14)$$

$$t^* = 0,\ t \le t_{inf};\, t^* = t - t_{inf},\ t > t_{inf}.$$

In (9.14), D represents infusion rate, t_{inf} is the duration of infusion (240 minutes for each subject in this case), and Cl and V represent clearance and volume of distribution, respectively. This equation describes the standard one-compartment model for drug disposition during and following a single intravenous infusion. A two-compartment model might be considered; however, data for most subjects did not support fitting this more complicated model. Accordingly, results reported below are based on (9.14). Positivity of estimates was ensured by parameterization in terms of $Cl^* = \log Cl$ and $V^* = \log V$. Thus, the nonlinear model $f(\boldsymbol{x}, \boldsymbol{\beta})$ for argatroban concentration y considered below has the form (9.14) with $\boldsymbol{x} = [D, t]'$ and $\boldsymbol{\beta} = [\beta_1, \beta_2]' = [Cl^*, V^*]'$.

For the analyses presented below, intra-subject errors were taken to be uncorrelated, with variance proportional to a power of the mean response. The following hierarchical model was used to describe the population PK of argatroban:

Stage 1 (within-subject variation)

$$y_{ij} = \frac{D_i}{e^{Cl_i^*}}\left\{\exp\left(-\frac{e^{Cl_i^*}}{e^{V_i^*}}t_{ij}^*\right) - \exp\left(-\frac{e^{Cl_i^*}}{e^{V_i^*}}t_{ij}\right)\right\} + e_{ij},$$

$$(\boldsymbol{e}_i|\boldsymbol{\beta}_i) \sim \{\boldsymbol{0}, \boldsymbol{R}_i(\boldsymbol{\beta}_i, \boldsymbol{\xi})\}, \tag{9.15}$$

where y_{ij} represents the jth concentration measurement taken at dose D_i and time t_{ij} for the ith subject, and $\boldsymbol{R}_i(\boldsymbol{\beta}_i, \boldsymbol{\xi}) = \sigma^2 \boldsymbol{G}_i(\boldsymbol{\beta}_i, \theta)$, where $\boldsymbol{G}_i(\boldsymbol{\beta}_i, \theta) = \text{diag}[f^{2\theta}(x_{i1}, \boldsymbol{\beta}_i), \ldots, f^{2\theta}(x_{in_i}, \boldsymbol{\beta}_i)]$, $\boldsymbol{x}_{ij} = [D_i, t_{ij}]'$.

Stage 2 (subject-to-subject variation)

$$\boldsymbol{\beta}_i = [\beta_{1i}, \beta_{2i}]' = [Cl_i^*, V_i^*]',$$

$$\boldsymbol{\beta}_i = \boldsymbol{\beta} + \boldsymbol{b}_i, \quad \boldsymbol{b}_i \sim (\boldsymbol{0}, \boldsymbol{D}), \tag{9.16}$$

where $\boldsymbol{\beta} = [\beta_1, \beta_2]'$ and $\boldsymbol{b}_i = [b_{1i}, b_{2i}]'$.

In the absence of covariate information, only the simplest population model, given in Stage 2, was assumed for the PK parameters.

For the concentration data, we performed population analysis using methods based on individual parameter estimates and linearization of the model. Specifically, we implemented the global two-stage (GTS) algorithm described in section 5.3.2, with individual parameter estimates obtained by generalized least squares (GLS); all computations were programmed in GAUSS. In the GLS procedure, variance function estimation of the within-subject variance parameters (σ, θ) was carried out by the pooled methods of section 5.2.2, using the PL, REML, and AR objective functions. Because all three techniques gave very similar answers, we present GTS results only for individual estimates obtained by GLS with pooled pseudolikelihood estimation. Inference based on linearization was accomplished by the conditional first-order procedure of Lindstrom and Bates (1990) using the NLME suite of Splus functions and the first-order maximum likelihood method (section 6.2.2) implemented in the NONMEM package. For NLME, covariance parameters were estimated using restricted maximum likelihood. To accommodate possible heterogeneity of intra-subject variance, a square-root transformation was applied to both sides of the model (9.15); the square-root transformation was adopted based on the use of HPLC methods for determining argatroban concentrations. NONMEM analyses appeared to support a constant within-subject variance structure; results below are based on this specification. In the GTS and NLME analyses, the random effects covariance matrix $\boldsymbol{D}$ was taken as unrestricted; in the NONMEM analyses, the default specification of diagonal $\boldsymbol{D}$ was used.

Pharmacokinetic analysis – results

Table 9.7 summarizes the results for the GTS, NLME, and NONMEM fits to the argatroban concentration-time data. With the exception of the off-diagonal element of $\boldsymbol{D}$ for the first two methods and the parameter σ, estimates by the methods are in fairly close agreement. Empirical Bayes estimates of the subject-specific parameters obtained by all methods (not shown) also agree well; those for $\log Cl$ are virtually identical, and the NLME estimates of $\log V$ are consistently smaller than those obtained by GTS by about 0.1 unit. Figure 9.7 shows the NLME empirical Bayes fits for four selected subjects.

Pharmacodynamic analysis – model and methods

For argatroban, the PD endpoint of interest is measured directly in the blood, so there is no rationale for using an 'effect compartment' to accommodate lag between concentrations in the circulation and at the effect site. In the notation of section 9.2.2, $C_p(t)$ and $C_e(t)$ must be the same for these data, whether or not argatroban

Table 9.7. *Fits of the hierarchical model (9.15), (9.16) to the argatroban concentration data by GTS, NLME, and NONMEM. The disparity among estimates for σ reflects the difference in response scales and within-subject variance specifications.*

	Fitting method		
	GTS	*NLME*	*NONMEM*
β_1	−5.43	−5.42	−5.56
(SE)	(0.062)	(0.064)	(0.072)
β_2	−1.93	−1.84	−1.86
(SE)	(0.026)	(0.024)	(0.037)
D_{11}	0.1374	0.1478	0.142
D_{12}	0.0060	0.0193	–
D_{22}	0.0062	0.0056	0.019
σ	23.86	2.07	92.14

Pooled PL estimate of $\theta = 0.22$

concentrations are at steady state (given the timing of PD measurements relative to the infusion period for these data, there is little ability to assess hysteresis empirically). Accordingly, we used all available argatroban concentrations in modeling the effect on aPTT. The plot of aPTT against predicted argatroban concentration for all subjects, shown in Figure 9.8, guided our modeling strategy. Figure 9.9 shows the same plot, separately for each of four selected subjects. These figures suggest use of the sigmoid E_{max} model in (9.1) to describe the pharmacodynamic relationship, although the need to go beyond a linear relationship over this concentration range is not overwhelming. We fitted the E_{max} model for purposes of illustration. It is clear from Figure 9.9 that fitting an overall E_{max} model by methods based on individual estimates is not an option for these data. As is commonly the case in PD studies, subjects in the lower dose groups do not generally achieve a broad enough range of concentrations to allow individual E_{max} model parameters to be determined sensibly.

Accordingly, we used methods based on linearization to fit the PD model to the aPTT responses. Specifically, the conditional first-order linearization method of Lindstrom and Bates (1990) and the first-order method of Beal and Sheiner (1992) in sections 6.3.2 and 6.2.2, respectively, as implemented in the NLME and NONMEM programs, were used. For each, two approaches were implemented:

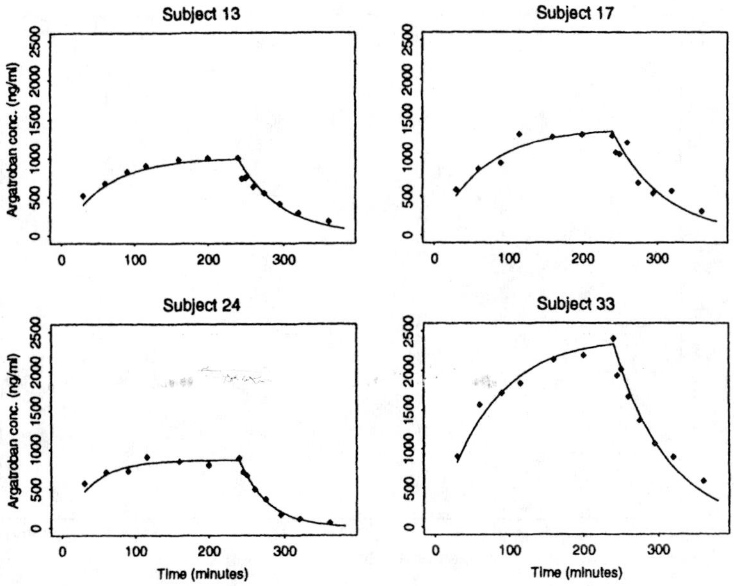

Figure 9.7. *Empirical Bayes profiles from the NLME fits of model (9.14) to argatroban concentration-time data for four subjects. Dose groups: subject 13, 2.5 μg/kg/min; subject 17, 3.0 μg/kg/min; subject 24, 3.5 μg/kg/min; subject 33, 5.0 μg/kg/min.*

(i) Sequential modeling: First, the PK model was fitted to the argatroban concentration-time data. For each subject, the empirical Bayes estimates of the individual-specific PK parameter vector, $\boldsymbol{\beta}_i$, were used to obtain predicted argatroban concentrations corresponding to the times at which PD measurements were taken. Use of predicted, rather than observed, concentrations may reduce attenuation bias in estimating the population PD parameters, as observed concentrations are measured with error, although predicted concentrations still contain error due to estimation. Obtaining reliable standard errors may be difficult in either case, as uncertainty in measured or predicted concentrations is not taken into account. For these analyses, intra-subject errors were taken to be uncorrelated and the following hierarchical model was used to characterize population PD:

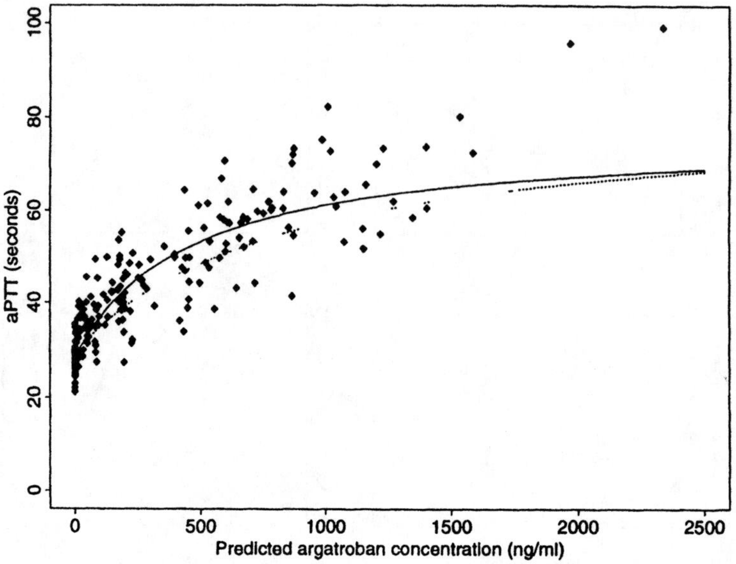

Figure 9.8. *aPTT versus predicted argatroban concentration for all subjects, argatroban pharmacodynamic study. Predicted aPTT values based on the estimated population mean fixed effects from the fits of (9.17), (9.18) by NLME (solid line) and NONMEM (dashed line) are superimposed.*

Stage 1 (within-subject variation)

$$y_{ij} = E_{0i} + \frac{E_{\max i} - E_{0i}}{1 + EC_{50i}/c_{ij}} + e_{ij},$$

$$(\boldsymbol{e}_i|\boldsymbol{\beta}_i) \sim \{\mathbf{0}, \boldsymbol{R}_i(\boldsymbol{\beta}_i, \boldsymbol{\xi})\}, \tag{9.17}$$

where now y_{ij} represents the jth aPTT measurement with associated predicted argatroban concentration c_{ij} for the ith subject, and $\boldsymbol{R}_i(\boldsymbol{\beta}_i, \boldsymbol{\xi})$ is an assumed within-subject covariance structure.

Stage 2 (subject-to-subject variation)

$$\boldsymbol{\beta}_i = [\beta_{1i}, \beta_{2i}, \beta_{3i}]' = [E_{0i}, E_{\max i}, EC_{50i}]',$$

$$\boldsymbol{\beta}_i = \boldsymbol{\beta} + \boldsymbol{b}_i, \quad \boldsymbol{b}_i \sim (\mathbf{0}, \boldsymbol{D}), \tag{9.18}$$

where $\boldsymbol{\beta} = [\beta_1, \beta_2, \beta_3]'$ and $\boldsymbol{b}_i = [b_{1i}, b_{2i}, b_{3i}]'$.

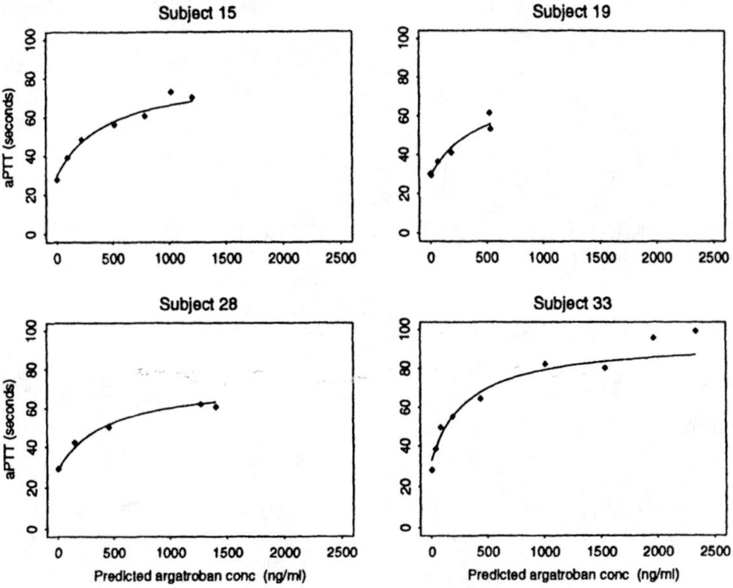

Figure 9.9. *aPTT versus argatroban concentration for four subjects. Empirical Bayes profiles from the NLME fits of models (9.17), (9.18) are superimposed. Dose groups: subject 15, 2.5 μg/kg/min; subject 19, 3.0 μg/kg/min; subject 28, 4.0 μg/kg/min; subject 33, 5.0 μg/kg/min.*

Thus, interest focuses on the PD fixed-effects parameter $\boldsymbol{\beta}$ representing the population mean values for E_0, $E_{\max}$, and EC_{50}.

In the NLME implementation of these fits, a transform both sides strategy was used to accommodate heterogeneous intra-individual variance in both the PK and PD measurements. Specifically, as discussed above, a square-root transformation was adopted in fitting the PK model; for the PD analysis, a log transformation was used. For the NONMEM fits, a constant within-individual variance was used for the pharmacokinetics, and an inter-subject multiplicative error structure was assumed for the pharmacodynamics.

(ii) Simultaneous modeling: This method involves simultaneous fitting of PK and PD models, proceeding as described in section 4.2.6. Specifically, let n_i now be the total number of PK and PD measurements on subject i, and let the $(n_i \times 1)$ vector $\boldsymbol{y}_i$ denote the 'stacked' PK and PD measurements taken on subject i, with jth element y_{ij}, so that j indexes position in the vector. Then one may specify a 'combined' hierarchical model based on (9.15)–(9.18) for the elements of $\boldsymbol{y}_i$ as follows:

Stage 1 (within-subject variation)

$$y_{ij} = C_p(t_{ij}) = \frac{D_i}{e^{Cl_i^*}}\left\{\exp\left(-\frac{e^{Cl_i^*}}{e^{V_i^*}}t_{ij}^*\right) - \exp\left(-\frac{e^{Cl_i^*}}{e^{V_i^*}}t_{ij}\right)\right\} + e_{ij}$$

if observation j on subject i is an argatroban concentration (PK) taken at time t_{ij}, and

$$y_{ij} = E_{0i} + \frac{E_{\max i} - E_{0i}}{1 + EC_{50i}/C_p(t_{ij})} + e_{ij}$$

if the jth observation on subject i at time t_{ij} is an aPTT (PD) value, and

$$(\boldsymbol{e}_i|\boldsymbol{\beta}_i) \sim \{\boldsymbol{0}, \boldsymbol{R}_i(\boldsymbol{\beta}_i, \boldsymbol{\xi})\}, \tag{9.19}$$

where $\boldsymbol{R}_i(\boldsymbol{\beta}_i, \boldsymbol{\xi})$ is an assumed within-subject covariance structure for the stacked error vector $\boldsymbol{e}_i$.

Stage 2 (subject-to-subject variation)

$$\boldsymbol{\beta}_i = [\beta_{1i}, \beta_{2i}, \beta_{3i}, \beta_{4i}, \beta_{5i}]' = [Cl_i^*, V_i^*, E_{0i}, E_{\max i}, EC_{50i}]',$$

$$\boldsymbol{\beta}_i = \boldsymbol{\beta} + \boldsymbol{b}_i, \quad \boldsymbol{b}_i \sim (\boldsymbol{0}, \boldsymbol{D}), \tag{9.20}$$

where $\boldsymbol{\beta} = [\beta_1, \beta_2, \beta_3, \beta_4, \beta_5]'$ and $\boldsymbol{b}_i = [b_{1i}, b_{2i}, b_{3i}, b_{4i}, b_{5i}]'$.

This model thus accommodates simultaneous assessment of PK and PD variability through Stage 2, where individual PK and PD parameters are assumed to be jointly distributed, with (5×5) covariance matrix $\boldsymbol{D}$. The specification for within-subject error, $\boldsymbol{R}_i(\boldsymbol{\beta}_i, \boldsymbol{\xi})$, should account for likely different response variance models for PK and PD components.

Fitting the simultaneous model (9.19), (9.20) was straightforward using the NONMEM software; the flexibility allowed in specifying the model for intra-individual variance makes it easy to accommodate a response vector for each subject with components corresponding to two different types of measurements. In contrast, a considerable amount of manipulation was needed to allow simultaneous modeling in NLME. Because the program assumes homoscedasticity of within-subject response, the presence of two differently scaled measurements poses a problem. To circumvent this, we rescaled the argatroban concentration measurements by a factor based on estimation of the within-subject variances from separate PK and PD fits, to make the estimation of a single intra-individual scale factor imposed by NLME reasonable. This rescaling followed the separate transformations used for the PK (square-root) and PD (log) measurements to achieve homogeneity of variance for each. This degree of response manipulation before fitting shows that NLME is not really designed to accommodate simultaneous fitting of different response types.

In both the PK and PD models for the sequential fit (i), and in the simultaneous PK/PD model (ii), estimates of the PD parameters E_0, $E_{\max}$, and EC_{50} were not constrained to be positive, although this could be accomplished by reparameterization of the $\mathrm{E_{max}}$ model.

Pharmacodynamic analysis – results

Table 9.8 summarizes results from sequential fits to the aPTT data based on the hierarchical model (9.17),(9.18), using NLME and NONMEM. Predicted aPTT values based on the population mean fit by NLME (solid line) and NONMEM (dashed line) are shown in Figure 9.8. We remark that parameter estimates obtained by fitting the aPTT data untransformed (NLME) or with constant within-subject variance (NONMEM), not reported, are extremely sensitive to inclusion of the two highest predicted concentration values (both from subject 33 in the highest dose group). Use of the log transformation or a multiplicative error structure, respectively, alleviates this sensitivity – estimates obtained with and without these two values are very similar.

Parameter estimates from the simultaneous PK/PD fit of (9.19), (9.20) are summarized in Table 9.9. Fixed-effect estimates are generally similar to those obtained from the separate fits for each method (compare Table 9.9 to Tables 9.7 and 9.8). Empirical Bayes prediction curves based on the fits to the joint PK/PD model were comparable to those in Figure 9.9, and are not reported.

Estimates of the EC_{50} parameter ranged from less than 100 to greater than 1000 ng/ml, depending on the method used, assumptions pertaining to within-individual and inter-individual error structures, and inclusion or exclusion of the two largest aPTT values. The instability in this particular parameter indicates that the data are insufficient to determine its value. Inspection of Figure 9.8 shows that this is not entirely unexpected; as mentioned previously, a linear relationship may well be adequate over this

Table 9.8. *Fits of the hierarchical model (9.17), (9.18) to the aPTT measurements by NLME and NONMEM, where* $\beta_1 = E_0$, $\beta_2 = E_{\max}$, $\beta_3 = EC_{50}$.

	Fitting method	
	NLME	*NONMEM*
β_1	28.4	29.4
(SE)	(0.53)	(0.57)
β_2	76.9	82.7
(SE)	(2.76)	(3.93)
β_3	472.4	89.3
(SE)	(51.9)	(14.9)
D_{11}	7.12	9.48
D_{12}	13.46	–
D_{13}	−254.1	–
D_{22}	52.77	56.30
D_{23}	−413.29	–
D_{33}	9227.6	1090.0
σ	0.081	0.083

range of concentrations. Table 9.9 shows other evidence of instability; for instance, the estimate of D_{44} in the NONMEM fit is implausible. In our experience, the fitting procedures discussed in this chapter are so complex as to have an inherent potential for this type of computational instability.

Discussion

These data exemplify many of the issues that arise commonly in PD modeling of early clinical data. In this case, the sequential and simultaneous modeling approaches gave similar conclusions; we do not have sufficient experience to comment on whether or not this is true in general. Lagged response was not an issue for these data, but is likely to be of concern for many pharmacodynamic responses. Further discussion of these issues may be found in the references cited in section 9.2.2.

9.6 Discussion

The examples presented in this chapter can only give a flavor for the applicability of population methods in PK/PD modeling.

Table 9.9. *Fits of the combined hierarchical model (9.19), (9.20) to the argatroban concentrations and aPTT measurements by NLME and NONMEM, where $\beta_1 = Cl^*$, $\beta_2 = V^*$ $\beta_3 = E_0$, $\beta_4 = E_{max}$, $\beta_5 = EC_{50}$. For brevity, only estimates of the diagonal elements of D are reported. Because of the approach taken to estimating the within-subject scale parameters, only an estimate for that corresponding to the PK data was obtained from NLME; for NONMEM, a separate scale parameter was estimated for PK and PD data.*

	Fitting method	
	NLME	*NONMEM*
β_1	−5.35	−5.58
(SE)	(0.066)	(0.072)
β_2	−1.77	−1.82
(SE)	(0.026)	(0.043)
β_3	28.4	30.1
(SE)	(0.54)	(0.75)
β_4	75.5	87.2
(SE)	(2.98)	(5.34)
β_5	456.0	822.0
(SE)	(57.0)	(126.0)
D_{11}	0.1567	0.137
D_{22}	0.0090	0.017
D_{33}	7.575	14.60
D_{44}	66.73	1.61×10^{-8}
D_{55}	14.329	8070.0
σ	0.083 (PK)	0.098 (PK)
	–	92.52 (PD)

Population pharmacokinetics, in particular, have seen widespread use within the last decade. Reports of population PK analyses appear regularly in the pharmacology literature in journals such as *Journal of Pharmacokinetics and Biopharmaceutics*, *Clinical Pharmacokinetics*, *Clinical Pharmacology and Therapeutics*, and *British Journal of Clinical Pharmacokinetics.* It would be impossible to list a comprehensive bibliography here, and we refer the reader to these journals for further examples.

The use of population PK analyses as a basis for individualization of dosage regimens is gaining widespread acceptance. An excellent overview of the principles involved, as well as practical issues of implementation, may be found in Peck and Rodman (1991).

Use of population PD modeling techniques is less well-developed. In part, this may be a reflection of the difficulties enumerated in section 9.2.2.

CHAPTER 10

Analysis of assay data

10.1 Introduction

In this chapter, we explore several statistical issues which arise in the collection and interpretation of assay data. The hierarchical nonlinear modeling techniques discussed in this book, particularly those in Chapter 5, provide a useful approach to these issues. In developing drugs for therapeutic use, an important challenge is the derivation of suitable assays for measuring drug levels in serum, plasma, or other biological matrices. For instance, availability of reliable assays is central to the conduct of pharmacokinetic studies of the type discussed in the previous chapter. For low molecular weight drugs, assay techniques such as high-performance liquid chromatography (HPLC) are often adequate. The growth of genetic engineering and the consequent ability to exploit recombinant techniques to manufacture macromolecules such as proteins for therapeutic use has presented new challenges for assay development. Traditional methods such as HPLC often lack the specificity and sensitivity required to detect proteins in complex biological matrices. Instead, methods used to measure protein levels frequently exploit the specificity of immunoassay techniques such as radioimmunoassay (RIA) or enzyme-linked immunosorbent assay (ELISA). Such techniques can often achieve sensitivity in the pg/ml range. Although immunoassays are extremely useful in quantifying protein levels, a common requirement is the development of an assay that provides a direct measure of the biological activity of the protein, such as a cell-based bioassay. Such an assay is usually required for regulatory purposes, and reflects the fact that the degree of antigen binding exhibited by a polypeptide such as an antibody may not always provide a good indication of its biological activity. The relaxin data, illustrated in Figure 1.3, provide an example of this type of assay.

Section 10.2 provides a brief description of the principles underlying RIA and ELISA methods. Quantifying drug or protein levels corresponds to the statistical problem of *calibration*, where one estimates analyte concentration in an unknown sample from an interpolating curve fitted to known standard concentrations. In section 10.3 we show that the hierarchical nonlinear model described in section 4.2 provides an appropriate framework for calibration inference. Each of the following four sections addresses a particular aspect of analyzing assay data. Section 10.4 discusses the evaluation of intra-assay precision, and the construction and uses of precision profiles. In section 10.5 we discuss evaluation of lower limits of assay reliability, in particular, the determination of the minimum detectable concentration, or MDC. The construction of accurate confidence intervals about calibrated concentrations is considered in section 10.6. Application of the empirical Bayes methods of section 5.3 to data from multiple assay runs is explored in section 10.7. This chapter closes with a discussion and with some brief bibliographic notes. This chapter is self-contained; readers who do not need to concern themselves with assay or clinical chemistry data may proceed to Chapter 11 without loss of continuity.

10.2 Background on assay methods

10.2.1 Radioimmunoassay

The calibration of a given analyte in an unknown sample involves measurement of a signal and exploiting the monotonic nature of the signal versus concentration profile to obtain a concentration estimate by interpolation. In radioimmunoassay, as the name implies, the signal corresponds to a radioactive count. Referring to the substance to be quantitated as the *antigen*, the typical experimental setup is as follows.

To each of several tubes, a fixed amount of radiolabeled ('hot') antigen is added, as well as a fixed amount of antibody which binds to the hot antigen. Several known concentrations of unlabeled ('cold') antigen are added to those tubes used to generate standard curve information. The cold antigen competes with the fixed amount of hot for binding sites on the antibody until equilibrium is reached. Thus tubes with higher concentrations of the cold standard have a lower amount of bound hot antigen at equilibrium. Once the system has reached equilibrium, a second antibody is added which precipitates the bound complex. The complex

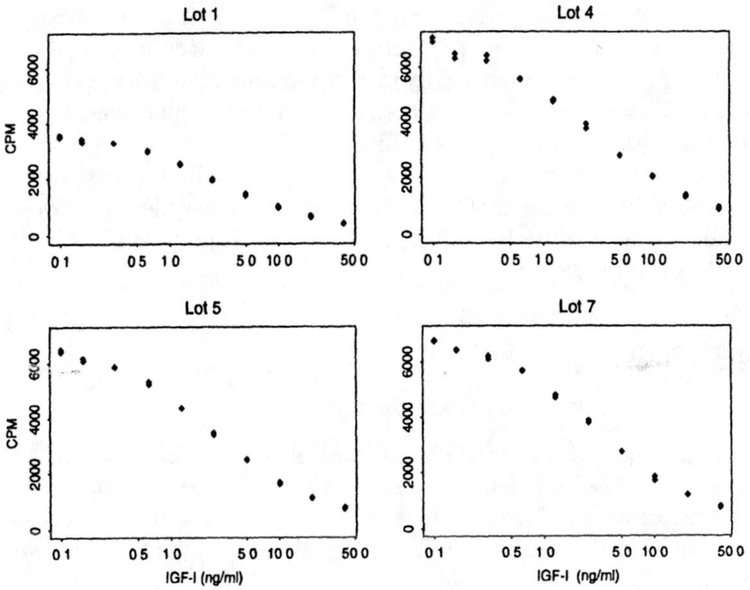

Figure 10.1. *Response-concentration profiles from four runs of a radioimmunoassay for the protein IGF-I.*

is counted on a gamma counter and the number of counts is inversely proportional to the amount of unlabeled antigen present in each tube. Figure 10.1 shows standard data from several runs of a competitive RIA for the protein IGF-I (insulin-like growth factor I).

10.2.2 Enzyme-linked immunosorbent assay

The main difference between RIA and ELISA is the labeling method used to generate the signal. ELISA techniques use an enzymatic labeling scheme wherein the amount of enzyme label present is measured by reacting with the appropriate substrate. A typical experimental protocol is as follows, where again the term 'antigen' designates the analyte to be measured.

The antigen to be quantitated is sandwiched between two antibodies. One of these is attached to the wells of a microtiter plate and the other is tagged with an enzyme label (frequently horseradish peroxidase). The more antigen present in a sample, the greater the degree of capture of the labeled antibody. At equilibrium, substrate is added, which effects a colorimetric reaction proportional to the amount of antigen present in each well. The intensity of reaction is assessed by measuring optical density (OD) in each well, a procedure which is easily automated by platereader. An example of data from a typical ELISA was seen in Figure 5.1, which shows standard curve data from several runs of an immunoassay for the recombinant protein DNase in rat serum. An excellent overview of the principles underlying ELISA techniques is provided by Clark and Engvall (1978).

10.3 Model framework

10.3.1 Model for the mean response

Suppose that data pairs (x_{ij}, y_{ij}) are available from m runs of the assay, where y_{ij} denotes the observed response at one of the concentrations x_{ij} for the ith assay run, $j = 1, \ldots, n_i$. A natural parametric framework for inference is the hierarchical nonlinear model

$$y_{ij} = f(x_{ij}, \boldsymbol{\beta}_i) + e_{ij}. \tag{10.1}$$

In this equation, the subscript i indexes run number within a series of assay runs, and j indexes measurements within a run. Replicate measurements usually occur at some or all of the concentrations for a given run; this is not explicitly highlighted in (10.1).

Response, y, relates to the known standard concentration, x, according to the nonlinear regression function f. For many assays, the standard curve exhibits a characteristic sigmoidal shape that is adequately described by the four-parameter logistic function

$$f(x, \boldsymbol{\beta}) = \beta_1 + (\beta_2 - \beta_1)/[1 + \exp\{\beta_4(\log x - \beta_3)\}]. \tag{10.2}$$

For example, this function may be used to fit the relaxin data plotted in Figure 1.3. It may also be used to characterize a decreasing S-shaped curve, as would be appropriate for the radioimmunoassay data shown in Figure 10.1. Choice of the four-parameter logistic standard curve is fairly typical in the context of immunoassay or bioassay. This choice may not always be adequate, however, and one may substitute any appropriate function f of a general p-dimensional parameter vector $\boldsymbol{\beta}$ in (10.1).

As an example, consider the data shown in Figure 10.2, representing concentration-response values for standards in the so-called Threshold assay, used to detect very low levels of DNA; these data

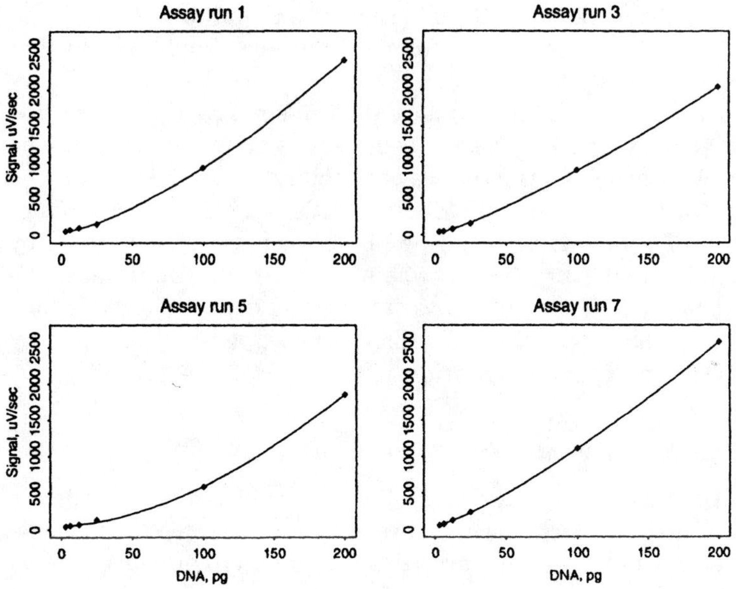

Figure 10.2. *Response-concentration profiles for the Threshold assay, used to measure very low levels of DNA.*

were kindly provided by Matt Field and Melissa Landi, Genentech, Inc. Measured values in the Threshold instrument are obtained by complexing single-stranded DNA with two binding proteins and labeling the complex with biotin. The biotinylated complex is placed in the Threshold reader and reacted with the substrate urea. This changes the local pH, resulting in a proportional change in potential. Voltage for each sample is measured at evenly-spaced time intervals and the rate of change (μV/sec), referred to as the signal, is an increasing function of the quantity of DNA in the sample.

Exploratory analysis of standard-curve data from several runs of this assay indicates that the four-parameter logistic family does not provide an adequate fit. Instead, a reasonable representation of these profiles is provided by the family of power curves

$$f(x, \boldsymbol{\beta}) = \beta_0 + \beta_1 x^{\beta_2}. \tag{10.3}$$

The parameter vector $\boldsymbol{\beta} = [\beta_0, \beta_1, \beta_2]'$ may vary from run to run. Fitted curves shown in Figure 10.2 were obtained using GLS with variance proportional to the mean response (iteratively-reweighted least squares with weights proportional to the reciprocal of the current fitted value). Determination of an appropriate weighting scheme based on assay data from several runs is addressed further in section 10.4.

It is clear from the relaxin data (Figure 1.3) that regression parameter values can change considerably from run to run. The model accommodates interassay variability by considering $\boldsymbol{\beta}_i$ as a random vector from a distribution with mean $\boldsymbol{\beta}$ and covariance matrix $\boldsymbol{D}$. This is accomplished in the context of the model by assuming that $\boldsymbol{\beta}_i = \boldsymbol{\beta} + \boldsymbol{b}_i$, where $\boldsymbol{b}_i \sim (\mathbf{0}, \boldsymbol{D})$. The usual specification for $\boldsymbol{b}_i$ is that of a p-variate normal distribution; this may be generalized to include distributions such as the lognormal by appropriate reparameterization.

10.3.2 Modeling intra-assay response variation

In (10.1) the intra-assay errors e_{ij} are assumed to be independently distributed with zero mean and variance $\sigma^2 g^2\{f(x_{ij}, \boldsymbol{\beta}_i), \boldsymbol{\theta}\}$, independently of the $\boldsymbol{b}_i$. The further assumption of normality is often reasonable. A routine phenomenon in assay data is heteroscedasticity in the intra-assay response. Typically, this is manifested as a systematic relationship between the intra-assay variability and the response level. This is taken into account by the scale parameter σ and a variance function g. As we did in Chapters 2 and 4, we usually assume that g depends on the conditional mean $f(x_{ij}, \boldsymbol{\beta}_i)$ and a q-dimensional parameter vector $\boldsymbol{\theta}$; that is:

$$\mathrm{E}(y_{ij}|\boldsymbol{\beta}_i) = f(x_{ij}, \boldsymbol{\beta}_i); \;\; \mathrm{Var}(y_{ij}|\boldsymbol{\beta}_i) = \sigma^2 g^2\{f(x_{ij}, \boldsymbol{\beta}_i), \boldsymbol{\theta}\}, \tag{10.4}$$

although g may be any general function of x_{ij}, $\boldsymbol{\beta}_i$, and $\boldsymbol{\theta}$. One chooses the variance function g to reflect the likely character of intra-assay response variability. In (10.4), g, σ, and $\boldsymbol{\theta}$ are assumed common across assay runs. This reflects a belief that the pattern of variability is consistent across assays, which seems reasonable provided assay procedures have stabilized. One should weigh the appropriateness of this assumption in any given situation; it is certainly possible conceptually to allow g, σ, and $\boldsymbol{\theta}$ to vary with i, though this will complicate inference on variance parameters; in particular, the pooled estimation methods described in section 5.2 will not apply.

In the case of assay data, one often models the variance as a function of the mean $\mu_{ij} = f(x_{ij}, \boldsymbol{\beta}_i)$; a common choice is the power model $g(\mu, \theta) = \mu^\theta$. Other possibilities have been discussed earlier; see section 2.2.3, in particular equations (2.6)–(2.8). The components of variance model (2.8) is particularly attractive in the context of assay data, where it is quite likely that a nonzero component of variance exists which dominates at low concentrations. We assume here that the form of g is known, but the analyst may be unable to specify an appropriate value for $\boldsymbol{\theta}$. As described in section 2.2.3, in developing a new assay, the variability may not be well understood; the power model may be reasonable, but common choices, such as $\theta = 0.5$ or 1.0 may be inappropriate (Finney, 1976). In quadratic or exponential models, it may be difficult to specify a value for $\boldsymbol{\theta}$ *a priori*. In such cases, a reasonable strategy is to estimate $\boldsymbol{\theta}$ from the data; we discuss methods for doing this in the next section.

10.4 Assessing intra-assay precision

In this section we discuss estimation of the variance parameters $(\sigma, \boldsymbol{\theta})$, given the model specifications described by (10.1) and (10.4). The numerical illustrations below involve the specific choice of f as the four-parameter logistic function. The methods apply for general choices of f, however, and throughout the exposition we continue to regard f as a general nonlinear function of x and of the regression parameter $\boldsymbol{\beta}_i$.

10.4.1 Estimation of variance parameters

In the assay context, standard concentration-response data are almost always available from a series of individual assay runs. This is fortunate; as discussed in Chapter 2, data from a single individual are rarely sufficient to provide a reliable characterization of intra-individual variability. In our development in this chapter, we shall assume that standard curve data are available from several runs of the assay under consideration. In practice, we have found that data from about 8 to 12 assay runs are generally adequate to provide a stable characterization of the pattern of intra-assay response variation.

Given the availability of data from several assay runs, the methods described in section 5.2 may be applied to estimate $(\sigma, \boldsymbol{\theta})$. Specifically, it is straightforward to use generalized least squares methods, where variance parameters are estimated based on pooled residuals across individual assay runs. As discussed in section 5.3, several variations of GLS are possible. Illustrations presented in this chapter are based on normal-theory pseudolikelihood estimation of $(\sigma,\boldsymbol{\theta})$, unless otherwise stated.

10.4.2 Intra-assay precision profiles

The main focus in analyzing assay data is *calibration*, the estimation of analyte concentration in an unknown sample, based on the level of response in replicates of the sample. Proper characterization of intra-assay variability in *response* is of interest only in as much as it affects evaluation of precision in calibrating unknown samples. The following development shows that assessment of intra-assay variability in y directly affects one's ability to evaluate the precision in estimating an unknown concentration, x.

A standard technique for evaluating the intra-assay precision associated with calibration of an unknown sample in a given experiment is to construct a *precision profile* (Ekins and Edwards, 1983). Such a profile is simply a plot of the estimated precision of an estimated concentration against concentration across the assay range. Several methods of constructing a precision profile are possible, depending on the measure of precision used (standard error or coefficient of variation) and the method used to calculate it. For concreteness, we focus here on one computational scheme, but the issues raised will also arise in the case of other calculation methods.

Approximate standard error for estimated concentration

Let $h(y,\ \boldsymbol{\beta})$ denote the inverse of the regression function f, specified in (10.1). In the particular case of the four-parameter logistic function, h has the form

$$x = h(y,\ \boldsymbol{\beta}) = \exp[\beta_3 + \log\{(\beta_2 - y)/(y - \beta_1)\}/\beta_4]. \qquad (10.5)$$

Let $h_y(y,\ \boldsymbol{\beta})$ and the p-variate vector-valued function $\boldsymbol{h}_\beta(y,\ \boldsymbol{\beta})$ represent the derivatives of h with respect to y and $\boldsymbol{\beta}$, respectively. Then if y_0 represents the average response for r_0 replicates of an unknown sample at concentration x_0 for the ith run, the estimated value of x_0 is given by $\hat{x}_0 = h(y_0, \hat{\boldsymbol{\beta}}_i)$, where $\hat{\boldsymbol{\beta}}_i$ is any estimate of $\boldsymbol{\beta}_i$ with estimated covariance matrix $\hat{\sigma}^2\hat{\boldsymbol{\Sigma}}_i$. We can obtain an estimated approximate variance for $\hat{x}_0$ by a standard Taylor series linearization, namely,

$$\mathrm{Var}(\hat{x}_0) = h_y^2(y_0, \hat{\boldsymbol{\beta}}_i)\hat{\sigma}^2 g^2(y_0, \hat{\boldsymbol{\theta}})/r_0 + \boldsymbol{h}_\beta'(y_0, \hat{\boldsymbol{\beta}}_i)\hat{\sigma}^2\hat{\boldsymbol{\Sigma}}_i\boldsymbol{h}_\beta(y_0, \hat{\boldsymbol{\beta}}_i), \qquad (10.6)$$

where $(\hat{\sigma}, \hat{\boldsymbol{\theta}})$ are estimates of the intra-assay variance parameters. The first term in (10.6) reflects uncertainty in the measurement of y_0. The derivative of the inverse function, h_y, converts error in the measured response to error in the estimated concentration. If the fitted standard curve is locally steep then the conversion factor h_y will be small. In contrast, in 'flat' portions of the standard curve, the inverse function h will have an effectively infinite derivative, thus reflecting the translation of even minor errors in y to huge differences in the estimate of x in this situation. In general, the first term of (10.6) is expected to dominate the second term, which corresponds to uncertainty in the fitted standard curve. For some assays, however, the contribution of the second term may be substantial; in such cases, the widespread practice in the assay and clinical chemistry literature of taking only the first term of (10.6) may result in highly optimistic estimates of intra-assay precision.

Importance of correct estimation of variance parameters

To illustrate the importance of a good understanding of the correct value of the variance parameter $\boldsymbol{\theta}$, consider the profiles shown in Figure 10.3. These represent precision profiles of $\mathrm{SE}(\hat{x}_0)/\hat{x}_0$ versus $\hat{x}_0$, based on run #1 of the relaxin bioassay, using the power variance model, with the standard error calculated as the square root (10.6) under the assumption that the exponent in the power variance model was $\theta = 0$, 0.5 and 1.0, respectively, and assuming triplicate response measurements for the unknown sample in all cases ($r_0 = 3$). The assessment of intra-assay precision in calibration diverges widely, according to our belief about the appropriate value of θ. Not all three can be correct; our point here is simply that accurate knowledge of θ is necessary to construct an accurate precision profile. Use of the pooled estimation techniques described in section 5.2 based on data from 8 to 12 assays generally allows reliable characterization of the precision of calibration.

For the relaxin bioassay data of Figure 1.4, pooled estimation of the power variance parameter, θ, based on nine assay runs, yields a value of 1.03, suggesting that a default value of $\theta = 1.0$ may be reasonable. This analysis addresses the issue of which of the three precision profiles illustrated in Figure 10.3 is most accurate; that for $\theta = 1.0$ is preferred.

The idea that pooled estimation of θ is superior to obtaining estimates based on data from only a single assay, given the validity of the assumption of a common pattern of variation across runs, is

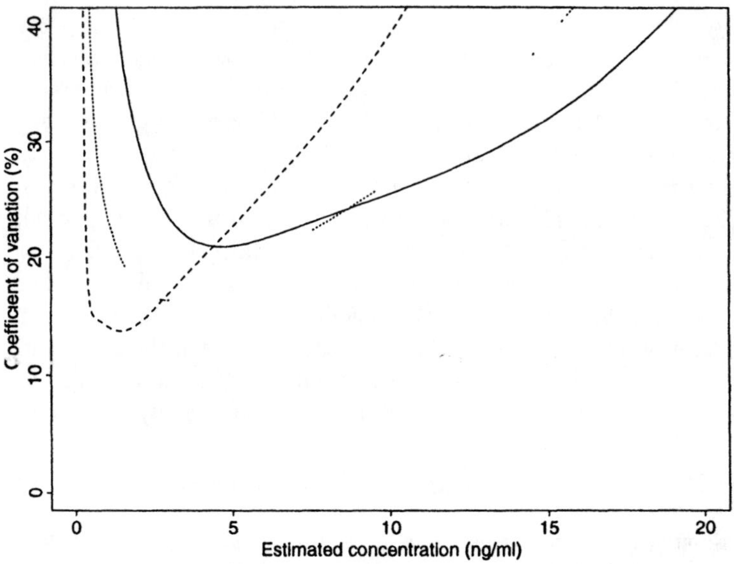

Figure 10.3. *Precision profiles for run #1 of the relaxin bioassay for different values of θ in the power variance model. Solid line, θ = 0.0; dotted line, θ = 0.5; dashed line, θ = 1.0.*

implicit throughout our discussion so far. This belief is vindicated by comparison of confidence intervals constructed for pooled and individual estimates of θ. For the relaxin assay, the approximate 95% confidence interval for θ obtained by the profile log-likelihood method (section 5.2) based on data from run #2 alone is (0.82, 1.48); that using pooled data from all nine runs of the bioassay is (0.90, 1.09). Pooling information across runs reduces the width of the confidence interval considerably.

Applications of precision profiles

Intra-assay precision profiles serve as a useful tool, especially during the development stage of an assay. The following examples illustrate three major uses:

Table 10.1. *Assay reporting ranges for the relaxin bioassay.*

Assay run	*Conc. range with CV < 20% (ng/ml)*	*EC20 to EC80 (ng/ml)*
1	0.50 to 4.9 (EC4 to EC50)	1.7 to 12.6
2	0.48 to 3.1 (EC3 to EC52)	1.7 to 10.7
3	0.38 to 6.8 (EC2 to EC 62)	1.85 to 12.0
4	0.65 to 4.0 (EC3 to EC46)	2.4 to 6.6
5	0.50 to 3.0 (EC3.5 to EC44)	1.7 to 6.0
6	0.30 to 4.9 (EC1 to EC53)	1.9 to 10.4
7	0.60 to 3.6 (EC2 to EC51)	1.85 to 7.0
8	0.50 to 5.4 (EC2.5 to EC64)	1.75 to 8.0
9	0.45 to 5.5 (EC2 to EC70)	1.7 to 6.5

1. Determination of assay range
2. Assessing the effect of varying the degree of replication on precision
3. Evaluating the effect of assay design changes on precision.

For the relaxin bioassay, assay users consider an intra-assay coefficient of variation of 20% or less acceptable. The precision profile provides an objective criterion for determining this 'usable' portion of the assay range. Table 10.1 summarizes assay ranges, based on separate precision profiles for each assay run, assuming $\theta = 1.0$ and $r_0 = 3$ in all cases. The rule of thumb, which had been in effect before formal precision analysis, was to take the EC20 to EC80 as the assay range; here, for instance, the EC20 is defined as that concentration eliciting a response 20% of the way between the minimum and maximum expected response. Inspection of the precision profiles shows this rule is flawed; its symmetry about the inflection point fails to account for the asymmetric nature of the response variability across the response range. The assay range derived from the precision profile, in contrast, is quite asymmetric about the EC50 value, typically corresponding to approximate EC3 to EC55 values. Given the extreme heteroscedasticity in the response, this represents a more credible assessment of the acceptable range.

The intra-assay precision profile may also be used to gauge the effect of changing the number of replicates per unknown sample on the final precision of calibration. For example, Figure 10.4 shows precision profiles for run #1 of the relaxin bioassay, assuming

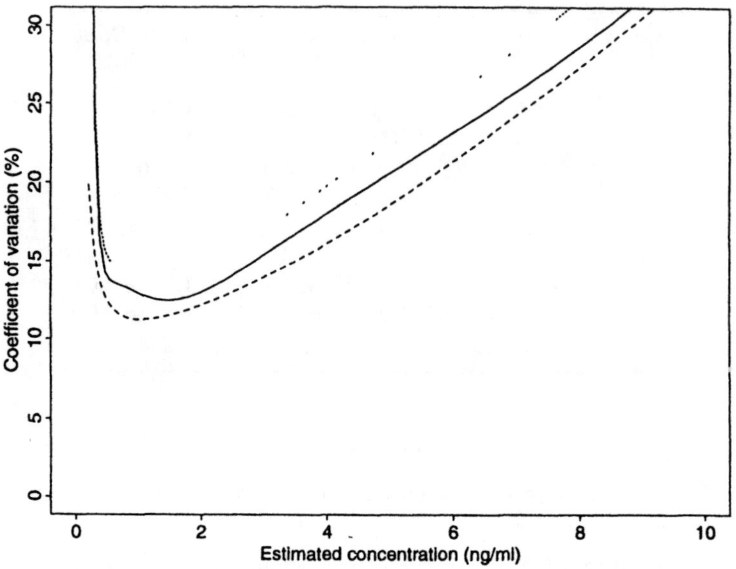

Figure 10.4. *Precision profiles for run #1 of the relaxin bioassay. Solid line, $r_0 = 4$; dotted line, $r_0 = 3$; dashed line, $r_0 = 3$ ignoring the second term in 10.6.*

$\theta = 1.0$, for $r_0 = 3$ and 4 replicates of unknown samples, respectively. Inspection of these profiles allows the analyst to assess the tradeoff between increased precision and diminished throughput when changing the degree of replication.

Figure 10.4 also shows a precision profile for run #1 of the relaxin bioassay, assuming $\theta = 1.0$ and $r_0 = 3$, but ignoring the second term in (10.6), the error due to fit. Comparison with the profile that correctly incorporates both terms in (10.6) shows that ignoring the error in fitting the standard curve can lead to considerable optimism in the assessment of precision.

Figure 10.5 shows standard curve data obtained for eight assay runs during the development stage of a cell-based assay for the immunoglobulin IgE; these data were kindly made available by David Fei and John Lowe, Genentech, Inc. These eight runs constitute a full 2^3 factorial design in the three factors incubation temperature (on ice or at room temperature), cell seeding density (20 or 40 thousand per well), and cell line (RBL48 or CHO3D10). For

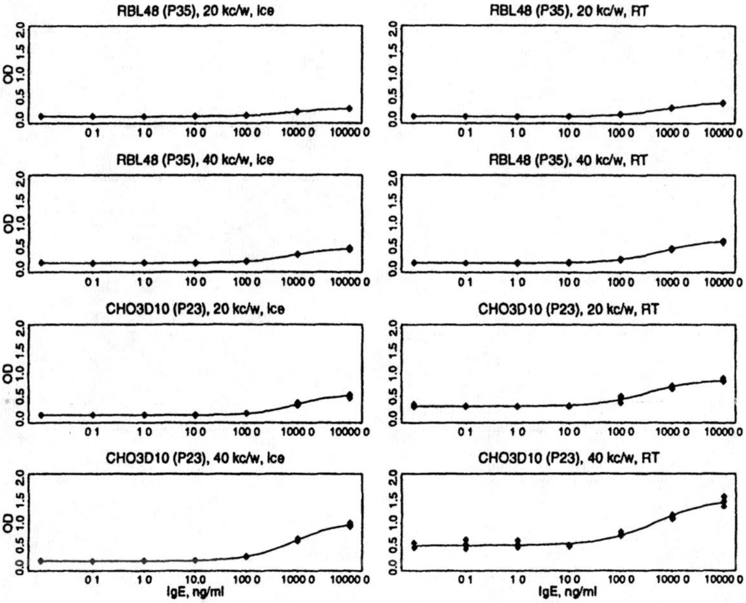

Figure 10.5. *Standard curves for eight runs of a cell-based assay for IgE.*

routine use of the assay, the best set of experimental conditions would maximize the range of the signal, and thus the steepness of the concentration- response curve, while minimizing variability among replicate responses. The overall effect on precision of varying the different experimental conditions is hard to gauge directly from Figure 10.5 alone. However, inspection of the corresponding precision profiles (based on $r_0 = 3$ replicates and a value of θ of 1.0), shown in Figure 10.6, indicates that the most promising set of experimental conditions appears to be the combination corresponding to use of CHO3D10 cells, seeded at the higher density, and incubated on ice. Obviously, further runs are necessary to confirm this result. The precision profile provides a useful intermediate tool in investigating the optimal set of experimental conditions before final qualification of the assay.

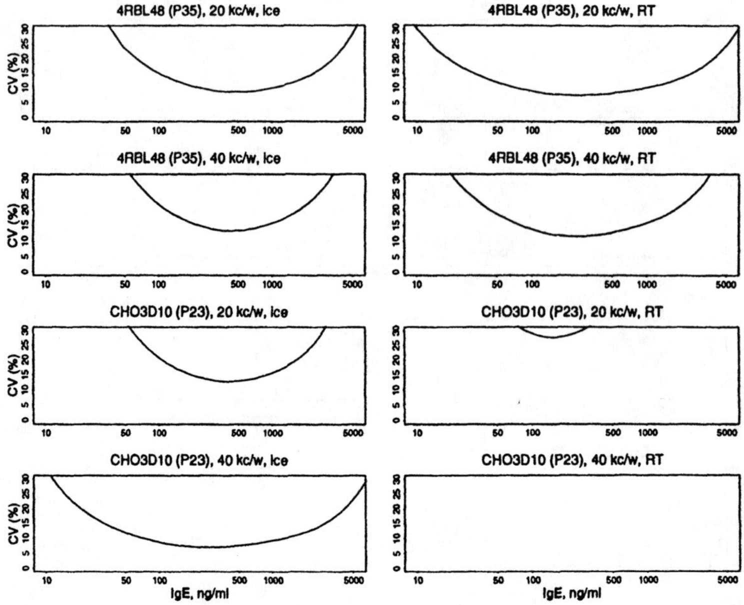

Figure 10.6. *Precision profiles for eight runs of an assay for IgE. The last panel, with cell line CHO3D10, the higher cell seeding density, and room temperature, is empty because the estimated CV never fell below 30%.*

10.5 Assay sensitivity and detection limits

The 'sensitivity' of an assay is commonly used to evaluate the limitations of a particular analytical technique, and may be used as a criterion to compare performance among several assays or laboratories. As Rodbard (1978) and others have pointed out, numerous definitions of sensitivity have been proposed; a common feature of these definitions is a pronounced dependence on assumptions about the variability in response at low concentrations. This is not surprising; proper characterization of the reliability with which low concentrations can be determined may be expected to require a good understanding of the pattern of intra-assay variation.

As a concrete illustration of the need for good characterization of the pattern of intra-assay variance in determining sensitivity, we consider estimation of the minimum detectable concentration (MDC) for the relaxin assays. Similar issues arise in the construction of other measures of sensitivity, such as the critical level, detection limit, or determination limit of Oppenheimer *et al.* (1983) We use the definition of MDC of Rodbard. Let $\bar{y}(x, r_0)$ be the mean response of r_0 future replicates at concentration x for the ith run, and let $\hat{\boldsymbol{\beta}}_i$ be an estimator of $\boldsymbol{\beta}_i$. The MDC x_{MDC} at level $(1-\alpha)$ for the ith assay is the smallest value x for which $\Pr\{\bar{y}(x, r_0) \geq f(0, \hat{\boldsymbol{\beta}}_i)\}$. If $t_{\alpha,df}$ is the $100(1-\alpha)$ percentile of the t distribution with *df* degrees of freedom, the usual estimate $\hat{x}_{MDC}$ satisfies

$$\{f(\hat{x}_{MDC}, \hat{\boldsymbol{\beta}}_i) - f(0, \hat{\boldsymbol{\beta}}_i)\}^2$$
$$= t^2_{\alpha,df}[\hat{\sigma}^2 g^2\{f(\hat{x}_{MDC}, \hat{\boldsymbol{\beta}}_i), \boldsymbol{\theta}\}/r_0 + v\{f(0, \hat{\boldsymbol{\beta}}_i)\}],$$

where $v\{f(0, \hat{\boldsymbol{\beta}}_i)\}$ estimates $\text{Var}\{f(0, \hat{\boldsymbol{\beta}}_i)\}$ and $\hat{\sigma}^2$ is the mean squared error from the weighted fit. A usual choice for $v\{f(0, \hat{\boldsymbol{\beta}}_i)\}$ is to substitute $x = 0$ in the expression $\boldsymbol{f}'_{\beta}(x, \hat{\boldsymbol{\beta}}_i)\hat{\sigma}^2 \hat{\boldsymbol{\Sigma}}_i \boldsymbol{f}_{\beta}(x, \hat{\boldsymbol{\beta}}_i)$, where $\boldsymbol{f}_{\beta}(x, \hat{\boldsymbol{\beta}}_i)$ is the $(n_i \times p)$ matrix of partial derivatives of f with respect to $\boldsymbol{\beta}_i$ evaluated at $\hat{\boldsymbol{\beta}}_i$ and $\hat{\sigma}^2 \hat{\boldsymbol{\Sigma}}_i$ is an estimate of the covariance matrix of $\hat{\boldsymbol{\beta}}_i$ based on asymptotic normal theory. The value for degrees of freedom *df* is based on the degrees of freedom associated with estimation of σ^2; for MDC estimation based on data from the ith assay only, $df = n_i - p$. For MDC estimation using $(\hat{\sigma}, \hat{\boldsymbol{\theta}})$ based on pooled data, $df = N - mp$; this value is generally sufficiently large to justify use of the $100(1-\alpha)$ percentile of the standard normal distribution.

Several authors have noted that MDC estimates are conservative if one ignores heteroscedasticity in the response (Rodbard, 1978; Oppenheimer *et al.*, 1983; Davidian, Carroll and Smith, 1988). In the notation of our model, with the power variance function and increasing mean, this amounts to a gross underestimate of the power θ (by setting it to zero), which results in overestimation of the response variation at low concentrations. The converse is also true, that overestimation of θ will lead to underestimation of the response variability at concentrations near zero, with a correspondingly optimistic estimate of the MDC.

To illustrate this point, consider estimation of MDC for runs #1 and #9 of the relaxin assay. For run #1, the estimate of θ obtained from that assay alone, 0.926 ng/ml, and the pooled estimate, 1.028 ng/ml, are similar; however, the corresponding MDC estimates based on the individual fit, 0.168 ng/ml, and based on the fit obtained from the pooled data, 0.137 ng/ml, differ in relative magnitude by 22%. The MDC estimates for run #9, where the estimated θ values differ fairly substantially, are 0.078 ng/ml based on the data for run #9 only and 0.259 ng/ml based on the pooled data, a discrepancy on the order of 70%. In this extreme case, the MDC estimate based on the data for the single assay is roughly two dilutions below the lowest standard, which is unrealistic. For both assay runs, the discrepancy in MDC estimates once again exemplifies the difference in evaluation of precision that can result from pooled versus individual-assay estimation of θ. For the reasons given above, we believe pooled estimation methods result in superior inference.

10.6 Confidence intervals for unknown concentrations

In section 10.4 we described how to calculate an approximate standard error for an estimated unknown concentration, $\hat{x}_0$, given in equation (10.6). Another assessment of the precision of calibration is by construction of a confidence interval for x_0. In this section we describe three methods of forming confidence intervals and briefly summarize empirical work comparing their performance characteristics. Our presentation is based on that in Belanger *et al.* (1995); the methods described below by no means exhaust possible approaches to the problem.

Assume that, for the ith assay run, the total of n_i observations is such that there are c distinct concentrations, with r_{ic} replicates at each, $\sum_{i=1}^{c} r_{ic} = n_i$.

10.6.1 Methods of forming confidence intervals

Wald intervals on the untransformed concentration scale.

Assuming that $\hat{x}_0$ is approximately asymptotically normally distributed with variance given in (10.6), form a $100(1-\alpha)\%$ calibration confidence interval according to the formula

$$\hat{x}_0 \pm c_\alpha \{\text{var}(\hat{x}_0)\}^{1/2}, \qquad (10.7)$$

where c_α is a suitably chosen critical value. Choices for c_α include $z_{\alpha/2}$, the $(1-\alpha/2)$ percentile of the standard normal distribution or $t_{df,\alpha/2}$, that for the t-distribution with df degrees of freedom; $df = n_i - p$ and $df = c - p$ are standard choices.

Wald intervals on the log concentration scale.

The Wald interval given in (10.7) is symmetric by construction. Standard concentrations are often chosen according to a geometric dilution series, so that a calibration interval that is symmetric on the log scale seems more natural. A supporting argument is that exact, Fieller-type intervals for the linear regression case are more nearly symmetric on the log scale than on the untransformed scale. Furthermore, for the four-parameter logistic function in (10.2), it is intuitively clear that confidence intervals at the low and high ends of the assay range ought to be asymmetric, in order to account for the asymptotic nature of the curve. This suggests forming intervals based on $\log(\hat{x}_0)$. An estimate, $\mathrm{Var}(\log \hat{x}_0)$ may be obtained by a Taylor series linearization analogous to that in section 10.4.2 above. An approximate $100(1-\alpha)\%$ for x_0 is then calculated by back-transformation:

$$\exp[\log \hat{x}_0 \pm c_\alpha \{\mathrm{var}(\log \hat{x_0})\}^{1/2}]. \tag{10.8}$$

Inversion of asymptotic prediction intervals.

An approximate $100(1-\alpha)\%$ confidence interval may be obtained by 'inverting' $100(1-\alpha)\%$ prediction intervals for the response (Carroll and Ruppert, 1988, section 2.9.3.) For the pair (x_0, y_0), the prediction interval for y_0 is given by

$$I(x_0) = \{\text{all } y \text{ in the interval} f(x_0, \hat{\boldsymbol{\beta}}_i) \pm c_\alpha \hat{\sigma} q(x_0, \hat{\boldsymbol{\beta}}_i, \hat{\boldsymbol{\theta}})\}, \tag{10.9}$$

$$q^2(x_0, \boldsymbol{\beta}_i, \boldsymbol{\theta}) = g^2\{f(x_0, \boldsymbol{\beta}_i), \boldsymbol{\theta}\}/r_0 + \boldsymbol{f}'_\beta(x_0, \boldsymbol{\beta}_i)\hat{\boldsymbol{\Sigma}}_i \boldsymbol{f}_\beta(x_0, \boldsymbol{\beta}), \tag{10.10}$$

where the p-variate vector-valued function $\boldsymbol{f}_\beta(x, \boldsymbol{\beta})$ is the derivative of f with respect to $\boldsymbol{\beta}$. The calibration interval is taken as

$$\{\text{all } x \text{ with } y_0 \in I(x)\}. \tag{10.11}$$

Uncertainty in y_0 and that due to the fit is taken into account through the term $q^2(x_0, \boldsymbol{\beta}, \boldsymbol{\theta})$. Note that this interval, which is symmetric on the response scale, is not, in general, symmetric on the concentration or log concentration scales, particularly at concentrations where the response is close to the upper or lower limits.

10.6.2 Examples

To investigate the performance of the three methods of confidence interval construction described above, we applied them to each of

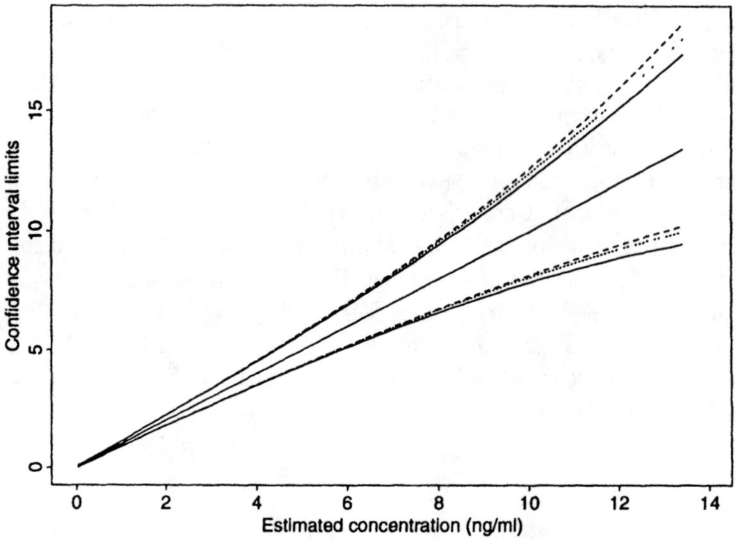

Figure 10.7. *Confidence intervals for run #10 of the DNase ELISA by three methods. Solid line, Wald intervals; dotted line, log Wald intervals; dashed line, inversion intervals.*

two datasets, the DNase ELISA data discussed in Chapter 5 and the relaxin bioassay data shown in Figure 1.4.

ELISA for DNase in rat serum. Recall from section 5.2 that pooled pseudolikelihood estimation of the intra-assay variance parameters in the power of the mean variance model, based on 11 assay runs, yields estimates of $(0.023, 0.50)$ for (σ, θ). Figure 10.7 shows 95% calibration confidence intervals based on a GLS fit of the four-parameter logistic function to the data of Figure 5.1 (run #10) using the pooled estimate $\hat{\theta} = 0.50$, by each of the three methods described above. For these data, the intervals are virtually indistinguishable across the working range of the assay.

Bioassay for relaxin. Figure 10.8 shows 95% calibration intervals based on a GLS fit to run #1 of the relaxin assay, using the value $\theta = 1.0$, for each of the three methods of interval construction. In

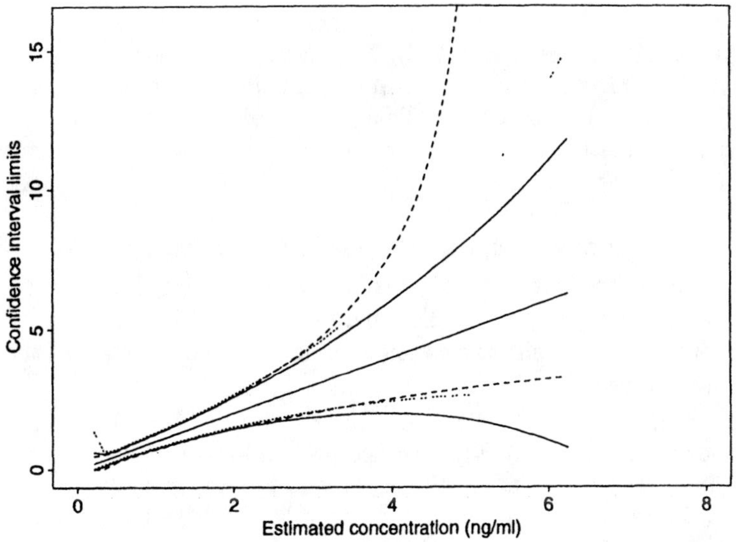

Figure 10.8. *Confidence intervals for run #1 of the relaxin bioassay by three methods. Solid line, Wald intervals; dotted line, log Wald intervals; dashed line, inversion intervals.*

contrast to the previous example, there is wide disparity among the intervals across a large segment of the concentration range.

The relatively good agreement among the three methods in the first example and the pronounced divergence in the second case raise several questions. Under what circumstances do the methods agree, in general? In the event that they diverge, which method, if any, gives 'best' results? All three methods are approximate; there is no reason *a priori* to believe that any one of them should exhibit uniform superiority. Indeed, there is no guarantee that any of the methods necessarily achieves the nominal coverage probability across the entire range of an assay. It is thus of interest to assess the operating characteristics of the different calibration intervals.

10.6.3 Simulation studies

We report very briefly on the conclusions from two simulation studies to investigate this question. Further details may be found in the paper by Belanger *et al.* (1995). These studies focused on estimation of variance parameters based on individual data only; pooled estimation is not considered. Thus, one would expect that the effect of estimation of variance parameters on calibration inference would be highlighted, as the estimates $(\sigma, \boldsymbol{\theta})$ will likely be of poor quality.

Simulation results supported the following conclusions:

- In situations where the degree of heteroscedasticity is moderate, all three interval construction methods gave similar results, with coverage probabilities close to nominal values over much of the assay range.
- For more severe variance heterogeneity (e.g. $\theta = 1.0$ in the power variance model), coverage probabilities can be substantially lower than nominal values. The quality of estimation of variance parameters plays a major role; the better the estimation of θ, the more accurate the coverage probability. This suggests that pooled estimation of variance parameters across assays should improve the quality of interval estimation.
- In general, performance of the intervals obtained by inversion of the prediction interval was best, particularly for the case of sever heteroscedasticity, coverage for both types of Wald intervals could be very poor, especially towards the extremes of the assay range. This appeared to stem from bias in the estimation of calibrated concentrations, as well as poor estimation of the standard errors. Behavior of Wald intervals on the log scale was particularly unimpressive; intervals were often extremely long, with no compensating improvement in coverage probability.

10.7 Bayesian calibration

The development and examples in sections 10.4–10.6 emphasized the advantages gained by estimating variance parameters based upon pooled data across several assay runs rather than data from only a single run. Given the hierarchical nature of the model for the regression parameters, it is natural to ask whether we can achieve similar gains in estimating the $\boldsymbol{\beta}_i$ by pooling data across assay runs in an appropriate Bayesian fashion. While implementation of Bayesian techniques is common in the population pharmacokinetic literature, they have been used rarely in the calibration context; an exception is the paper by Unadkat *et al.* (1986). In this section, we briefly describe one method for Bayesian calibration, illustrate its use for the relaxin data, and report on its performance based on limited simulation work.

In the calibration context, there are generally enough concentration-response data available for the standard to allow for individual estimates $\boldsymbol{\beta}_i^*$ for the regression parameters. In the numerical illustrations below for the relaxin assay we consider the specific case where $\boldsymbol{\beta}_i^*$ is a GLS estimator with weights based on the pooled pseudolikelihood estimates of parameters in the power variance, but any reasonable competing estimator could be used. Given the availability of such assay-specific estimates, $\boldsymbol{\beta}_i^*$, one way of implementing Bayesian calibration is to use the GTS method described in section 5.3.2. Specifically, one may implement the iterative EM algorithm defined in (5.12)–(5.14). This EM scheme involves construction of refined individual estimates (5.14), which are approximate current Bayes estimates, borrowing information across the sample.

Table 10.2 summarizes results of the GTS scheme for the relaxin bioassay data. Regression parameter estimates, $\boldsymbol{\beta}_i^*$, obtained for each assay using the pooled pseudolikelihood estimates of σ and θ and the refined GTS versions of these estimates, $\hat{\boldsymbol{\beta}}_i$, appear in Table 10.2(a). To illustrate the effect of the Bayesian refinement on calibration, Table 10.2(b) shows calibrated concentrations for responses at several levels based on the unrefined estimates $\boldsymbol{\beta}_i^*$ and refined estimates $\hat{\boldsymbol{\beta}}_i$ for assay #9. Differences of up to 15% of calibrated values are observed.

Giltinan and Davidian (1994) report on simulation work to investigate the effect of different methods of estimation of the standard curve for a particular assay on subsequent calibration inference. Results from that work suggest that a marked increase in efficiency may be realized by using refined estimates both for parameter estimation and for calibration. This reflects the standard Bayesian view that one may gain efficiency by borrowing information across assays. One must, however, weigh this gain against the practicality of implementing a Bayesian calibration scheme in an environment where hundreds of thousands of samples are processed in several hundred different assays annually, typically the case for an established biotechnology company. Furthermore, regulatory issues may mitigate against routine use of Bayesian calibration methods;

Table 10.2. *(a) Estimation of β based on pooled GLS and refined GTS estimation for the relaxin bioassay. (b) Calibration based on pooled GLS and refined GTS estimation, experiment 9, relaxin bioassay.*

(a)

	Pooled GLS-PL				Refined GTS			
Run	β_1	β_2	β_3	β_4	β_1	β_2	β_3	β_4
1	127.32	1.76	1.57	1.43	120.47	1.89	1.47	1.52
2	105.56	1.90	1.23	1.59	119.60	1.94	1.47	1.51
3	101.83	1.91	1.69	1.38	91.09	1.84	1.45	1.49
4	121.58	1.76	1.66	1.49	106.06	1.65	1.45	1.54
5	62.13	1.06	1.30	1.64	65.72	1.16	1.40	1.59
6	164.12	1.38	1.56	1.58	145.17	1.54	1.47	1.61
7	121.39	1.34	1.47	1.68	113.89	1.29	1.44	1.62
8	93.83	2.02	1.56	1.44	89.89	1.85	1.44	1.49
9	98.45	2.56	1.37	1.64	103.35	1.92	1.46	1.50

(b)

Response level (pmoles/ml)	*GLS-PL estimated conc. (ng/ml)*	*GTS estimated conc. (ng/ml)*
3.5	0.238	0.269
4.0	0.309	0.324
6.5	0.579	0.558
8.0	0.712	0.681
20.0	1.579	1.544
60.0	5.038	5.219

where regulatory requirements stipulate strict guidelines for sample tracking, audit trails, and software validation, the practical drawbacks to routine implementation of Bayesian techniques may outweigh potential benefits.

10.8 Discussion

Our investigation of the performance of nonlinear calibration methods indicates that the pattern of response variability and the quality of variance function estimation are key issues in assessment of assay precision and reliability of calibration inference. When response heteroscedasticity is not too severe and can be well characterized, approximate large-sample calibration confidence intervals are valid, and different methods often yield similar results. Intervals obtained by inversion of prediction intervals appear most reliable. Wald-type methods may break down at the extremes of the assay range, although precision of calibration may not be high enough in this range for calibrated values to be useful in general. The interval constructed on the log concentration scale is particularly troublesome. When heterogeneity of variance is severe or information for estimation of variance parameters is scarce, usefulness or reliability of any of the methods is suspect.

The implication is that the common practice of setting variance parameters to fixed values without adequate investigation may lead to erroneous calibration inference. Good estimation of intra-assay variance parameters is as important for forming reliable confidence intervals as it was for constructing accurate precision profiles. As we have mentioned before, this is unlikely to be possible based on data from a single run only, unless the degree of replication is unusually high. The utility of the pooled estimation methods described in section 5.2, which combine information from several assay runs, is evident in this context.

A common alternative to variance modeling and estimation is transformation of the response, or of both the response and regression model. We note that this approach will be subject to similar problems. Accurate characterization of the transformation is essential, and the relationship between transformations and variance modeling when σ is 'small' (Carroll and Ruppert, 1988, Chapter 4) suggests that there will be a penalty for estimation or misspecification of the transformation.

Earlier in this chapter, we remarked that the model $g^2(\mu, \theta) = \theta_1 + \mu^{\theta_2}$ represents an appealing choice for the intra-assay variance function, since it incorporates a nonzero variance component that dominates at low response levels. In the absence of a considerable degree of replication at zero concentration, as in the case of the examples in this chapter, reliable characterization of the variance parameters in this model is often problematic. The power variance function may be a reasonable approximation to the components of variance structure, but this suggests that caution should be exercised in the estimation of constructs such as the MDC; if sufficient data are not collected at low concentrations, estimates may be undesirably model-dependent.

10.9 Bibliographic notes

References on the analysis of assay data are contained both in the clinical chemistry and the statistics literature. Two early papers by Finney (1976, 1978) discuss modeling and computational aspects. In the clinical chemistry literature, a series of papers by Rodbard and co-workers (Rodbard and Hutt, 1974; Rodbard *et al.*, 1976; Rodbard, 1978) addresses modeling issues, both for mean response and variability. Variance function estimation is considered by Raab (1981), Sadler and Smith (1985), and Davidian and Carroll (1987). Ekins and Edwards (1983) and Dudley *et al.* (1985) discuss construction and implementation of precision profiles. The importance of correct modeling and estimation of variance functions for setting lower assay limits is stressed by several authors (Rodbard, 1978; Oppenheimer *et al.*, 1983; Davidian *et al.*, 1988). Bayesian calibration is discussed in Unadkat *et al.* (1986). Construction of calibration confidence intervals is treated by Schwenke and Milliken (1991) and Belanger *et al.* (1995). Recent expository papers include those by Davidian and Haaland (1992), O'Connell *et al.* (1993), and Giltinan and Davidian (1994).

CHAPTER 11

Further applications

11.1 Introduction

In Chapters 9 and 10, we focused on the use of hierarchical nonlinear modeling in two specific areas of application, pharmacokinetics and pharmacodynamics and assay analysis. The methods in this book are potentially useful in a wide variety of problems. We illustrate this in this chapter by presenting three self-contained case studies drawn from quite distinct subject-matter areas: crop science, forestry, and seismology.

11.2 Comparison of soybean growth patterns

Data description

Data from an experiment to compare growth patterns of two genotypes of soybean were introduced in section 1.1.3. These data were kindly provided by Colleen Hudak and are part of a study conducted by researchers in the Department of Crop Science at North Carolina State University.

In each of three consecutive years (1988–1990), 16 plots were planted with seeds, eight with the genotype Forrest (F), a commercial variety, and eight with Plant Introduction #416937 (P), an experimental strain. The experiment took place at different sites in each year, for a total of $m = 48$ plots over the entire three year period. At approximate weekly intervals during the growing season, six plants were randomly selected from each plot, the leaves from these plants were aggregated and weighed, and average leaf weight per plant, y, was calculated for the plot. Sampling began roughly two weeks after planting, and continued until 10 to 12 weeks after planting, for a planned total of eight to ten measurements per plot. A slightly different sampling scheme was followed in each year, and some values were not recorded for some plots for reasons unrelated to year or genotype. Weather patterns differed from year to year: 1988 was unusually dry, 1989 was wet, and conditions in 1990 were normal.

Profiles of average leaf weight values over time, x, for selected plots are shown in Figure 1.2. The data for each plot thus consist of the pairs (y_{ij}, x_{ij}), where y_{ij} is the average leaf weight per plant calculated for the ith plot at sampling time x_{ij} (in units of days after planting), $i = 1, \ldots, 48$, $j = 1, \ldots, n_i$, where n_i is between 8 and 10; genotype and year information for each plot is also available. A complete data listing is given in Davidian and Giltinan (1993b).

Objective and model formulation

A main goal of the study was to compare growth patterns over the growing season for the two soybean genotypes; more specifically, to contrast initial average leaf weight, growth rate, final limiting average leaf weight attained, and the time at which 50% of limiting growth is realized. Because the three study years differed markedly in terms of precipitation, representing extremes as well as typical conditions, comparison of these growth characteristics across and within years was also of interest, both to provide insight into the effect of weather patterns on soybean growth and to identify possible differences in response to weather patterns between genotypes. Thus, in the analyses reported here, we consider 'year' to be a fixed factor; we discuss an alternative approach at the end of this section. Under this assumption, we may view the data as arising from six groups determined by the six genotype-year combinations.

As discussed in section 1.1.3, the logistic growth model

$$y = \frac{\beta_1}{1 + \beta_2 \exp(\beta_3 x)}, \quad \beta_1, \beta_2 > 0, \; \beta_3 < 0, \qquad (11.1)$$

provides an adequate characterization of the growth pattern for an individual plot. To allow for inter-individual variation, we may consider the 48 plots as arising from a population of plots and assume that each has its own individual-specific vector of random parameters $\boldsymbol{\beta}_i = [\beta_{1i}, \beta_{2i}, \beta_{3i}]'$. Because plot-to-plot variation in growth patterns may be attributable in part to genotype (F or P) or weather condition (1988, 1989, 1990), and because investigation of this issue is the primary objective, we began with a model for $\boldsymbol{\beta}_i$ that incorporates information on the systematic effects of genotype and year. Following the discussion in section 4.2.3, and indexing the six possible combination groups by g, where $g = 1$ corresponds to genotype F in 1988, $g = 2$ corresponds to genotype P in 1988, and so on, we consider the following hierarchical model:

Stage 1. (within-plot variation)

$$y_{ij} = f(x_{ij}, \boldsymbol{\beta}_i) = \frac{\beta_{1i}}{1 + \beta_{2i}\exp(\beta_{3i}x_{ij})} + e_{ij},$$

$$(\boldsymbol{e}_i|\boldsymbol{\beta}_i) \sim \{\boldsymbol{0}, \boldsymbol{R}_i(\boldsymbol{\beta}_i, \boldsymbol{\xi})\}, \qquad (11.2)$$

where $\boldsymbol{R}_i(\boldsymbol{\beta}_i, \boldsymbol{\xi})$ is an assumed within-plot covariance structure.

Stage 2. (plot-to-plot variation)

$$\boldsymbol{\beta}_i = \boldsymbol{A}_i\boldsymbol{\beta} + \boldsymbol{b}_i, \quad \boldsymbol{b}_i \sim (\boldsymbol{0}, \boldsymbol{D}), \qquad (11.3)$$

where $\boldsymbol{b}_i$ is a (3×1) vector of random effects, $\boldsymbol{\beta} = [\boldsymbol{\beta}^{(1)\prime}, \ldots, \boldsymbol{\beta}^{(6)\prime}]'$, an (18×1) vector consisting of the six group mean vectors $\boldsymbol{\beta}^{(g)} = [\beta_1^{(g)}, \beta_2^{(g)}, \beta_3^{(g)}]'$ stacked vertically, $g = 1, \ldots, 6$; and $\boldsymbol{A}_i$ is a (3×18) 'design' matrix such that

$$\boldsymbol{A}_i = [\boldsymbol{I}_3|\boldsymbol{0}_{3\times3}|\cdots|\boldsymbol{0}_{3\times3}]$$

if $g = 1$,

$$\boldsymbol{A}_i = [\boldsymbol{0}_{3\times3}|\boldsymbol{I}_3|\cdots|\boldsymbol{0}_{3\times3}]$$

if $g = 2$, and so on, where $\boldsymbol{0}_{3\times3}$ is a (3×3) matrix of zeroes.

The specification (11.3) implies that the regression parameter for plot i appearing in the gth genotype-year combination group is given by

$$\boldsymbol{\beta}_i = \boldsymbol{\beta}^{(g)} + \boldsymbol{b}_i. \qquad (11.4)$$

The parameters of the logistic growth model have interpretations that pertain directly to the study objectives. The function (11.1) may be derived by assuming that rate of growth relative to current size is in constant proportion, by a factor β_3, to remaining growth to be attained (e.g. Draper and Smith, 1981). Consequently, β_3 plays the role of a growth rate constant, so that comparison of growth rate among the soybean genotypes and weather patterns may be based on this parameter. From the form of (11.1), β_1 represents the limiting growth value. Other quantities of interest, such as initial growth value representing 'seed' leaf mass, may be obtained as a function of the three parameters. Thus, the main objectives of the study may be addressed by investigating appropriate functions of the elements of $\boldsymbol{\beta}^{(g)}$.

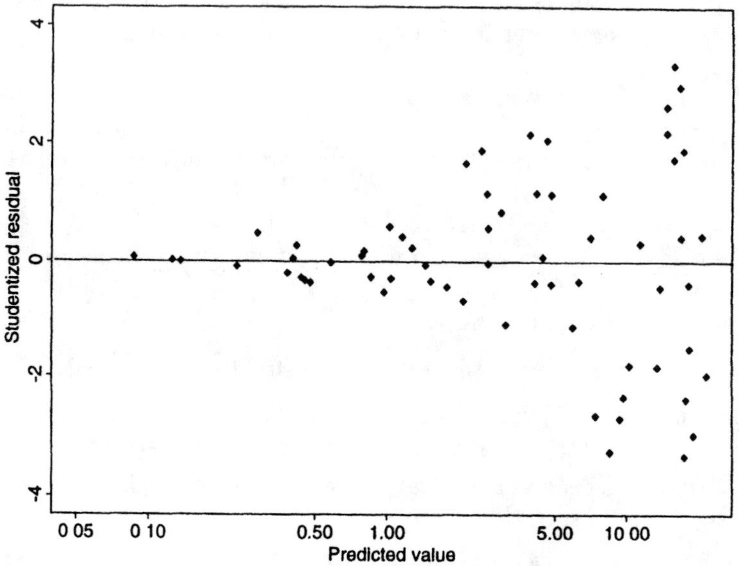

Figure 11.1. *Studentized residuals versus predicted values from individual ordinary least squares fits of model (11.1) for eight plots of genotype P, 1990.*

Methods

Because both the number of plots and the number of observations per plot are large, methods for population inference based on individual estimates may be used; thus, these data allow a direct comparison between these techniques and other approaches, such as those based on linearization. As discussed in section 1.1.3, for this type of growth data, intra-plot variance that increases systematically with average leaf weight is likely. This is confirmed in Figure 11.1, which shows studentized residuals (Carroll and Ruppert, 1988, section 2.7) from separate ordinary least squares fits of the logistic model (11.1) plotted against predicted values for the plots in one combination group. Moreover, because of the serial nature of data collection, it is reasonable to expect autocorrelation among observations within the same plot. Based on these considerations, we fit the hierarchical model to the soybean data using two approaches that allow straightforward accommodation of these features: (i) GLS estimation of individual regression parameters incorporating pooled estimation of within-plot covariance parameters (section 5.2.2), followed by inference on population parameters based on the global two-stage (GTS) algorithm (section 5.3.2); and (ii) GLS estimation based on the first-order linearization using the method of Vonesh and Carter (1992), modified for estimation of common intra-individual covariance parameters (section 6.2.3). Both procedures were implemented in the GAUSS matrix programming language.

We report on the fits of two models. In all cases, the second stage was as in (11.3); the first stage of each model represents a different assumption about intra-plot random variation:

Model 1. (intra-plot variance proportional to a power of mean response, uncorrelated errors)

$$(\boldsymbol{e}_i|\boldsymbol{\beta}_i) \sim \{\boldsymbol{0}, \boldsymbol{R}_i(\boldsymbol{\beta}_i, \boldsymbol{\xi})\}, \quad \boldsymbol{R}_i(\boldsymbol{\beta}_i, \boldsymbol{\xi}) = \sigma^2 \boldsymbol{G}_i(\boldsymbol{\beta}_i, \theta),$$

where $\boldsymbol{G}_i(\boldsymbol{\beta}_i, \theta) = \mathrm{diag}[f^{2\theta}(x_{i1}, \boldsymbol{\beta}_i), \ldots, f^{2\theta}(x_{in_i}, \boldsymbol{\beta}_i)]$.

Model 2. (intra-plot variance proportional to a power of mean response, correlated errors)

$$(\boldsymbol{e}_i|\boldsymbol{\beta}_i) \sim \{\boldsymbol{0}, \boldsymbol{R}_i(\boldsymbol{\beta}_i, \boldsymbol{\xi})\},$$

$$\boldsymbol{R}_i(\boldsymbol{\beta}_i, \boldsymbol{\xi}) = \sigma^2 \boldsymbol{G}_i^{1/2}(\boldsymbol{\beta}_i, \theta)\boldsymbol{\Gamma}_i(\alpha)\boldsymbol{G}_i^{1/2}(\boldsymbol{\beta}_i, \theta),$$

where $\boldsymbol{G}_i(\boldsymbol{\beta}_i, \theta)$ is as in Model 1, and $\boldsymbol{\Gamma}_i(\alpha)$ is the $(n_i \times n_i)$ correlation matrix corresponding to an autoregressive process of order one as in (2.10):

$$\boldsymbol{\Gamma}_i(\alpha) = \begin{bmatrix} 1 & \alpha & \alpha^2 & \cdots & \alpha^{n_i-1} \\ & 1 & \alpha & \cdots & \alpha^{n_i-2} \\ & & \ddots & \ddots & \vdots \\ & & & \ddots & \alpha \\ & & & & 1 \end{bmatrix}.$$

For each model, we used the appropriate pseudolikelihood objective function, given in sections 5.2.2 and 6.2.3, respectively, to estimate the intra-plot covariance parameters σ, θ, and, in the case of Model 2, α.

Results

Although normality of the random effects, $\boldsymbol{b}_i$, is not a requirement for use of either of the fitting methods employed, the reliability of inference is likely to be improved if the $\boldsymbol{b}_i$ do indeed arise from a distribution that is close to normal or at least symmetric. To investigate the form of the distribution of the random effects, we adopted an ***ad hoc*** approach: from the fit of Model 1 using individual estimates and the GTS algorithm (i), we examined the empirical distributions of the components of plot 'residuals' $(\boldsymbol{\beta}_i^* - \boldsymbol{A}_i\hat{\boldsymbol{\beta}}_{GTS})$. The distributions appeared normal with the exception of that associated with β_2 in (11.1), which was markedly skewed. This preliminary investigation led us to reparameterize (11.1) as

$$f(x, \boldsymbol{\beta}) = \frac{\beta_1}{1 + \exp\{\beta_3(x - \beta_2)\}}, \quad \beta_1, \beta_2 > 0, \; \beta_3 < 0. \tag{11.5}$$

This form of the logistic growth model has the added advantage that the parameter β_2 represents soybean 'half-life,' the time at which 50% of limiting growth is achieved, with the parameters β_1 and β_3 maintaining their previous interpretations. In all subsequent analyses, we thus replaced the regression model in Stage 1 (11.2) of the basic hierarchy by

$$y_{ij} = f(x_{ij}, \boldsymbol{\beta}_i) = \frac{\beta_{1i}}{1 + \exp\{\beta_{3i}(x_{ij} - \beta_{2i})\}} + e_{ij}.$$

Table 11.1 summarizes the results of the fits to Models 1 and 2 using each of the inferential approaches. Estimates of the fixed effects $\boldsymbol{\beta}$ and the within-plot covariance parameters σ, θ, and, in the case of Model 2, α, are similar for both estimation techniques and models. The estimates of θ suggest that the standard assumption of a constant coefficient of variation in growth analysis may be a bit too severe for these data.

Under the assumption of an AR(1) intra-plot correlation structure in Model 2, both procedures (i) and (ii) yield a negative estimate for the lag-one correlation parameter α. Examination of lagged residual plots from the final GLS estimates of individual regression parameters from the pooled fit of Model 1, as in the analysis of the sweetgum data described in section 2.5.2, supports such an estimate. One possible explanation is that the sampling strategy for each plot may have favored larger plants, so that selection of the largest plants at a given sampling time might have led to a large average leaf weight and left smaller plants remaining.

Table 11.1. *Fits of the soybean data to Models 1 and 2: (i) based on individual GLS estimates obtained by pooling information on within-plot variation using the GTS algorithm; (ii) based on first-order linearization of the model using the GLS method of Vonesh and Carter (1992), modified for estimation of intra-plot covariance parameters. Each grouping in the fixed-effects portion of the table represents the estimates of* $\beta^{(g)}$ *in the order* $\beta_1^{(g)}$, $\beta_2^{(g)}$, $\beta_3^{(g)}$ *for the mean model (11.5) corresponding to the gth genotype-year combination (with estimated standard errors in parentheses).*

	Model/Method			
	1(i)	*1(ii)*	*2(i)*	*2(ii)*
		Fixed effects		
F-1988	19.04	19.64	19.42	19.72
($g = 1$)	(1.04)	(1.26)	(0.89)	(1.11)
	54.62	55.14	55.05	55.30
	(1.12)	(1.37)	(1.09)	(1.17)
	−0.124	−0.122	−0.122	−0.121
	(0.004)	(0.005)	(0.004)	(0.004)
P-1988	20.81	21.78	21.34	21.64
($g = 2$)	(1.04)	(1.27)	(0.89)	(1.11)
	53.65	54.30	54.11	54.36
	(1.07)	(1.34)	(1.07)	(1.11)
	−0.123	−0.122	−0.122	−0.121
	(0.004)	(0.005)	(0.004)	(0.004)
F-1989	10.33	10.37	10.78	10.79
($g = 3$)	(0.58)	(0.88)	(0.71)	(0.83)
	52.74	52.81	53.20	52.98
	(1.02)	(1.33)	(1.07)	(1.16)
	−0.137	−0.138	−0.137	−0.134
	(0.004)	(0.005)	(0.004)	(0.004)
P-1989	18.01	18.48	18.33	18.41
($g = 4$)	(0.87)	(1.11)	(0.82)	(1.11)
	51.95	52.36	52.34	52.33
	(1.02)	(1.30)	(1.06)	(1.11)
	−0.137	−0.135	−0.135	−0.134
	(0.004)	(0.005)	(0.004)	(0.004)

Table 11.1. *Continued.*

	Model/Method			
	1(i)	*1(ii)*	*2(i)*	*2(ii)*
		Fixed effects		
F-1990	15.67	16.75	16.24	16.69
($g = 5$)	(0.85)	(1.11)	(0.83)	(0.90)
	50.11	51.09	50.92	51.52
	(1.03)	(1.32)	(1.08)	(1.09)
	−0.134	−0.131	−0.130	−0.126
	(0.004)	(0.005)	(0.004)	(0.004)
P-1990	17.78	17.90	18.27	17.76
($g = 6$)	(0.90)	(1.13)	(0.84)	(0.91)
	49.47	49.41	50.25	49.43
	(1.01)	(1.31)	(1.07)	(1.10)
	−0.129	−0.130	−0.126	−0.128
	(0.004)	(0.005)	(0.004)	(0.004)
		Covariance parameters		
σ	0.23	0.24	0.18	0.19
θ	0.89	0.87	0.86	0.84
α	–	–	−0.34	−0.34
D_{11}	0.65	3.39	3.16	4.46
D_{12}	0.17	3.09	2.91	3.80
D_{13}	0.0003	0.007	0.008	0.008
D_{22}	2.15	6.28	6.35	7.15
D_{23}	0.008	0.019	0.021	0.020
D_{33} [1]	2.81	7.12	7.10	7.17

[1] Values should be multiplied by 10^{-5}

At the following sampling time, the measurement would then be smaller than expected, until size differentials once again became apparent, yielding larger-than-expected measurements at the subsequent sampling time. We remark, however, that estimation of within-individual correlation structure, even with a moderate sample size ($m = 48$, $n_i = 8$ to 10), is inherently difficult.

A notable feature of the results in Table 11.1 is the sensitivity of estimation of the random effects covariance matrix $\boldsymbol{D}$ to the assumption about intra-plot covariance structure for methods based on individual estimates; a similar phenomenon is not seen in the results of the linearization-based fits. The large discrepancy between GTS estimates of $\boldsymbol{D}$, in a situation where replication both within and among individuals is good, is surprising. This may well reflect the fundamental difficulty in estimating second-moment parameters or may indicate a particular sensitivity of methods based on individual estimates to the choice of within-plot covariance model for these data.

Inspection of the estimates of the elements of $\boldsymbol{\beta}$ corresponding to each genotype-year combination for the fits of both models implies that, within each year, the typical value of limiting average leaf weight is lower for the commercial variety F than for the experimental strain P. Regardless of genotype, limiting growth seems greatest in the dry year, 1988, and smallest in the wet year, 1989, while soybean half-life, averaged across genotypes, is ordered from largest to smallest from 1988 to 1990.

A formal investigation of the study objectives may be undertaken by construction of appropriate contrasts of the components of $\boldsymbol{\beta}$. For instance, to make inference on the main effect of genotype with respect to limiting growth value, the appropriate set of hypotheses is

$$H_0 : \boldsymbol{a}'\boldsymbol{\beta} = \mathbf{0} \quad H_1 : \boldsymbol{a}'\boldsymbol{\beta} \neq \mathbf{0},$$

$$\boldsymbol{a} = [-1/3, 0, 0, 1/3, 0, 0, -1/3, 0, 0, 1/3, 0, 0, -1/3, 0, 0, 1/3, 0, 0]'.$$

From (11.5), initial leaf weight is given by $\beta_1/\{1 + \exp(-\beta_3\beta_2)\}$, a nonlinear function of $\boldsymbol{\beta}$; a set of hypotheses of the form

$$H_0 : a(\boldsymbol{\beta}) = \mathbf{0} \quad H_1 : a(\boldsymbol{\beta}) \neq \mathbf{0},$$

where the $a(\boldsymbol{\beta})$ represents the appropriate contrast of the expression for initial leaf weight, may be constructed. d-dimensional hypotheses, for example, for year or interaction effects ($d = 2$), may be specified by the appropriate set of d contrasts applied to the relevant parameter or function of model parameters.

To conduct these hypothesis tests, we carried out approximate Wald-type inference using the methods described in section 5.3.3 for the GTS algorithm and in section 6.2.3 for the GLS linearization procedure. The qualitative conclusions of all tests were in agreement for both fitting methods and model specifications; accordingly, we limit presentation of the results to those from the linearization fit to Model 2. Table 11.2 presents values of the test statistics T^2 for main effects and genotype-year interaction for each

Table 11.2. *Test statistics for main effects and interactions for growth characteristics of soybean plants based on the GLS linearization fit of Model 2.*

Effect	*Degrees of freedom* (d)	T^2
	Limiting growth value	
Year	2	28.94
Genotype	1	15.84
Interaction	2	16.35
	Initial growth value	
Year	2	36.83
Genotype	1	8.15
Interaction	2	0.12
	Growth rate constant	
Year	2	9.35
Genotype	1	0.18
Interaction	2	0.10
	Time to reach 50% limiting growth	
Year	2	11.28
Genotype	1	0.83
Interaction	2	0.23

quantity of interest; as discussed in section 6.2.3, the approximate test may be conducted by referring the statistic to the χ^2_d distribution, where d is the degrees of freedom for the associated set of contrasts. As suggested by Table 11.1, the significant interaction for limiting growth value is attributable to differences in magnitude, not direction, of the difference. The data provide evidence to support the researcher's contention that, although the two genotypes have different initial and limiting leaf mass, growth patterns over the growing season are comparable; there is not sufficient evidence to suggest that growth rate or soybean half-life differ between them.

For comparison, we also carried out analyses using both inferential techniques under the assumption that the intra-plot covariance matrix $\boldsymbol{R}_i(\boldsymbol{\beta}_i, \boldsymbol{\xi}) = \sigma^2 \boldsymbol{I}_{n_i}$, thus ignoring within-individual heterogeneity of variance and possible serial correlation. Results of hypothesis tests analogous to those in Table 11.2 suggest no differences in initial or limiting leaf weights but do imply differing growth rates across the two genotypes. These conclusions are directly at odds with the preceding analyses and with the researcher's expectations, indicating that accurate characterization of the pattern of intra-individual variability is critical for sensible inference.

Examination of the estimates of $\boldsymbol{D}$ in Table 11.1 reveals that the estimated correlation between the random effects associated with β_2 and β_3 is greater than 0.90 in every case. This near-singularity of the estimated population covariance matrix may indicate a real association in the population or may reflect a model misspecification, for example, the need to treat one of the parameters in the logistic model as fixed across plots. The latter possibility could be accommodated by a refinement of the Stage 2 model by a modification of the matrix $\boldsymbol{A}_i$. Other modifications to the model, such as reduction of the dimension of the fixed effects vector $\boldsymbol{\beta}$ to allow growth parameters common to, for instance, year or genotype based on the results of the tests above, might also be considered.

Discussion

This example illustrates how the hierarchical nonlinear model provides an appealing alternative to more traditional approaches to testing main effects and interactions among treatment factors in the repeated measurement setting, such as univariate repeated measures analysis of variance. The traditional analysis cannot take full advantage of apparent functional relationship between average leaf weight and time, which appears to be well-approximated by the logistic growth model; moreover, it does not accommodate straightforward examination of the quantities of primary interest, such as limiting growth value, and requires techniques such as data transformation to account for nonconstant intra-plot variance.

A warning message to be taken from this example concerns the possible sensitivity of analyses based on the hierarchical model to specification of model components. For these data, estimation of the population covariance structure is highly dependent on the assumption about within-plot variability when methods based on individual estimates are used. Further investigation of this issue is required; in the absence of general guidelines, our experience with these data suggests that caution should be exercised in the interpretation of results and that it is worthwhile to investigate robustness of conclusions to modeling assumptions.

In the analyses reported here, we treated year as a fixed factor on the grounds that differences among years are likely to be mainly due to disparate weather patterns. Alternatively, one could specify a model in which year was taken to be random, and extend the hierarchical model to accommodate a further level of random effects. We do not discuss this here; see Goldstein (1991) for one approach to this kind of model.

11.3 Prediction of bole volume of sweetgum trees

Background

An important issue in the area of forest inventory and planning is the derivation of accurate models to predict the volume of trees without felling the trees. Typically, it is of interest to predict cumulative bole volume up to a specified diameter, above which the bole is too small to convert into merchantable product. To be useful, prediction should depend only on measurements easily obtained from standing trees; common practice in the forestry literature is to model cumulative bole volume, y, to an upper stem diameter, d, in terms of tree height H, and basal diameter, D (diameter of the tree at breast height). Given a prediction equation for y as a function $f(\boldsymbol{x}, \boldsymbol{\beta})$, for $\boldsymbol{x} = [d, D, H]'$, where the form of f is known up to the vector of regression parameters, $\boldsymbol{\beta}$, the goal of inference is to estimate $\boldsymbol{\beta}$. Inference is based on measurements from felled trees; usually, the following measurements are available for the ith tree, $i = 1, \ldots, m$: D_i, diameter at breast height; H_i, total height; and the pairs of values (d_{ij}, y_{ij}), $j = 1, \ldots, n_i$, sequential stem diameters and corresponding cumulative bole volume measurements within the tree.

Data of this type fall into the class of hierarchical models considered in this book. The functional form of the cumulative bole volume function, f, is typically nonlinear in $\boldsymbol{\beta}$, repeated measurements are taken on each of a number of trees, and some degree of correlation may be expected among measurements on the same tree. In our terminology, 'individuals' refers to trees in this context.

Data description and objective

We apply methods described in earlier chapters to the problem of obtaining bole prediction equations for sweetgum (*Liquidambar styraciflua* L.) trees. Data for $m = 39$ sweetgum trees were kindly provided by Timothy G. Gregoire, Virginia Polytechnic Institute and State University. Table 11.3 shows a partial data listing. The variables are as described above: basal diameter (D) at breast height (4.5 ft), total height (H) for each tree, and outer-bark diameter (d) and cumulative bole volume measurements y taken sequentially at three foot intervals within each tree. An average of 26 (d, y) pairs were available for each tree, for a total of $N = 986$ measurements. An analysis of these data may be found in Gregoire and Schabenberger (1994).

The analysis presented here differs from that in Gregoire and Schabenberger (1994) in two key respects. We have analyzed bole volume in terms of *outer*-bark diameter and volume measurements only. The original data also contained information on *inner*-bark diameter and volume. Conceptually, one could model cumulative bole volume as a function of both inner- and outer-bark measurements, accounting for correlation appropriately; this approach was taken by Gregoire and Schabenberger. For simplicity of exposition, we focus only on outer-bark measurements. A second difference in our analysis is that we have chosen to model cumulative bole volume beginning at breast height (4.5 ft), rather than from stump height. Total merchantable bole volume would, of course, be calculated from stump height. However, because the height above ground at which a tree is felled may vary, proper accounting for volume between stump and breast height would require inclusion of stump height as a variable in the regression model. This would compromise the utility of the model for predicting the volume of standing trees. One approach, suggested by Walters, Hann and Clyde (1985), is to use separate functions to estimate volume below and above breast height. We adopt this strategy, focusing for illustrative purposes on modeling cumulative bole volume above breast height in terms of diameter at breast height (D), total height (H), and stem diameter, d, for $d \leq D$.

Model formulation

Figure 11.2 plots the response, y, cumulative bole volume from breast height, against the difference between stem diameter d and basal diameter D for four trees. Plotting y against $D - d$ in this way gives an increasing function constrained to pass through the origin. Inspection of the shape of these profiles, and the obvious parallel to growth curve analysis, suggests use of a logistic function

Table 11.3. *Data for two sweetgum trees.*

Tree 2		*Tree 18*	
d (in)	y (cu ft)	d (in)	y (cu ft)
13.8	2.81	5.6	0.36
12.8	5.71	5.5	0.87
12.3	8.29	5.2	1.34
12.2	10.75	5.1	1.77
11.8	13.10	4.9	2.18
11.6	15.34	4.8	2.57
11.4	17.51	4.6	2.93
11.0	19.56	4.5	3.27
10.9	21.52	4.3	3.58
10.7	23.43	4.2	3.88
11.0	25.35	3.7	4.13
10.6	27.26	3.9	4.37
10.2	29.03	3.4	4.59
8.5	31.91	3.0	4.76
8.1	33.04	2.9	4.90
7.9	34.09	2.5	5.02
7.8	35.09	0.0	5.10
7.4	36.04		
6.6	36.84		
6.4	37.54		
5.8	38.15		
5.7	38.69		
4.0	39.08		
4.0	39.34		
0.0	39.78		
$D = 14.4$ in		$D = 5.8$ in	
$H = 94.4$ ft		$H = 59.0$ ft	

to model the relationship:

$$f(x, \beta) = \frac{\beta_1}{1 + \exp\{e^{\beta_3}(\log x - \beta_2)\}}, \quad x = D - d, \ 0 \le d \le D. \tag{11.6}$$

Parameterization to ensure positivity of the multiplier e^{β_3} improved stability of the fitting procedures.

We fitted a sequence of models, each based on (11.6), allowing different degrees of complexity in characterizing the variation among trees. Denoting the tree-specific parameter values by

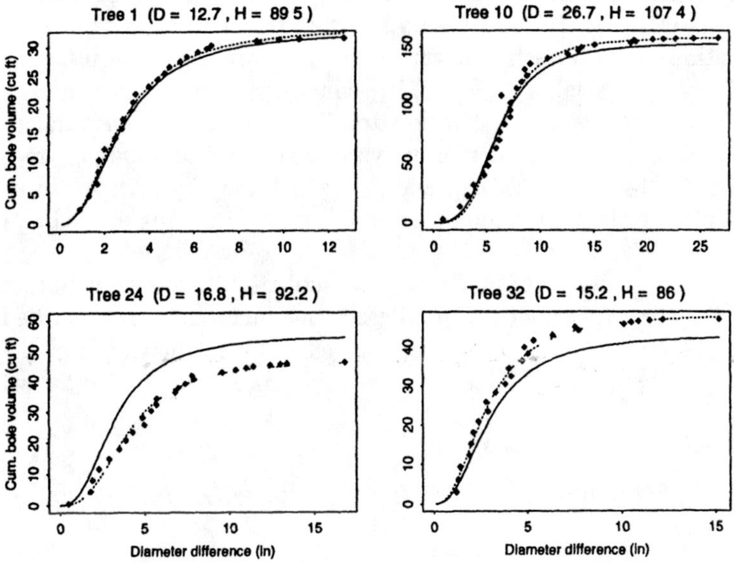

Figure 11.2. *Cumulative bole volume data for four sweetgum trees. Fits of model (11.6) are superimposed: Solid line, fit of the final population equation; dashed line, individual empirical Bayes fits using NLME.*

$\boldsymbol{\beta}_i = [\beta_{1i}, \beta_{2i}, \beta_{3i}]'$, the models may all be written in the hierarchical framework:

Stage 1. (within-tree variation)

$$y_{ij} = f(x_{ij}, \boldsymbol{\beta}_i) = \frac{\beta_{1i}}{1 + \exp\{e^{\beta_{3i}}(\log x_{ij} - \beta_{2i})\}} + e_{ij},$$

$$(\boldsymbol{e}_i | \boldsymbol{\beta}_i) \sim \{\boldsymbol{0}, \boldsymbol{R}_i(\boldsymbol{\beta}_i, \boldsymbol{\xi})\}, \qquad (11.7)$$

where $\boldsymbol{R}_i(\boldsymbol{\beta}_i, \boldsymbol{\xi})$ is an assumed within-plot covariance structure.

Stage 2. (tree-to-tree variation)

$$\boldsymbol{\beta}_i = \boldsymbol{A}_i\boldsymbol{\beta} + \boldsymbol{b}_i, \quad \boldsymbol{b}_i \sim (\boldsymbol{0}, \boldsymbol{D}), \qquad (11.8)$$

where $\boldsymbol{b}_i$ is a (3×1) vector of random effects, $\boldsymbol{\beta}$ is a $(r \times 1)$ vector of fixed effects, and $\boldsymbol{A}_i$ is a $(3 \times r)$ 'design' matrix.

Choice of the design matrix $\boldsymbol{A}_i$ allows flexibility in modeling the regression parameters $\boldsymbol{\beta}_i$. Tree-specific covariates of interest are tree height, H_i, and basal diameter, D_i. We followed the convention in the forestry literature of incorporating these in terms of a single variable $X_i = D_i^2 H_i$, which may be thought of as a crude overall measure of tree size; X_i would be proportional to volume if trees were perfectly cylindrical. Variation among trees in the parameters β_{1i}, β_{2i}, and β_{3i} (which may be interpreted as coefficients governing the shape and degree of tapering of the bole) was modeled linearly in the variable X_i. We report below on the following sequence of models to characterize tree-to-tree variation in the shape parameters. In all cases, the hierarchy was described by (11.7) and (11.8) above; the models differ in the choice of the matrix $\boldsymbol{A}_i$ in Stage 2. Specification of $\boldsymbol{R}_i(\boldsymbol{\beta}_i, \boldsymbol{\xi})$ is discussed below.

Model 1. (no covariates)

$$\beta_{1i} = \beta_1 + b_{1i}, \quad \beta_{2i} = \beta_2 + b_{2i}, \quad \beta_{3i} = \beta_3 + b_{3i},$$

$$\boldsymbol{\beta} = [\beta_1, \beta_2, \beta_3]'$$

Model 2. (X_i as a covariate for β_{1i})

$$\beta_{1i} = \beta_1 + \beta_4 X_i + b_{1i}, \quad \beta_{2i} = \beta_2 + b_{2i}, \quad \beta_{3i} = \beta_3 + b_{3i},$$

$$\boldsymbol{\beta} = [\beta_1, \beta_2, \beta_3, \beta_4]'$$

Model 3. (X_i as a covariate for β_{1i} and β_{2i})

$$\beta_{1i} = \beta_1 + \beta_4 X_i + b_{1i}, \quad \beta_{2i} = \beta_2 + \beta_5 X_i + b_{2i}, \quad \beta_{3i} = \beta_3 + b_{3i},$$

$$\boldsymbol{\beta} = [\beta_1, \beta_2, \beta_3, \beta_4, \beta_5]'$$

Model 4. (X_i as a covariate for β_{1i}, β_{2i}, and β_{3i})

$$\beta_{1i} = \beta_1 + \beta_4 X_i + b_{1i}, \quad \beta_{2i} = \beta_2 + \beta_5 X_i + b_{2i},$$

$$\beta_{3i} = \beta_3 + \beta_6 X_i + b_{3i},$$

$$\boldsymbol{\beta} = [\beta_1, \beta_2, \beta_3, \beta_4, \beta_5, \beta_6]'.$$

Model 1 corresponds to choice of $\boldsymbol{A}_i$ as the (3×3) identity matrix; in Model 4, for example, $\boldsymbol{A}_i$ is the (3×6) matrix

$$\boldsymbol{A}_i = \begin{bmatrix} 1 & 0 & 0 & X_i & 0 & 0 \\ 0 & 1 & 0 & 0 & X_i & 0 \\ 0 & 0 & 1 & 0 & 0 & X_i \end{bmatrix}.$$

We considered other intermediate possibilities; for instance, eliminating the intercept term β_1 from Model 2. In general, however, the intercept terms appeared to be needed. For brevity, we present results only for Models 1–4 below.

Inspection of profiles for individual trees did not suggest heterogeneity of intra-individual variance. However, it seems reasonable to suppose that the equally spaced measurements taken along the same bole are spatially correlated. Figure 11.3 shows plots of lagged residuals from individual ordinary least squares fits of the model (11.6), as discussed in section 2.5; the four panels of the figure show the residuals for all 39 trees plotted against themselves, lagged by one to four measurements within each tree. Inspection of these plots indicates that, although lag-1 residuals may be correlated (sample correlation = 0.30), this correlation does not extend to larger lags. This seems plausible; measurements were taken at three foot intervals, so one might speculate that immediately adjacent values would be correlated but that irregularities in bole shape would eliminate correlation among measurements that are further apart. Accordingly, we carried out fitting assuming a banded intra-individual correlation structure $\boldsymbol{R}_i(\boldsymbol{\beta}_i, \boldsymbol{\xi}) = \sigma^2 \boldsymbol{\Gamma}_i(\alpha)$, where

$$\boldsymbol{\Gamma}_i(\alpha) = \begin{bmatrix} 1 & \alpha & 0 & \cdots & 0 \\ & 1 & \alpha & \cdots & 0 \\ & & \ddots & \ddots & \vdots \\ & & & \ddots & \alpha \\ & & & & 1 \end{bmatrix}.$$

This correlation structure is that of a moving average process of order one, see Seber and Wild (1989, chapter 6).

We also considered the richer AR(1) correlation structure (2.10). Attempts to fit this correlation structure were unsuccessful; efforts to accommodate correlation between non-adjacent measurements implied by this model interfered with estimation of the mean function, reflecting the complicated interplay between mean model and

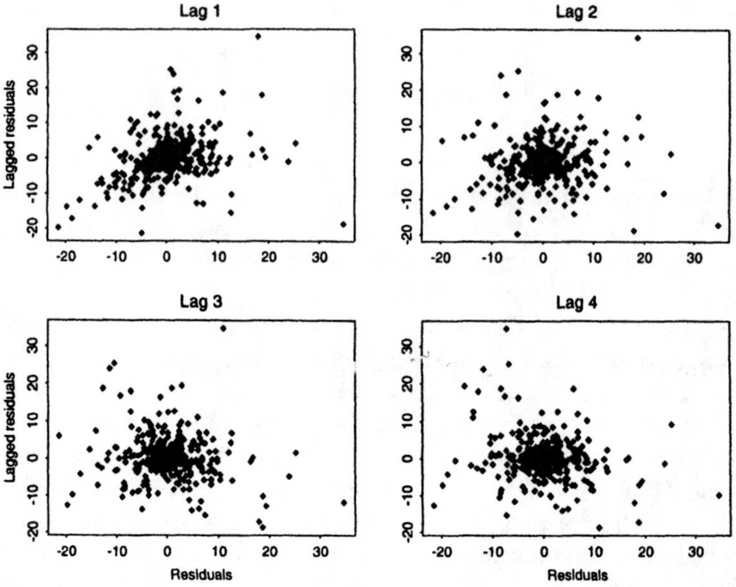

Figure 11.3. *Lagged OLS residuals for all trees based on individual fits to the sweetgum data. (a) residuals vs. lag-1 residuals; (b) residuals vs. lag-2 residuals; (c) residuals vs. lag-3 residuals; (d) residuals vs. lag-4 residuals.*

assumptions about associations among measurements discussed in section 5.2.4.

Methods

Given the large number of observations per tree, techniques based on individual estimates are a natural choice for these data. To obtain individual estimates, we used the GLS method of section 5.2 with the banded intra-individual correlation structure described above, estimated by pooled pseudolikelihood. We obtained also OLS estimates assuming no within-tree correlation, with pooled estimation of σ. The pooled estimate of the correlation parameter α was 0.25; however, individual estimates obtained with and without the assumption of within-individual correlation were virtually indistinguishable. Adopting the principle of parsimony, in what follows, we report results only for the model without correlation.

The tree-specific OLS estimates were input to the GTS scheme of section 5.3.2 to obtain estimates of $\boldsymbol{\beta}$, $\boldsymbol{D}$, and 'refined' estimates of the $\boldsymbol{\beta}_i$. All inference based on individual estimates was implemented using the GAUSS matrix programming language. Inference on $\boldsymbol{\beta}$ and $\boldsymbol{D}$, and empirical Bayes estimates of the $\boldsymbol{\beta}_i$ was also carried out using the conditional first-order linearization method of Lindstrom and Bates (1990) using the NLME software. Of necessity, the intra-individual covariance structure $\boldsymbol{R}_i(\boldsymbol{\beta}_i, \boldsymbol{\xi}) = \sigma^2 \boldsymbol{I}_{n_i}$ was assumed for this procedure.

Results

For these data, the GTS and NLME fits gave indistinguishable results. Accordingly, we present results for the NLME fit only, with maximum likelihood estimation of the covariance components. Table 11.4 shows these results. For brevity, we report estimates of only the diagonal elements of $\boldsymbol{D}$.

The preferred model, according to both the AIC and loglikelihood criteria, is Model 4. That is, the variable X ($= D^2H$) is an important covariate in explaining tree-to-tree variation in all 'shape' parameters. This is especially obvious in the case of β_1 (which roughly corresponds to maximum cumulative bole volume attained). The model with no covariates attributes all variation in β_1 to be random; this variation is considerable ($D_{11} = 2715.0$). Allowing X to enter the model as a covariate to explain the systematic among-tree variability in β_1 reduces the random component considerably ($D_{11} = 55$ to 64 in Models 2–4); the presence of X in β_2 and β_3 also explains much of the variation in these parameter values, as reflected by the small relative magnitudes of the estimates for D_{22} and D_{33} for Model 4. Intuitively, choice of Model 4 seems reasonable; it is plausible that tree-to-tree variation in parameters governing shape and degree of tapering depend on a measure of overall tree size.

The final mean equation for prediction of cumulative outer-bark bole volume, then, is based on the fit of Model 4:

$$y = \frac{4.46 + 0.00195X}{1 + \exp[e^* + \{\log(D - d) - (0.856 + 0.000125X)\}}',$$

where $e^* = \exp(0.758 + 0.00000575X)$ and $X = D^2H$.

The fit of this equation to selected trees is shown by the solid line in Figure 11.2. One may also obtain individualized fits based on empirical Bayes estimates of the model parameters (dashed line).

Table 11.4. *NLME fits to Models 1–4 to the sweetgum data.*

	Model			
	1	*2*	*3*	*4*
β_1	60.4	1.60	3.34	4.46
(SE)	(8.4)	(1.96)	(1.83)	(1.86)
β_2	1.20	1.24	0.901	0.856
(SE)	(0.06)	(0.06)	(0.05)	(0.01)
β_3	0.933	0.923	0.995	0.758
(SE)	(0.045)	(0.045)	(0.046)	(0.075)
β_4	–	2.06×10^{-3}	1.98×10^{-3}	1.95×10^{-3}
(SE)	–	(4.72×10^{-5})	(4.63×10^{-5})	(4.69×10^{-5})
β_5	–	–	1.08×10^{-5}	1.25×10^{-5}
(SE)	–	–	(9.91×10^{-7})	(1.10×10^{-6})
β_6	–	–	–	5.75×10^{-6}
(SE)	–	–	–	(1.46×10^{-6})
D_{11}	2715.0	63.98	55.54	55.64
D_{22}	0.147	0.135	0.025	0.023
D_{33}	0.053	0.052	0.046	0.033
σ	4.10	4.11	4.11	4.10
Log-like.	−2096.9	−2045.0	−2022.6	−2014.8
AIC	4213.8	4111.9	4069.3	4055.5

For the data at hand, these fits are very close to the fit of the population model, reflecting the small tree-to-tree variation in the parameters once X is taken into account.

Discussion

This example shows that methods based on conditional linearization and those based on individual estimates may lead to very similar conclusions in the case where the number of observations per individual is large. When this is the case, one may pursue refinements of the model using the most convenient approach. We have found NLME to be particularly well-suited to investigating a variety of model specifications for inter-individual variation in the regression parameters.

11.4 Prediction of strong motion of earthquakes

Data description

Brillinger (1987) summarizes an analysis of seismology data listed in Joyner and Boore (1981). The data consist of 182 maximum horizontal acceleration values recorded at available seismometer locations for 23 large earthquakes in western North America between 1940 and 1980. Table 11.5 shows a partial data listing. The key variables are a_{ij}, maximum horizontal acceleration (in g units) taken at the jth measurement station pertaining to the ith earthquake; d_{ij}, the distance in kilometers of the jth measuring station for the ith event from the event epicenter; and M_i, the magnitude of the ith quake on the Richter scale. In addition, information on site condition, classified as either 'soil' or 'rock,' was available for 17 of the 23 events. The number of measurements per event, n_i, ranged from one (for six quakes), to 38, for one event, at Imperial Valley, California in 1979.

Objective and model formulation

It is of interest to obtain an expression to predict maximum horizontal acceleration at a specified location during a large earthquake in terms of magnitude of the quake and distance from the epicenter. This information is useful for understanding the nature of damage caused by a particular quake and for determining the location of sensitive installations. The task may be formulated in terms of a regression problem: we wish to fit a model $f(\boldsymbol{x}, \boldsymbol{\beta})$ to the response, where $\boldsymbol{x} = [d, M]'$, the functional form of f is known, and the vector $\boldsymbol{\beta}$ represents a set of regression parameters to be estimated from the data (a_{ij}, d_{ij}, M_i), $i = 1, \ldots, 23$, $j = 1, \ldots, n_i$.

Joyner and Boore (1981) propose an attenuation law of the form

$$y = \log_{10} a = \beta_1 + \beta_2 M - \log_{10}(\sqrt{d^2 + \beta_3^2}) - \beta_4 \sqrt{d^2 + \beta_3^2}, \quad (11.9)$$

based partly on physical, partly on empirical considerations. This model for mean response is nonlinear in the parameter vector $\boldsymbol{\beta} = [\beta_1, \beta_2, \beta_3, \beta_4]'$, and thus the data may be viewed as nonlinear repeated measurements, where the unit of repeated measurement (the 'individual,' in the terminology of this book) is the earthquake. To accommodate the inter-individual variation, we might consider the 23 earthquakes in this data set as coming from a population of earthquakes, and thus regard $\boldsymbol{\beta}_i$, the vector of parameters in (11.9)

Table 11.5. *Strong motion data for two earthquakes.*

Station	*Distance (km)*	*Max. hor. accel. (g)*	*Soil type*
	Kern County, 1952		
1	148.0	0.014	soil
2	42.0	0.196	soil
3	85.0	0.135	soil
4	107.0	0.062	soil
5	109.0	0.054	soil
6	156.0	0.014	soil
7	224.0	0.018	soil
8	293.0	0.010	soil
9	359.0	0.004	soil
10	370.0	0.004	soil
	San Fernando, 1971		
1	17.0	0.374	rock
2	19.6	0.200	rock
3	20.2	0.147	rock
4	21.1	0.188	rock
5	21.9	0.204	rock
6	24.2	0.335	rock
7	66.0	0.057	rock
8	87.0	0.021	rock
9	23.4	0.152	soil
10	24.6	0.217	soil
11	25.7	0.114	soil
12	28.6	0.150	soil
13	37.4	0.148	soil
14	46.7	0.112	soil
15	56.9	0.043	soil
16	60.7	0.057	soil
17	61.4	0.030	soil
18	62.0	0.027	soil
19	64.0	0.028	soil
20	82.0	0.034	soil
21	88.0	0.030	soil
22	91.0	0.039	soil

for the ith event, as random. Because earthquake magnitude, M_i, is constant for a given individual, β_2, the regression coefficient in M for the ith quake, cannot sensibly be regarded as varying with i. To do so would overspecify the model; intuitively, with a single M_i value per quake, one cannot estimate a separate 'slope' for each quake. It might reasonably be expected that incorrect or inadvertent overspecification of a model, e.g. by treating a necessarily fixed parameter as random, should lead to computational difficulties. The reader should beware, however; warning signs from the program used may not be obvious. For instance, in 'stress-testing' some of the software for sensitivity to this type of nonidentifiability, we were able to achieve convergence in several cases. Signs of trouble were more subtle, e.g. the estimated covariance matrix of the random effects suggested perfect correlation among the random effect components. This should not be interpreted as criticism of the software; rather, it should serve as merely a reminder that the responsibility to specify a sensible model rests primarily with the analyst, not the program.

In the fits reported here, we considered β_3 and β_4 to be random and β_1 and β_2 to be fixed. As discussed above, treating β_2 as random would render the model unidentifiable; preliminary investigation indicated that regarding β_1 as fixed was reasonable. Omitting the covariate information on soil type for the moment, the basic hierarchical modeling structure may be stated in the notation of Chapter 4 as follows:

Stage 1. (within-earthquake variation)

$$y_{ij} = \log_{10} a_{ij} = \beta_{1i} + \beta_{2i} M_i$$
$$- \log_{10}\left(\sqrt{d_{ij}^2 + \beta_{3i}^2}\right) - \beta_{4i}\sqrt{d_{ij}^2 + \beta_{3i}^2} + e_{ij},$$

$$(\boldsymbol{e}_i | \boldsymbol{\beta}_i) \sim \{\boldsymbol{0}, \boldsymbol{R}_i(\boldsymbol{\beta}_i, \boldsymbol{\xi})\}, \tag{11.10}$$

where $\boldsymbol{R}_i(\boldsymbol{\beta}_i, \boldsymbol{\xi})$ is an assumed within-earthquake covariance structure.

Stage 2. (quake-to-quake variation)

$$\boldsymbol{\beta}_i = \boldsymbol{A}_i\boldsymbol{\beta} + \boldsymbol{B}_i\boldsymbol{b}_i, \quad \boldsymbol{b}_i \sim (\boldsymbol{0}, \boldsymbol{D}), \tag{11.11}$$

where $\boldsymbol{A}_i$ is the (4×4) identity matrix, $\boldsymbol{\beta} = [\beta_1, \beta_2, \beta_3, \beta_4]'$, $\boldsymbol{b}_i$ is a (2×1) vector of random effects corresponding to the parameters regarded as random, and the (4×2) design matrix $\boldsymbol{B}_i$ associated with the random effect is given by

$$\boldsymbol{B}_i = \begin{bmatrix} 0 & 0 \\ 0 & 0 \\ 1 & 0 \\ 0 & 1 \end{bmatrix}.$$

The specification (11.11) implies the following assumptions about the individual components of $\boldsymbol{\beta}_i$:

$$\beta_{1i} = \beta_1, \quad \beta_{2i} = \beta_2, \quad \beta_{3i} = \beta_3 + b_{3i}, \quad \beta_{4i} = \beta_4 + b_{4i},$$

where $\boldsymbol{b}_i = [b_{3i}, b_{4i}]'$.

The inferential challenge is thus to estimate the population parameters $\boldsymbol{\beta}$ and $\boldsymbol{D}$, incorporating the differential amounts of data available for different events, and correctly accounting for within- and among-earthquake variation. Empirical Bayes fits for individual events may be obtained as a byproduct of this population inference. Refinements to the basic model, such as inclusion of covariate information to improve characterization of inter-quake variation, may be undertaken in the analysis. Estimation of maximum horizontal acceleration for a particular location for a future earthquake of given magnitude may then be based on the final fitted model.

Methods

Given the sparsity of measurements for several of the earthquakes, inferential techniques based on individual estimates may not be used for these data. Accordingly, we fit the hierarchical model to the data using methods based on linearization, described in Chapter 6.

To determine an appropriate intra-individual covariance specification $\boldsymbol{R}_i(\boldsymbol{\beta}_i, \boldsymbol{\xi})$, we performed a preliminary analysis under the basic hierarchy (11.10), (11.11), using the GLS algorithm of Vonesh and Carter (1992), modified to allow estimation of within-individual covariance parameters (section 6.2.3); the results are not reported here. To investigate the possibility of heterogeneous intra-quake variation in $\log_{10}$ maximum horizontal acceleration values, we assumed measurements within an earthquake were uncorrelated, with variance proportional to a power, θ, of the mean response. The estimate of θ obtained from the fit of this model was close to zero; thus, in the analyses below, we took $\boldsymbol{R}_i(\boldsymbol{\beta}_i, \boldsymbol{\xi}) = \sigma^2 \boldsymbol{I}_{n_i}$.

The following sequence of three models was fitted using the conditional linearization method of Lindstrom and Bates (1990) (section 6.3.2), as implemented in the NLME software. In all cases, the first stage of each model was as in (11.10); the second stage of each model considered was as follows:

Model 1. (β_{1i}, β_{2i} fixed; β_{3i}, β_{4i} random)

$$\beta_{1i} = \beta_1, \quad \beta_{2i} = \beta_2, \quad \beta_{3i} = \beta_3 + b_{3i}, \quad \beta_{4i} = \beta_4 + b_{4i},$$

$$\boldsymbol{\beta} = [\beta_1, \beta_2, \beta_3, \beta_4]', \quad \boldsymbol{b}_i = [b_{3i}, b_{4i}]', \quad \boldsymbol{b}_i \sim (\boldsymbol{0}, \boldsymbol{D})$$

Model 2. ($\beta_{1i}, \beta_{2i}, \beta_{3i}$ fixed; β_{4i} random)

$$\beta_{1i} = \beta_1, \quad \beta_{2i} = \beta_2, \quad \beta_{3i} = \beta_3, \quad \beta_{4i} = \beta_4 + b_{4i},$$

$$\boldsymbol{\beta} = [\beta_1, \beta_2, \beta_3, \beta_4]', \quad b_i = b_{4i}, \quad b_i \sim (0, D)$$

Model 3. ($\beta_{1i}, \beta_{2i}, \beta_{3i}$ fixed; β_{4i} random, soil type as a covariate in β_{4i})

$$\beta_{1i} = \beta_1, \quad \beta_{2i} = \beta_2, \quad \beta_{3i} = \beta_3, \quad \beta_{4ij} = \beta_4 + \beta_5 s_{ij} + b_{4i},$$

$$\boldsymbol{\beta} = [\beta_1, \beta_2, \beta_3, \beta_4, \beta_5]', \quad b_i = b_{4i}, \quad b_i \sim (0, D),$$

where s_{ij} is an indicator variable describing the conditions at the jth site for the ith earthquake ($s_{ij} = 0$ for 'rock,' $s_{ij} = 1$ for 'soil'), and we have written β_{4ij} to indicate that the value of this parameter may change within a given earthquake.

Incorporation of soil type as a covariate in this example is analogous to the use of time-dependent covariates (section 4.2.5) in earlier examples, as the value of s_{ij} may vary within an individual in these data.

Model 1 was fitted to data from all 23 earthquakes, Model 3 was fitted only to those 17 events for which values of s_{ij} were recorded, and Model 2 was fitted both to the full and reduced data sets. The form of the random effects covariance matrix $\boldsymbol{D}$ for Model 1 was unconstrained; this quantity is a scalar for Models 2 and 3. Inference was carried out using both maximum likelihood and restricted maximum likelihood estimation of the covariance components σ and $\boldsymbol{D}$ or D. Results were comparable; only those for restricted maximum likelihood are reported below.

Table 11.6. *Fits to Models 1 and 2, full data set.*

	Model	
	1	*2*
β_1	-0.764	-0.763
(SE)	(0.257)	(0.258)
β_2	0.218	0.218
(SE)	(0.040)	(0.040)
β_3	8.365	8.380
(SE)	(1.511)	(1.550)
β_4	5.657×10^{-3}	5.657×10^{-3}
(SE)	(1.659×10^{-3})	(1.657×10^{-3})
D_{11}	4.437×10^{-3}	–
D_{12}	-1.454×10^{-4}	–
D_{22}[1]	2.171×10^{-5}	2.165×10^{-5}
σ	0.218	0.201

[1] *D* for Model 2

Results

Table 11.6 summarizes results for fitting Models 1 and 2 to the full data. It is evident from the fit of Model 1 that inter-event random variation in the β_{3i} values about their mean is negligible (CV = 0.7%). Accordingly, it seems reasonable to favor Model 2, which fixes β_3. This choice is supported by comparison of AIC values computed based on the final maximized restricted likelihood for each fit (Model 1, -297, versus Model 2, -301; smaller values are preferred). Table 11.7 shows results for fitting Models 2 and 3 to those events where soil type was recorded. The coefficient for soil type, β_5, has an associated Wald statistic of -2.07; comparison to an approximate reference standard normal distribution suggests that knowledge of soil type at a particular site may add some explanatory power to the overall model, although the evidence is not overwhelming.

Estimation of $\log_{10}$ maximum horizontal acceleration that would occur for a future earthquake at a particular location some given distance from the epicenter may be carried out by evaluating (11.9) at the final population estimates of the parameters. These analyses suggest the following final model from the fit of Model 3:

Table 11.7. *Fits to Models 2 and 3, reduced data set.*

	Model	
	2	3
β_1	-1.336	-1.355
(SE)	(0.367)	(0.373)
β_2	0.309	0.316
(SE)	(0.056)	(0.056)
β_3	8.141	8.629
(SE)	(1.573)	(1.629)
β_4	6.209×10^{-3}	7.671×10^{-3}
(SE)	(1.871×10^{-3})	(2.022×10^{-3})
β_5	-	-1.243×10^{-3}
(SE)	-	(6.000×10^{-4})
D	2.321×10^{-5}	2.496×10^{-5}
σ	0.205	0.201

$$\log_{10} a = -1.355 + 0.316M - \log_{10}(\sqrt{d^2 + 8.629^2})$$
$$-(0.00767 - 0.00124s)\sqrt{d^2 + 8.629^2}, \qquad (11.12)$$

where s is the soil type indicator. In the absence of information on s at a given location, the final fit to Model 2 may be used in place of (11.12).

Predicted values using individual empirical Bayes estimates and using the population mean fixed effects estimates from the fit of Model 2 are shown for four selected events in the data set in Figure 11.4. These plots provide a dramatic illustration of the way in which empirical Bayes estimation allows one to 'borrow strength' across events. The validity of doing so rests strongly on the assumption that the earthquakes in question may be regarded as coming from a common population.

Discussion

This example highlights the importance of a clear understanding of the limitations of the hierarchical model framework on the part of the data analyst. Specification of a parameter as random is predicated on the availability of information required to determine a separate value for each individual; it is intuitively evident that, with a single earthquake magnitude value for each earthquake, there is insufficient information to determine a separate coefficient

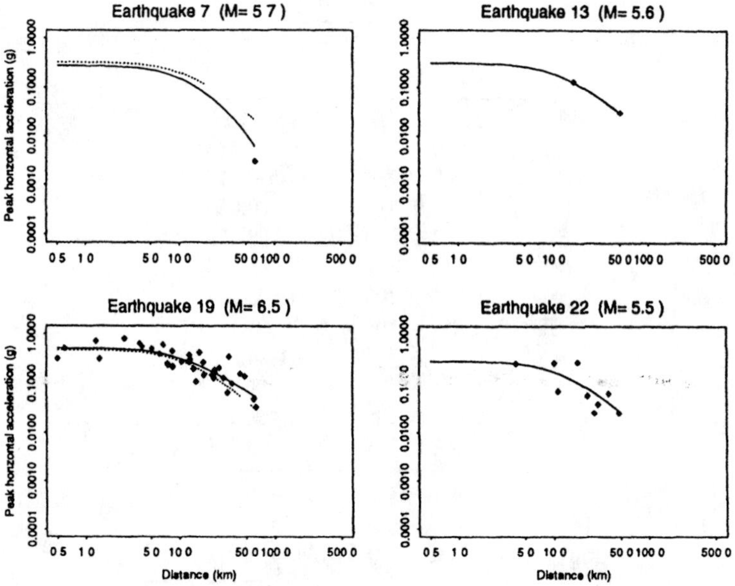

Figure 11.4. *Data for earthquakes 7 (Santa Rosa, 1969), 13 (Point Mugu, 1973), 19 (Imperial Valley, 1972), and 22 (Livermore Valley, 1980). Solid line, empirical Bayes fit of Model 2; dashed line, predicted values based on estimated population mean parameters for Model 2.*

for magnitude for each. Moreover, for the attenuation law (11.9), identifiability of the model is a critical issue. It is essential for the data analyst to contemplate such features prior to proceeding with an analysis, as the consequences of a flawed model specification may not always be obvious from the results.

For these data, examination of maximum horizontal acceleration measurements within a quake with several recording stations reveals that, for stations approximately the same distance from the epicenter, these values may be quite variable. In the analysis presented here, based on (11.9), this within-quake variation is attributed to random phenomena (through the intra-individual error e_{ij}). It seems reasonable to postulate that knowledge of the spatial orientation of a recording site relative to the epicenter may contain additional explanatory value beyond that provided by distance alone and might be incorporated as covariate information in (11.9) within the hierarchical modeling framework.

CHAPTER 12

Open problems and discussion

Hierarchical modeling provides a very flexible approach to the analysis of nonlinear repeated measurement data, with applications in a wide variety of fields. Chapters 9–11 showed examples from several specific areas of application; the methods described in this book obviously have potential utility in other areas as well. We hope that the exposition in this book may lead to their wider use.

We conclude by sketching some open problems and making some general comments.

Software development

A recurrent theme throughout this book has been the inherent computational difficulty associated with all of the inferential methods presented. This is not surprising; inference for hierarchical nonlinear models inherits the computational problems associated with the areas of nonlinear regression and hierarchical linear models, each of which can be treacherous in its own right. There is a great need for stable, well-documented software of inference in hierarchical nonlinear models. This comment applies, to one degree or another, to all the methods described in this book.

Accuracy of approximations and asymptotics

Many of the techniques for population inference considered in the preceding chapters derive validity from large-sample arguments; within the context of hierarchical models, this requires that the number of individuals be large. Methods based on individual estimates, discussed in Chapter 5, carry the further requirement of large within-individual sample size. For all methods, with moderate sample sizes, the implication is that inference is approximate. In addition, several of the procedures explicitly involve a further level of approximation to facilitate inference. Different types of approximations are possible, and operating characteristics of each are not well-understood. Our experience has been that differences among methods in results pertaining to second moments are quite common. Further experience in critical use of these techniques is needed.

Diagnostics

There is an urgent need for diagnostics to complement all of the methods discussed in this book. Even for the hierarchical linear model, the literature on diagnostic procedures is sparse; an example of a proposal in this case is that of Lange and Ryan (1989). In the nonlinear case, there is an additional level of complication in fitting, and this complexity may obscure key features of the data. In particular, methods for outlier and case influence analysis at each level of the modeling hierarchy would be extremely useful. Development of robust and resistant fitting procedures should also be of interest; such methods often provide useful diagnostic information. Estimation of covariance parameters can be particularly susceptible to influence of a small subset of data. Development of suitable diagnostics poses obvious computational challenges, but is an essential step if the fitting tools described in this book are to be adopted for widespread use.

Measurement error

It is often the case in practice that covariate values collected on individuals are measured with non-negligible error. We touched on this issue briefly in our discussion of pharmacokinetic and pharmacodynamic modeling in Chapter 9. The potential for bias in parameter estimates obtained ignoring measurement error is obvious. Estimation methods that reduce attenuation bias due to measurement error are not well-developed in the context of hierarchical nonlinear models.

Censored and missing data

Methods described in the preceding chapters allow one to deal with missing or incomplete repeated measurement data. However, a key assumption underlying the validity of the proposed techniques is that the data are missing at random. The cases of informative censoring and missing data are not covered by the methods described in this monograph.

Multiple levels of nesting

Throughout this book, we have considered explicitly only the case of a single level of nesting. Measurements on an individual are repeated at only one level so that only within- and inter-individual components of variance are of concern. Extension to higher levels of nesting within the hierarchy is straightforward in principle, although notational and computational aspects are more complicated; see Lindstrom and Bates (1988) for an example in the linear case. In pharmacokinetic applications, an instance where such an extension is useful is that where profiles are obtained on repeated occasions on each of a number of subjects, for example, on several occasions during a prolonged treatment course with an experimental drug. Analyses of this kind of data should account for two levels of variation within subject, inter- and intra-occasion varibility; see Karlsson and Sheiner (1994). Longford (1993, chapter 6) considers multiple levels of nesting for hierarchical linear models and gives an example of application to a three-level data set involving repeated measurement of hearing ability in each ear on each of a number of subjects. Multilevel models are also considered by Goldstein (1986, 1991); software for inference in these models is available in the package ML3 (Prosser *et al.*, 1991).

Modeling correlation structures

We have considered models that allow considerable flexibility in the specification of intra-individual correlation structures. The examples considered in this book provide mixed messages on the utility of this approach, and further experience is needed. One point of view is that complicated modeling of within-individual correlation is unnecessary, as the random effects assumption induces an overall correlation structure in the marginal covariance. We believe that this induced correlation may not provide a sufficiently rich structure in all cases. However, specifying a detailed model for intra-individual, although justifiable in theory, may not always be supported by the data. As Longford (1993, section 4.4.2) writes, 'in general, it is difficult to make an informed decision about how complex a description for the variation part of a model can be supported by a given data set.' Second moment behavior is inherently difficult to characterize, and this is especially true for correlation parameters. Accordingly, we recommend specifying intra-individual correlation structures as an exploratory tool, to be used with caution.

Modeling intra-individual variance

The first stage of the hierarchy developed in Chapter 4 also allows specification of a model for heterogeneous within-individual variance. One could take the same perspective as for intra-individual correlation above and conclude that the induced marginal covariance will be adequate to characterize the overall pattern of variation at the marginal level, so that such modeling is not really required. In many of the areas of application considered in this book, intra-individual heterogeneity of variance is a widely acknowledged feature of the data, and there is a tradition of modeling this variation. Although the benefit of modeling within-individual correlation is unclear, failure to take heterogeneous intra-individual variance into account when it is evident or expected may be deleterious for inference both on fixed effects and other covariance components; this was seen in the soybean growth example in Chapter 11. Davidian and Giltinan (1993a) present simulation evidence that shows the dramatic effect, in terms of inefficiency, of improper modeling of intra-individual variation in the context of hierarchical nonlinear modeling.

Subject-specific modeling of discrete data

As discussed in Chapter 4, in the terminology of Zeger *et al.* (1988), our approach to modeling has been *subject-specific*, beginning from the conditional distribution at Stage 1 of the hierarchy. For the applications in this book, *subject-specific* modeling seems most natural. Throughout, we have considered only continuous responses. For discrete, especially binary responses, for which common models are inherently nonlinear, there has been much recent interest in both *population-averaged* and subject-specific modeling. In the case of the latter approach, Stiratelli *et al.* (1984), Zeger and Karim (1991), Longford (1993, chapter 8), and Diggle *et al.* (1994, chapters 7 and 9) describe inferential strategies for such models. Many of these techniques are based on the same principles as those presented in Chapters 6 and 8 of this book. Other methods described in this book, such as the semiparametric method described in Chapter 7, are also applicable to inference under subject-specific models for discrete data.

Performance of nonparametric and semiparametric methods

The lack or mildness of distributional assumptions underlying the methods of Chapter 7 is appealing. We repeat the caveat expressed in that chapter: further experience is needed with these methods. The associated high-dimensional optimization problems are intrinsically difficult. Only repeated use of these methods in a variety of problems will allow proper evaluation of their operating characteristics.

Bayesian inference

The Bayesian methods of Chapter 8 provide a natural framework for general hierarchical models, not just normal response data. Computational advances such as the Gibbs sampler show great promise for the future of the Bayesian approach; again, further experience is needed. While the Gibbs sampler is appealing due to its relative conceptual simplicity, it seems likely that future work in this area will lead to more sophisticated, and possibly more efficient Markov chain Monte Carlo techniques.

Conclusion

The preceding list of open issues, which is not exhaustive, indicates that modeling of nonlinear repeated measurement data, although useful, is not an enterprise that should be undertaken lightly. There is considerable current interest in this rapidly developing field, and we look forward to continued progress in the near future.

References

Amemiya, T. (1973) Regression analysis when the variance of the dependent variable is proportional to the square of its expectation. *Journal of the American Statistical Association*, **68**, 928–934.

Aziz, N.S., Gambertoglio, J.G., Lin, E.T., Grausz, H. and Benet, L.Z. (1978) Pharmacokinetics of cefamandole using a HPLC assay. *Journal of Pharmacokinetics and Biopharmaceutics*, **6**, 153–164.

Bates, D.M. and Watts, D.G. (1988) *Nonlinear Regression Analysis and its Applications.* Wiley, New York.

Beal, S.L. (1984a) Population pharmacokinetic and parameter estimation based on their first two statistical moments. *Drug Metabolism Reviews*, **15**, 173–193.

Beal, S.L. (1984b) Asymptotic properties of optimization estimators for the independent and not identically distributed case with application to extended least squares. Technical report, Division of Clinical Pharmacology, University of California, San Francisco.

Beal, S.L. and Sheiner, L.B. (1982) Estimating population kinetics. *CRC Critical Reviews in Biomedical Engineering*, **8**, 195–222.

Beal, S.L. and Sheiner, L.B. (1985) Methodology of population pharmacokinetics. In *Drug Fate and Metabolism–Methods and Techniques* (eds. E.R. Garrett and J.L. Hirtz), Marcel Dekker, New York.

Beal, S.L. and Sheiner, L.B. (1988) Heteroscedastic nonlinear regression. *Technometrics*, **30**, 327–338.

Beal, S.L. and Sheiner, L.B. (1992) *NONMEM User's Guides.* NONMEM Project Group, University of California, San Francisco.

Belanger, B.A., Davidian, M. and Giltinan, D.M. (1995) The effect of variance function estimation on nonlinear calibration inference in immunoassay. *Biometrics*, to appear.

Berkey, C.S. (1982) Bayesian approach for a nonlinear growth model. *Biometrics*, **38**, 953–961.

BMDP Statistical Software (1990) University of California Press, Berkeley.

Boeckmann, A.J., Sheiner, L.B. and Beal, S.L. (1992) *NONMEM User's Guide, Part V, Introductory Guide.* University of California, San Francisco.

Box, G.E.P. and Cox, D.R. (1964) An analysis of transformations. *Journal of the Royal Statistical Society, Series B*, **26**, 211–246.

Box, G.E.P. and Hill, W.J. (1974) Correcting inhomogeneity of variance with power transformation weighting. *Technometrics*, **16**, 385–389.

Box, G.E.P. and Meyer, R.D. (1986a) An analysis for unreplicated fractional factorials. *Technometrics*, **28**, 11–18.

Box, G.E.P. and Meyer, R.D. (1986b) Dispersion effects from fractional designs. *Technometrics*, **28**, 19–27.

Breslow, N.E. and Clayton, D.G. (1993) Approximate inference in generalized linear mixed models. *Journal of the American Statistical Association*, **88**, 9–25.

Brillinger, D.R. (1987) Comment on paper by C.R. Rao. *Statistical Science*, **2**, 448–450.

Burnett, R.T., Ross, W.H. and Krewksi, D. (1995) Nonlinear random effects regression models. *Environmetrics*, to appear.

Carroll, R.J. and Cline, D. (1988) An asymptotic theory for weighted least squares with weights estimated by replication. *Biometrika*, **75**, 35–43.

Carroll, R.J. and Ruppert, D. (1982a) Robust estimation in heteroscedastic linear models. *Annals of Statistics*, **10**,429–441.

Carroll, R.J. and Ruppert, D. (1982b) A comparison between maximum likelihood and generalized least squares in a heteroscedastic linear model. *Journal of the American Statistical Association*, **77**, 878–882.

Carroll, R.J. and Ruppert, D. (1984) Power transformations when fitting theoretical models to data. *Journal of the American Statistical Association*, **79**, 321–328.

Carroll, R.J. and Ruppert, D. (1988) *Transformations and Weighting in Regression*. Chapman & Hall, New York.

Carroll, R.J., Wu, C.F.J. and Ruppert, D. (1988) The effect of estimating weights in weighted least squares. *Journal of the American Statistical Association*, **83**, 1045–1054.

Carter, R.L. and Yang, M.C.K. (1986) Large-sample inference in random coefficient regression models. *Communications in Statistics – Theory and Methods*, **8**, 2507–2526.

Casella, G. and George, E.I. (1992) Explaining the Gibbs sampler. *The American Statistician*, **46**, 167–174.

Chambers, J.M. and Hastie, T.J. (1992) *Statistical Models in S.* Wadsworth and Brooks/Cole, Pacific Grove, California.

Chi, E.M. and Reinsel, G.C. (1989) Models for longitudinal data with random effects and AR(1) errors. *Journal of the American Statistical Association*, **84**, 452–459.

Clark, B.R. and Engvall, E. (1978) Enzyme-linked immunosorbent assay (ELISA): Theoretical and practical aspects. In *Enzyme Immunoassay* (ed. E.T. Corrie), CRC Press, Boston.

Corbeil, R.R. and Searle, S.R. (1976) Restricted maximum likelihood

(REML) estimation of variance components in the mixed model. *Technometrics*, **18**, 31–38.

Crowder, M.J. (1978) Beta-binomial ANOVA for proportions. *Applied Statistics*, **27**, 34–37.

Crowder, M.J. and Hand, D.J. (1990) *Analysis of Repeated Measures.* Chapman & Hall, New York.

D'Argenio, D.Z. and Schumitzky, A. (1992) Adapt II User's Guide. Biomedical Simulations Resource, University of Southern California, Los Angeles.

Davidian, M. (1990) Estimation of variance functions in assays with possibly unequal replication and nonnormal data. *Biometrika*, **77**, 43–54.

Davidian, M. and Carroll, R.J. (1987) Variance function estimation. *Journal of the American Statistical Association*, **82**, 1079–1091.

Davidian, M. and Carroll, R.J. (1988) A note on extended quasilikelihood. *Journal of the Royal Statistical Society, Series B*, **50**, 74–82.

Davidian, M., Carroll, R.J. and Smith, W. (1988) Variance functions and the minimum detectable concentration in assays. *Biometrika*, **75**, 549–556.

Davidian, M. and Gallant, A.R. (1992) Smooth nonparametric maximum likelihood estimation for population pharmacokinetics, with application to quinidine. *Journal of Pharmacokinetics and Biopharmaceutics*, **20**, 529–556.

Davidian, M. and Gallant, A.R. (1993) The nonlinear mixed effects model with a smooth random effects density. *Biometrika*, **80**, 475–488.

Davidian, M. and Gallant, A.R. (1994) Nlmix: a program for maximum likelihood estimation of the nonlinear mixed effects model with a smooth random effects density. Unpublished technical report.

Davidian, M. and Giltinan, D.M. (1993a) Some simple methods for estimating intra-individual variability in nonlinear mixed effects models. *Biometrics*, **49**, 59–73.

Davidian, M. and Giltinan, D.M. (1993b) Some general estimation methods for nonlinear mixed effects models. *Journal of Biopharmaceutical Statistics*, **3**, 23–55.

Davidian, M. and Giltinan, D.M. (1993c) Analysis of repeated measurement data using the nonlinear mixed effects model. *Chemometrics and Intelligent Laboratory Systems*, **20**, 1–24.

Davidian, M. and Haaland, P.D. (1990) Regression and calibration with nonconstant error variance. *Chemometrics and Intelligent Laboratory Systems*, **9**, 231–248.

Dempster, A.P., Laird, N.M. and Rubin, D.B. (1977) Maximum likelihood from incomplete data via the EM algorithm (with Discussion). *Journal of the Royal Statistical Society, Series B*, **39**, 1–38.

Dempster, A.P., Rubin, D.B. and Tsutakawa, R.K. (1981) Estimation in covariance components models. *Journal of the American Statistical Association*, **76**, 341–353.

Dempster, A.P., Selwyn, M.R., Patel, C.M. and Roth, A.J. (1984) Statistical and computational aspects of mixed model analysis. *Applied Statistics*, **33**, 203–214.

Diggle, P.J. (1988) An approach to the analysis of repeated measurements. *Biometrics*, **44**, 959–971.

Diggle, P.J. and Kenward, M.G. (1994) Informative dropout in longitudinal data analysis (with Discussion). *Applied Statistics*, **43**, 49–93.

Diggle, P.J., Liang, K.-Y. and Zeger, S.L. (1994) *Analysis of Longitudinal Data.* Oxford University Press, Oxford.

Donaldson, J.R. and Schnabel, R.B. (1987) Computational experiences with confidence regions and confidence intervals for nonlinear least squares. *Technometrics*, **29**, 67–82.

Draper, N.R. and Smith, H. (1981) *Applied Regression Analysis.* Wiley, New York.

Dudley, R.A., Edwards, P., Ekins, R.P., Finney, D.J., McKenzie, I.G.M., Raab, G.M., Rodbard, D. and Rodgers, R.P.C. (1985) Guidelines for immunoassay data processing. *Clinical Chemistry* **31**, 1264–1271.

Eastwood, B.J. (1991) Asymptotic normality and consistency of semi-nonparametric regression estimators using an upward F test truncation rule. *Journal of Econometrics*, **48**, 151–182.

Eaves, D.M. (1983) On Bayesian nonlinear regression with an enzyme example. *Biometrika*, **70**, 373–379.

Efron, B. (1986) Double exponential families and their use in generalized linear regression. *Journal of the American Statistical Association*, **81**, 709–721.

Ekins, R.P. and Edwards, P.R. (1983) The precision profile: Its use in assay design, assessment, and quality control. In *Immunoassays for Clinical Chemistry* (eds. W.M. Hunter and J.E.T. Corrie), Churchill Livingston, Edinburgh.

Epicenter Software (1994) *Epilog Plus, Statistics Package for Epidemiology and Clinical Trials.* Pasadena, California.

Escobar, M. and West, M. (1992) Computing Bayesian nonparametric hierarchical models. ISDS Discussion Paper 92-A20, Duke University.

Fattinger, K., Vozeh, S., Ha, H.R., Borner, M. and Follath, F. (1991) Population pharmacokinetics of quinidine. *British Journal of Clinical Pharmacokinetics*, **31**, 279–286.

Fattinger Bachmann, K.E., Sheiner, L.B. and Verotta, D. (1993) A new method to explore the distribution of interindividual random effects in nonlinear mixed effects models. Technical report, Departments of Laboratory Medicine and Pharmacy, University of California, San Francisco.

Fedorov, V.V. (1972) *Theory of Optimal Experiments.* Academic Press, New York.

Fei, D.T.W., Gross, M.C., Lofgren, J., Mora-Worms, M. and Chen, A.B. (1990) Cyclic AMP response to recombinant human relaxin by cultured endometrial cells – a specific and high throughput in-vitro bioassay. *Biochemical and Biophysical Research Communications*, **170**, 214–222.

Finney, D.J. (1976) Radioligand assay. *Biometrics*, **32**, 721–740.

Finney, D.J. (1978) *Statistical Method in Bioassay* (3rd edn). Griffin, High Wycombe, Bucks.

Fuller, W.A. and Rao, J.N.K. (1978) Estimation for a linear regression model with unknown diagonal covariance matrix. *Annals of Statistics*, **6**, 1149–1158.

Gallant, A.R. (1975) Nonlinear regression. *The American Statistician*, **29**, 73–81.

Gallant, A.R. (1987) *Nonlinear Statistical Models*, Wiley, New York.

Gallant, A.R. and Goebel, J.J. (1976) Nonlinear regression with autocorrelated errors. *Journal of the American Statistical Association*, **71**, 961–967.

Gallant, A.R. and Nychka, D.W. (1987) Seminonparametric maximum likelihood estimation. *Econometrica*, **55**, 363–390.

Gallant, A.R. and Tauchen, G.E. (1989) Seminonparametric estimation of conditionally constrained heterogeneous processes: asset pricing applications. *Econometrica*, **57**, 1091–1120.

Gelfand, A.E., Hills, S.E., Racine-Poon, A. and Smith, A.F.M. (1990) Illustration of Bayesian inference in normal data models using Gibbs sampling. *Journal of the American Statistical Association*, **85**, 972–985.

Gelfand, A.E. and Smith, A.F.M. (1990) Sampling-based approaches to calculating marginal densities. *Journal of the American Statistical Association*, **85**, 398–409.

Gelman, A. and Rubin, D.B. (1992) Inferences from iterative simulation using multiple sequences. *Statistical Science*, **7**, 457–472.

Geman, S. and Geman, D. (1984) Stochastic relaxation, Gibbs distributions, and the Bayesian restoration of images. *IEEE Transactions on Pattern Analysis and Machine Intelligence*, **6**, 721–741.

Geyer, C. (1992) Practical Markov chain Monte Carlo. *Statistical Science*, **7**, 473–483.

Gibaldi, M. and Perrier, D. (1982) *Pharmacokinetics*. Marcel Dekker, New York.

Gilks, W.R., Clayton, D.G., Spiegelhalter, D., Best, N.G., McNeil, A.J., Sharples, L.D. and Kirby, A.J. (1993) Modelling complexity: Applications of Gibbs sampling in medicine. *Journal of the Royal Statistical Society, Series B*, **55**, 39–52.

Gilks, W.R. and Wild, P. (1992) Adaptive rejection sampling for Gibbs sampling. *Applied Statistics*, **41**, 337–348.

Giltinan, D.M., Carroll, R.J. and Ruppert, D. (1986) Some new methods for weighted regression when there are possible outliers. *Technometrics*, **28**, 219–230.

Giltinan, D.M. and Davidian, M. (1994) Assays for recombinant proteins: A problem in nonlinear calibration. *Statistics in Medicine*, **13**, 1165–1179.

Giltinan, D.M. and Ruppert, D. (1989) Fitting heteroscedastic regression models to individual pharmacokinetic data using standard statistical software. *Journal of Pharmacokinetics and Biopharmaceutics*, **17**, 601–614.

Glejser, H. (1969) A new test for heteroscedasticity. *Journal of the American Statistical Association*, **64**, 316–323.

Goldstein, H. (1986) Multilevel mixed linear model analysis using iterative generalized least squares. *Biometrika*, **73**, 43–56.

Goldstein, H (1991) Nonlinear multilevel models, with an application to discrete response data. *Biometrika*, **78**, 45–51.

Grasela, T.H., Jr. and Donn, S.M. (1985) Neonatal population pharmacokinetics of phenobarbital derived from routine clinical data. *Developmental Pharmacology and Therapeutics*, **8**, 374–383.

Green, P.J. (1984) Iteratively reweighted least squares for maximum likelihood estimation, and some robust and resistant alternatives (with Discussion). *Journal of the Royal Statistical Society, Series B*, **46**, 149–192.

Green, P.J. (1987) Penalized likelihood for general semi-parametric regression models. *International Statistical Review*, **55**, 245–259.

Gregoire, T.G. and Schabenberger, O. (1994) Fitting bole-volume equations to spatially correlated within-tree data. *Proceedings of the 6th Annual Conference on Applied Statistics in Agriculture, Manhattan, Kansas.*

Gumpertz, M.L. and Pantula, S.G. (1989) A simple approach to inference in random coefficient models. *The American Statistician*, **43**, 203–210.

Gumpertz, M.L. and Pantula, S.G. (1992) Nonlinear regression with variance components. *Journal of the American Statistical Association*, **87**, 201–209.

Hartley, H.O. and Rao, J.N.K. (1967) Maximum likelihood estimation for mixed analysis of variance models. *Biometrika*, **54**, 93–108.

Harvey, A.C. (1976) Estimating regression models with multiplicative heteroscedasticity. *Econometrics*, **44**, 461–465.

Harville, D. (1974) Bayesian inference for variance components using only error contrasts. *Biometrika*, **61**, 383–385.

Harville, D. (1976) Extension of the Gauss-Markov theorem to include the estimation of random effects. *The Annals of Statistics*, **2**, 384–395.

Harville, D. (1977) Maximum likelihood approaches to variance component estimation and to related problems. *Journal of the American Statistical Association*, **72**, 320–340.

Harville, D. (1990) BLUP (Best Linear Unbiased Prediction) and beyond. In *Advances in Statistical Methods for Genetic Improvement of Livestock*. Springer-Verlag, New York.

Hemmerle, W.J. and Hartley, H.O. (1973) Computing maximum likelihood estimates for the mixed A.O.V. model using the W-transformation. *Technometrics*, **15**, 819–831.

Henderson, C.R. (1953) Estimation of variance and covariance components. *Biometrics*, **9**, 226–252.

Henderson, C.R. (1975) Best linear unbiased estimation and prediction under a selection model. *Biometrics*, **31**, 423–447.

Henderson, C.R. (1984) Applications of linear models to animal breeding. University of Guelph technical report.

Henderson, C.R. (1990) Statistical methods in animal improvement: Historical overview. In *Advances in Statistical Methods for Genetic Improvement of Livestock*. Springer-Verlag, New York.

Higgins, K.M., Davidian, M. and Giltinan, D.M. (1995) A two-stage approach to measurement error in time-dependent covariates in nonlinear mixed effects models, with application to IGF-I pharmacokinetics. Unpublished manuscript.

Holford, N.H.G. and Sheiner, L.B. (1981) Understanding the dose-effect relationship: Clinical application of pharmacokinetic-pharmacodynamic models. *Clinical Pharmacokinetics*, **6**, 429–453.

Huber, P.J. (1981) *Robust Statistics*. Wiley, New York.

Hui, S.L. and Berger, J.O. (1983) Empirical Bayes estimation of rates in longitudinal studies. *Journal of the American Statistical Association*, **78**, 753–761.

Jacquez, J.A., Mather, F.J. and Crawford, C.R. (1968) Linear regression with non-constant, unknown error variances: sampling experiments with least squares and maximum likelihood estimators. *Biometrics*, **24**, 607–626.

James, W. and Stein, C. (1961) Estimation with quadratic loss. *Proceedings of the Fourth Berkeley Symposium on Mathematical Statistics and Probability*, Vol. 1, 361–379.

Jelliffe, R.W., Gomis, P. and Schumitzky, A. (1990) A population model of gentamicin made with a new nonparametric EM algorithm. Technical report 90-4, Laboratory of Applied Pharmacokinetics, University of Southern California.

Jelliffe, R.W., Iglesias, T., Hurst, A., Foo, K. and Rodriguez, J. (1991) Individualizing gentamicin dosage regimens: A comparative review of selected models, data fitting methods, and monitoring strategies. *Clinical Pharmacokinetics*, **21**, 461–478.

Jelliffe, R.W., Schumitzky, A. and Van Guilder, M. (1994) User manual for version 10.0 of the USC*PACK collection of PC programs. Laboratory of Applied Pharmacokinetics, University of Southern California.

Jennrich, R.I. (1969) Asymptotic properties of nonlinear least squares estimation. *The Annals of Mathematical Statistics*, **40**, 633–643.

Jennrich, R.I. and Sampson, P.F. (1976) Newton-Raphson and related algorithms for maximum likelihood variance component estimation. *Technometrics*, **18**, 11–17.

Jennrich, R.I. and Schluchter, M.D. (1986) Unbalanced repeated measures models with structured covariance matrices. *Biometrics*, **42**, 805–820.

Jobson, J.D. and Fuller, W.A. (1980) Least squares estimation when the covariance matrix and parameter vector are functionally related. *Journal of the American Statistical Association*, **75**, 176–181.

Johnson, N.L. and Kotz, S. (1972) *Distributions in Statistics: Continuous Multivariate*. Wiley, New York.

Jones, R.H. (1993) *Longitudinal Data With Serial Correlation: A State-space Approach*. Chapman & Hall, New York.

Joyner, W.B. and Boore, D.M. (1981) Peak horizontal acceleration and velocity from strong-motion records including records from the 1979 Imperial Valley, California, earthquake. *Bulletin of the Seismological Society of America*, **71**, 2011–2038.

Judge, G.G., Griffiths, W.E., Hill, R.C., Lütkepohl, H. and Lee, T.-C. (1985) *The Theory and Practice of Econometrics* (2nd edn). Wiley, New York.

Karlsson, M.O. and Sheiner, L.B. (1994) The importance of modeling inter-occasion variation in population pharmacokinetics. *Journal of Pharmacokinetics and Biopharmaceutics*, **22**, 735–750.

Kass, R.E. and Steffey, D. (1989) Approximate Bayesian inference in conditionally independent hierarchical models (parametric empirical Bayes models). *Journal of the American Statistical Association*, **84**, 717–726.

Katz, D., Azen, S.P. and Schumitzky, A. (1981) Bayesian approach to the analysis of nonlinear models: Implementation and evaluation. *Biometrics*, **37**, 137–142.

Kennedy, W.J., Jr. and Gentle, J.E. (1980) *Statistical Computing*. Marcel Dekker, New York.

Kiefer, J. and Wolfowitz, J. (1960) The equivalence of two extremum problems. *Canadian Journal of Mathematics* **12**, 363–366.

Kreft, I.G.G., de Leeuw, J. and Van Der Leeden, R. (1995) Review of five multi-level analysis programs: BMDP-5V, GENMOD, HLM, ML3, and VARCL. *The American Statistician*, **48**, 324–335.

Kwan, K.C., Breault, G.O., Umbenhauer, E.R., McMahon, F.G. and Duggan, D.E. (1976) Kinetics of indomethacin absorption, elimination, and enterohepatic circulation in man. *Journal of Pharmacokinetics and Biopharmaceutics*, **4**, 255–280.

Laird, N.M. (1978) Nonparametric maximum likelihood estimation of a mixing distribution. *Journal of the American Statistical Association*, **73**, 805–813.

Laird, N., Lange, N. and Stram, D. (1987) Maximum likelihood computations with repeated measures: Application of the EM algorithm. *Journal of the American Statistical Association*, **82**, 97–105.

Laird, N.M. and Ware, J.H. (1982) Random effects models for longitudinal data. *Biometrics*, **38**, 963–974.

Lamotte, L.R. (1973) Quadratic estimation of variance components. *Biometrics*, **29**, 311–330.

Lange, N. and Ryan, L. (1989) Assessing normality in random effects models. *Annals of Statistics*, **17**, 624–642.

Liang, K.-Y. and Zeger, S.L. (1986) Longitudinal data analysis using generalized linear models. *Biometrika*, **73**, 13–22.

Liang, K.-Y., Zeger, S.L. and Qaqish, B. (1992) Multivariate regression analyses for categorical data (with Discussion). *Journal of the Royal Statistical Society, Series B*, **54**, 3–40.

Lindley, D.V. and Smith, A.F.M. (1972) Bayes estimates for the linear model (with Discussion). *Journal of the Royal Statistical Society, Series B*, **34**, 1–42.

Lindsay, B.G. (1983) The geometry of mixture likelihoods: a general theory. *Annals of Statistics*, **11**, 86–94.

Lindsey, J.K. (1993) *Models for Repeated Measurments.* Oxford University Press, Oxford.

Lindstrom, M.J. and Bates, D.M. (1988) Newton-Raphson and EM algorithms for linear mixed-effects models for repeated-measures data. *Journal of the American Statistical Association*, **83**, 1014–1022.

Lindstrom, M.J. and Bates, D.M. (1990) Nonlinear mixed effects models for repeated measures data. *Biometrics*, **46**, 673–687.

Lindstrom, F.T. and Birkes, D.S. (1984) Estimation of population pharamcokinetic parameters using destructively obtained experimental data: A simulation study of the one-compartment open model. *Drug Metabolism Reviews*, **15**, 195–264.

Little, R.J.A. (1993) Pattern-mixture models for multivariate incomplete data. *Journal of the American Statistical Association*, **88**, 125–134.

Little, R.J.A. and Rubin, D.B. (1987) *Statistical Analysis with Missing Data.* Wiley, New York.

Longford, N.T. (1993) *Random Coefficient Models.* Oxford University Press, Oxford.

Louis, T.A. (1991) Using empirical Bayes methods in biopharmaceutical research. *Statistics in Medicine*, **10**, 811–829.

Magder, L. (1994). Estimating mixing distributions using mixtures of Gaussians. Unpublished Ph.D. dissertation, Johns Hopkins University.

Maitre, P.O., Buhrer, M., Thomson, D. and Stanski, D.R. (1991) A three-step approach combining Bayesian regression and NONMEM population analysis: application to midazolam. *Journal of Pharma-*

cokinetics and Biopharmaceutics, **19**, 377–384.

Mallet, A. (1986) A maximum likelihood estimation method for random coefficient regression models. *Biometrika*, **73**, 645–656.

Mallet, A. (1992) A nonparametric method for population pharmacokinetics: a case study. *Proceedings of the Biopharmaceutical Section of the American Statistical Association, Joint Statistical Meetings.*

Mallet, A., Mentré, F., Gilles, J., Kelman, A.W., Thomson, A.H., Bryson, S.M. and Whiting, B. (1988a) Handling covariates in population pharmacokinetics, with an application to gentamycin. *Biomedical Measurement and Information Control*, **2**, 138–146.

Mallet, A., Mentré, F., Steimer, J.-L. and Lokiec, F. (1988b) Nonparametric maximum likelihood estimation for population pharmacokinetics, with application to cyclosporine. *Journal of Pharmacokinetics and Biopharmaceutics*, **16**, 311–327.

Mandema, J.W., Verotta, D. and Sheiner, L.B. (1992) Building population pharmacokinetic/pharmacodynamic models. *Journal of Pharmacokinetics and Biopharmaceutics*, **20**, 511–529.

Marquardt, D.W. (1963) An algorithm for least-squares estimation of nonlinear parameters. *Journal of the Society for Industrial and Applied Mathematics*, **11**, 431–441.

McCullagh, P. (1983) Quasi-likelihood functions. *The Annals of Statistics*, **11**, 59–67.

McCullagh, P. and Nelder, J.A. (1989) *Generalized Linear Models* (2nd edn). Chapman & Hall, New York.

Mentré, F. and Mallet, A. (1994) Handling covariates in population pharmacokinetics. *International Journal of Bio-Medical Computing*, **36**, 25–33.

Moore, D.F. and Tsiatis, A. (1991) Robust estimation of the variance in moment methods for extra-binomial and extra-Poisson variation. *Biometrics*, **47**, 383–401.

Nelder, J.A. and Pregibon, D. (1987) An extended quasi-likelihood function. *Biometrika*, **74**, 221–232.

Ochs, H.R., Greenblatt, D.J. and Woo, E. (1980) Clinical pharmacokinetics of quinidine. *Clinical Pharmacokinetics*, **5**, 150–168.

O'Connell, M., Belanger, B.A. and Haaland, P.D. (1993) Calibration and assay development using the four-parameter logistic model. *Chemometrics and Intelligent Laboratory Systems*, **20**, 97–114.

Odell, P.L. and Fieveson, A.H. (1966) A numerical procedure to generate a sample covariance matrix. *Journal of the American Statistical Association*, **61**, 198–203.

Oppenheimer, L., Capizzi, T.P., Weppelman, R.M. and Mehta, H. (1983) Determining the lowest limit of reliable assay measurement. *Analytical Chemistry*, **55**, 638–646.

Paik, M.C. (1992) Parametric variance function estimation for nonnormal repeated measurement data. *Biometrics*, **48**, 19–30.

Patterson, H D. and Thompson, R. (1971) Recovery of inter-block information when block sizes are unequal. *Biometrika*, **58**, 545–554.

Peck, C.C. (1990) The randomized concentration-controlled trial: An information-rich alternative to the randomized placebo-controlled trial. *Clinical Pharmacokinetics and Therapeutics*, **47**, 126.

Peck, C.C., Beal, S.L., Sheiner, L.B. and Nichols, A.I. (1984) Extended least squares nonlinear regression: A possible solution to the choice of weights problem in analysis of individual pharmacokinetic data. *Journal of Pharmacokinetics and Biopharmaceutics*, **12**, 545–558.

Peck, C.C. and Rodman, J. H. (1991) Analysis of clinical pharmacokinetic data for individualizing drug dosage regimens. In *Applied Pharmacokinetics* (3rd edn) (eds. W.E. Evans *et al.*), Applied Therapeutics, Inc., Vancouver, WA.

Pinheiro, J.C. and Bates, D.M. (1995) Approximations to the loglikelihood function in the nonlinear mixed effects model. *Journal of Computational and Graphical Statistics*, **4**, 12–35.

Pinheiro, J.C., Bates, D.M. and Lindstrom, M. (1993) Nonlinear mixed effects classes and methods for S. Technical report No. 906, Department of Statistics, University of Wisconsin.

Pinheiro, J.C., Bates, D.M. and Lindstrom, M. (1994) Model building for nonlinear mixed effects models. Technical report No. 931, Department of Statistics, University of Wisconsin.

Potthoff, R.F. and Roy, S.N. (1964) A generalized multivariate analysis of variance useful especially for growth curve problems. *Biometrika*, **51**, 313–326.

Prentice, R.L. (1988) Correlated binary regression with covariates specific to each binary observation. *Biometrics*, **44**, 1033–1048.

Prentice, R.L. and Zhao, L.P. (1991) Estimating equations for parameters in means and covariance of multivariate discrete and and continuous responses. *Biometrics*, **47**, 825–840.

Prevost, G. (1977) Estimation of a normal probability density function from samples measured with nonnegligible and nonconstant dispersion. Internal report 6-77, Adersa-Gerbois, 2 avenue de le Mai, F-91120, Palaiseau.

Pritchard, D.J., Downie, J. and Bacon, D.W. (1977) Further consideration of heteroscedasticity in fitting kinetic models. *Technometrics*, **19**, 227–236.

Prosser, R., Rasbash, J. and Goldstein, H. (1991) *ML3 Software for Three-level Analysis: Users Guide for V.2.* Institute of Education, London.

Raab, G.M. (1981) Estimation of a variance function, with application to radioimmunoassay. *Applied Statistics*, **30**, 32–40.

Racine, A., Grieve, A.P., Flühler, H. and Smith, A.F.M. (1986) Bayesian methods in practice: experiences in the pharmaceutical industry (with Discussion). *Applied Statistics*, **35**. 93–150.

Racine-Poon, A. (1985) A Bayesian approach to nonlinear random effects models. *Biometrics*, **41**, 1015–1023.

Racine-Poon, A. and Smith, A.F.M. (1990) Population models. In *Statistical Methodology in the Pharmaceutical Sciences* (ed. D.A. Berry), Marcel Dekker, New York.

Ramos, R.Q. (1993) Estimation for nonlinear mixed effects and random coefficient models. Unpublished Ph.D. dissertation, Department of Statistics, North Carolina State University, Raleigh.

Rao, C.R. (1973) *Linear Statistical Inference and its Applications* (2nd edn). Wiley, New York.

Ratkowsky, D.A. (1983) *Nonlinear Regression Modeling.* Marcel Dekker, New York.

Reinsel, G.C. (1982) Multivariate repeated-measurement or growth curve models with multivariate random-effects covariance structure. *Journal of the American Statistical Association*, **77**, 190–195.

Reinsel, G.C. (1984) Estimation and prediction in a multivariate random-effects generalized linear model. *Journal of the American Statistical Association*, **79**, 406–414.

Reinsel, G.C. (1985) Mean squared error properties of empirical Bayes estimators in a multivariate random effects general linear model. *Journal of the American Statistical Association*, **80**, 642–650.

Robinson, G.K. (1991) That BLUP is a good thing – The estimation of random effects. *Statistical Science*, **6**, 15–51.

Rodbard, D. (1978) Statistical estimation of the minimal detectable concentration ('sensitivity') for radioligand assays. *Analytical Biochemistry,* **90**, 1–12.

Rodbard, D. and Frazier, G.R. (1975) Statistical analysis of radioligand assay data. *Methods of Enzymology*, **37**, 3–22.

Rodbard, D. and Hutt, D.M. (1974) Statistical analysis of radioimmunoassays and immunoradiometric (labelled antibody) assays: A generalized weighted, iterative, least squares method for logistic curve fitting. In *Radioimmunoassay and Related Procedures in Medicine, I,* International Atomic Energy Agency, Vienna, 165–192.

Rodbard D., Lennox R.H., Wray H.L., Ramseth D. (1976) Statistical characterization of the random errors in the radioimmunoassay dose-response variable. *Clinical Chemistry*, **22**, 350-358.

Rodman, J., Jelliffe, R., Kolb, E., Tuey, D., de Guzman, M., Wagers, P.W. and Haywood, J. (1984) Clinical studies with computer-assisted initial lidocaine infusion therapy. *Archives of Internal Medicine*, **144**, 703–709.

Ross, G.J.S. (1990) *Nonlinear Estimation.* Springer-Verlag, New York.

Rotnitzky, A. and Jewell, N.P. (1990) Hypothesis testing of regression parameters in semiparametric generalized linear models for cluster correlated data. *Biometrika*, **77**, 484–497.

Rowland, M., Sheiner, L.B. and Steimer, J.-L., eds. (1985) *Variability in Drug Therapy: Description, Estimation, and Control.* Raven Press, New York.

Rubin, D.B. (1976) Inference and missing data (with Discussion). *Biometrika*, **63**, 581–592.

Rubin, D.B. (1980) Using empirical Bayes techniques in the law school validity studies. *Journal of the American Statistical Association*, **75**, 801–816.

Ruppert, D., Cressie, N. and Carroll, R.J. (1989) A transformation/weighting model for estimating Michaelis-Menten parameters. *Biometrics*, **45**, 637–656.

Sadler, W.A. and Smith, M.H. (1985) Estimation of the response-error relationship in immunoassay. *Clinical Chemistry*, **31(11)**, 1–37.

SAS Institute, Inc. (1992) SAS Technical Report P-229, *SAS/STAT Software: Changes and Enhancements, Release 6.07.* Chapter 16, The MIXED Procedure, Cary, NC.

Schabenberger, O. (1994) Nonlinear mixed effects growth models for repeated measures in ecology. *Proceedings of the Environmental Section of the American Statistical Association, Joint Statistical Meetings.*

Schall, R. (1991) Estimation in generalized linear models with random effects. *Biometrika*, **78**, 719–727.

Schumitzky, A. (1991a) Nonparametric EM algorithms for estimating prior distributions. *Applied Mathematics and Computation*, **45**, 141–157.

Schumitzky, A. (1991b) Applications of stochastic control theory to optimal design of dosage regimens. In *Advanced Methods of Pharmacokinetic and Pharmacodynamic Systems Analysis* (ed. D.Z. D'Argenio), Plenum Press, New York.

Schumitzky, A. (1993) The nonparametric maximum likelihood approach to pharmacokinetic population analysis. *Proceedings of the Western Simulation Multiconference-Simulation in Health Care, Society for Computer Simulation.*

Schwenke, J.R. and Milliken, G. (1991) On the calibration problem extended to nonlinear models. *Biometrics*, **47**, 563–574.

Searle, S.R., Casella, G. and McCulloch, C.E. (1992) *Variance Components.* Wiley, New York.

Seber, G.A.F. and Wild, C.J. (1989) *Nonlinear Regression.* Wiley, New York.

Sheiner, L.B. (1985) Modeling pharmacodynamics: Parametric and nonparametric approaches. *Variability in Drug Therapy: Description, Estimation, and Control* (eds. M. Rowland, L.B. Sheiner and J.-L. Steimer). Raven Press, New York.

Sheiner, L.B. and Beal, S.L. (1980) Evaluation of methods for estimating population pharmacokinetic parameters. I. Michaelis-Menten Model: Routine clinical pharmacokinetic data. *Journal of Pharmacokinetics and Biopharmaceutics*, **8**, 553–571.

Sheiner, L.B. and Beal, S.L. (1981) Evaluation of methods for estimating population pharmacokinetic parameters. II. Biexponential model and experimental pharmacokinetic data. *Journal of Pharmacokinetics and Biopharmaceutics*, **9**, 635–651.

Sheiner, L.B. and Beal, S.L. (1983) Evaluation of methods for estimating population pharmacokinetic parameters. III. Monoexponential model: Routine clinical pharmacokinetic data. *Journal of Pharmacokinetics and Biopharmaceutics*, **11**, 303–319.

Sheiner, L.B. and Ludden, T.M. (1992) Population pharmacokinetics/pharmacodynamics. *Annual Review of Pharmacological Toxicology*, **32**, 185–209.

Sheiner, L.B., Rosenberg, B. and Marathe, V.V. (1977) Estimation of population characteristics of pharmacokinetic parameters from routine clinical data. *Journal of Pharmacokinetics and Biopharmaceutics*, **8**, 635–651.

Sheiner, L.B., Rosenberg, B. and Melmon, K.L. (1972) Modeling of individual pharmacokinetics for computer-aided drug dosing. *Computers and Biomedical Research*, **5**, 441–459.

Sheiner, L.B., Stanski, D.R., Vozeh, S., Miller, R.D. and Ham, J. (1979) Simultaneous modeling of pharmacokinetics and pharmacodynamics: Application to *d*-tubocurarine. *Clinical Pharmacology and Therapeutics*, **25**, 358–371.

Silverman, B.W. (1986) *Density Estimation for Statistics and Data Analysis.* Chapman & Hall, New York.

Silvey, S.D. (1980) *Optimal Design.* Chapman & Hall, New York.

Smith, A.F.M. and Gelfand, A.E. (1992) Bayesian statistics without tears: A sampling-resampling perspective. *The American Statistician*, **46**, 84–88.

Smith, A.F.M. and Roberts, G.O. (1993) Bayesian computation via the Gibbs sampler and related Markov chain Monte Carlo methods. *Journal of the Royal Statistical Society, Series B*, **55**, 3–23.

Solomon, P.J. and Cox, D.R. (1992) Nonlinear component of variance models. *Biometrika*, **79**, 1–11.

Spiegelhalter, D., Thomas, A. and Gilks, W. (1993) *BUGS Manual 0.20*, MRC Biostatistics Unit, Cambridge.

Statistical Consultants, Inc. (1986) PCNONLIN and NONLIN84: Software for the statistical analysis of nonlinear models. *The American Statistician*, **40**, 52.

Steimer, J.-L., Mallet, A., Golmard, J.L. and Boisvieux, J.F. (1984) Alternative approaches to estimation of population pharmacokinetic parameters: Comparison with the nonlinear mixed effect model. *Drug Metabolism Reviews*, **15**, 265–292.

Steimer, J.-L., Mallet, A. and Mentré, F. (1985) Estimating interindividual pharmacokinetic variability. In *Variability in Drug Therapy* (eds. M. Rowland *et al.*). Raven Press, New York.

Stiratelli, R., Laird, N.M., and Ware, J. (1984) Random effects models for serial observations with binary responses. *Biometrics*, **40**, 961–971.

Swamy, P.A.V.B. (1971) *Statistical Inference in Random Coefficient Regression Models.* Springer Verlag, Berlin.

Tanner, M.A. (1993) *Tools for Statistical Inference: Methods for the Exploration of Posterior Distributions and Likelihood Functions* (2nd edn). Springer-Verlag, New York.

Thall, P.F. and Vail, S.C. (1990) Some covariance models for longitudinal count data with overdispersion. *Biometrics*, **46**, 657–671.

Tierney, L. (1995) Markov chains for exploring posterior distributions. *Annals of Statistics*, to appear.

Unadkat, J.D., Beal, S.L. and Sheiner, L.B. (1986) Bayesian calibration. *Analytica Chimica Acta*, **181**, 27–36.

van Houwelingen, J.C. (1988) Use and abuse of variance models in regression. *Biometrics*, **43**, 1073–1081.

Verme, C.N., Ludden, T.M., Clementi, W.A. and Harris, S.C. (1992) Pharmacokinetics of quinidine in male patients: A population analysis. *Clinical Pharmacokinetics*, **22**, 468–480.

Vonesh, E.F. (1992) Nonlinear models for the analysis of longitudinal data. *Statistics in Medicine*, **11**, 1929–1954.

Vonesh, E.F. (1993) MIXNLIN: A SAS macro. User's Guide.

Vonesh, E.F. and Carter, R.L. (1987) Efficient inference for a random coefficient growth curve model with unbalanced data. *Biometrics*, **43**, 617–628.

Vonesh, E.F. and Carter, R.L. (1992) Mixed effects nonlinear regression for unbalanced repeated measures. *Biometrics*, **48**, 1–18.

Waclawiw, M.A. and Liang, K.-Y. (1993) Prediction of random effects in the generalized linear model. *Journal of the American Statistical Association*, **88**, 171–178.

Wakefield, J.C. (1995) The Bayesian analysis of population pharmacokinetic models. *Journal of the American Statistical Association*, to appear.

Wakefield, J.C., Smith, A.F.M., Racine-Poon, A. and Gelfand, A.E. (1994) Bayesian analysis of linear and nonlinear population models using the Gibbs sampler. *Applied Statistics*, **43**, 201–222.

Wakefield, J.C. and Walker, S. (1994) Population models with a nonparametric random coefficient distribution. Technical report, Department of Mathematics, Imperial College.

Walters, D.K., Hann, D.W. and Clyde, M.A. (1985) Equations and tables predicting gross total stem volumes in cubic feet for six major conifers of southwest Oregon. Research Bulletin 50, Forest Research Laboratory, Oregon State University.

Ware, J.H. (1985) Linear models for the analysis of longitudinal studies. *The American Statistician*, **39**, 95–101.

Wedderburn, R.W.M. (1974) Quasi-likelihood functions, generalized linear models, and the Gauss-Newton method. *Biometrika*, **61**, 439–447.

Williams, J.S. (1975) The analysis of binary responses from toxicological experiments involving reproduction and teratogenicity. *Biometrics*, **31**, 949–952.

Wolfinger, R. (1993) Laplace's approximation for nonlinear mixed models. *Biometrika*, **80**, 791–795.

Wolfinger, R. and O'Connell, M. (1993) Generalized linear mixed models: A pseudo-likelihood approach. *Journal of Statistical Computation and Simulation*, **48**, 233–243.

Wu, M.C. and Bailey, K.R. (1989) Estimation and comparison of changes in the presence of informative right censoring: Conditional linear model. *Biometrics*, **45**, 939–955.

Wu, M.C. and Carroll, R.J. (1988) Estimation and comparison of changes in the presence of informative right censoring by modeling the censoring process. *Biometrics*, **44**, 175–188.

Ye, K. and Berger, J.O. (1991) Noninformative priors for inference in exponential regression models. *Biometrika*, **78**, 645–656.

Zeger, S.L. and Karim, M.R. (1991) Generalized linear models with random effects: A Gibbs sampling approach. *Journal of the American Statistical Association*, **86**, 79–86.

Zeger, S.L. and Liang, K.-Y. (1986) Longitudinal data analysis for discrete and continuous outcomes. *Biometrics*, **42**, 121–130.

Zeger, S.L., Liang, K. and Albert, P. (1988) Models for longitudinal data: A generalized estimating equation approach. *Biometrics*, **44**, 1049–1060.

Zhao, L.P. and Prentice, R.L. (1990) Correlated binary responses using a quadratic exponential model. *Biometrika*, **77**, 642–648.

Author index

Subject index